An Introduction to Financial Mathematics

Option Valuation

Second Edition

CHAPMAN & HALL/CRC
Financial Mathematics Series

Aims and scope:
The field of financial mathematics forms an ever-expanding slice of the financial sector. This series aims to capture new developments and summarize what is known over the whole spectrum of this field. It will include a broad range of textbooks, reference works and handbooks that are meant to appeal to both academics and practitioners. The inclusion of numerical code and concrete real-world examples is highly encouraged.

High-Performance Computing in Finance
Problems, Methods, and Solutions
M.A.H. Dempster, Juho Kanniainen, John Keane, Erik Vynckier

An Introduction to Computational Risk Management of Equity-Linked Insurance
Runhuan Feng

Derivative Pricing
A Problem-Based Primer
Ambrose Lo

Portfolio Rebalancing
Edward E. Qian

Interest Rate Modeling
Theory and Practice, 2nd Edition
Lixin Wu

An Introduction to Financial Mathematics
Option Valuation, Second Edition
Hugo D. Junghenn

For more information about this series please visit: *https://www.crcpress.com/Chapman-and-HallCRC-Financial-Mathematics-Series/book-series/CHFINANCMTH*

An Introduction to Financial Mathematics

Option Valuation

Second Edition

Hugo D. Junghenn

CRC Press is an imprint of the
Taylor & Francis Group, an **informa** business
A CHAPMAN & HALL BOOK

CRC Press
Taylor & Francis Group
6000 Broken Sound Parkway NW, Suite 300
Boca Raton, FL 33487-2742

Printed on acid-free paper
Version Date: 20190207

International Standard Book Number-13: 978-0-367-20882-0 (Hardback)

Visit the Taylor & Francis Web site at
http://www.taylorandfrancis.com

and the CRC Press Web site at
http://www.crcpress.com

TO MY FAMILY

Mary, Katie, Patrick, Sadie

Contents

Preface

This text is intended as an introduction to the mathematics and models used in the valuation of financial derivatives. It is designed for an audience with a background in standard multivariable calculus. Otherwise, the book is essentially self-contained: the requisite probability theory is developed from first principles and introduced as needed, and finance theory is explained in detail under the assumption that the reader has no background in the subject.

The book is an outgrowth of a set of notes developed for an undergraduate course in financial mathematics offered at The George Washington University. The course serves mainly majors in mathematics, economics, or finance, and is intended to provide a straightforward account of the principles of option pricing. The primary goal of the text is to examine these principles in detail via the standard discrete and stochastic calculus models. Of course, a rigorous exposition of such models requires a systematic development of the requisite mathematical background, and it is an equally important goal to provide this background in a careful manner consistent with the scope of the text. Indeed, it is hoped that the text may serve as an introduction to applied probability (through the lens of mathematical finance).

The book consists of fifteen chapters, the first ten of which develop option valuation techniques in discrete time, the last five describing the theory in continuous time. The emphasis is on two models, the (discrete time) *binomial* model and the (continuous time) *Black-Scholes-Merton* model. The binomial model serves two purposes: First, it provides a practical way to price options using relatively elementary mathematical tools. Second, it allows a straightforward concrete exposition of fundamental principles of finance, such as arbitrage and hedging, without the possible distraction of complex mathematical constructs. Many of the ideas that arise in the binomial model foreshadow notions inherent in the more mathematically sophisticated Black-Scholes-Merton model.

Chapter 1 gives an elementary account of the basic principles of finance. Here, the focus is on risk-free investments, such money market accounts and bonds, whose values are determined by an interest rate. Investments of this type provide a way to measure the value of a risky asset, such as a stock or commodity, and mathematical descriptions of such investments form an important component of option pricing techniques.

Chapters 2, 3, 6, and 8 form the core of the general probability portion of the text. The exposition is essentially self-contained and uses only basic combinatorics and elementary calculus. Appendix A provides a brief overview of the elementary set theory and combinatorics used in these chapters. Readers with a good background in probability may safely give this part of the text a cursory reading. While our approach is largely standard, the more sophisticated notions of event σ-field and filtration are introduced early to prepare the reader for the martingale theory developed in later chapters. We have avoided using Lebesgue integration by considering only discrete and continuous random variables.

Chapter 4 describes the most common types of financial derivatives and emphasizes the role of arbitrage in finance theory. The assumption of an arbitrage-free market, that is, one that allows no "free lunch," is crucial in developing useful pricing models. An important consequence of this assumption is the put-call parity formula, which relates the cost of a standard call option to that of the corresponding put.

Discrete-time stochastic processes are introduced in Chapter 5 to provide a rigorous mathematical framework for the notion of a self-financing portfolio. The chapter describes how such portfolios may be used to replicate options in an arbitrage-free market.

Chapter 7 introduces the reader to the binomial model. The main result is the construction of a replicating, self-financing portfolio for a general European claim. An important consequence is the Cox-Ross-Rubinstein formula for the price of a call option. Chapter 9 considers the binomial model from the vantage point of discrete-time martingale theory.

Chapter 10 takes up the more difficult problems of pricing and hedging an American claim. Solutions to the problems are based on the notions of stopping times and supermartingales.

Chapter 11 gives an overview of Brownian motion, constructs the Ito integral for processes with continuous paths, and uses Ito's formula to solve various stochastic differential equations. Our approach to stochastic calculus builds on the reader's knowledge of classical calculus and emphasizes the similarities and differences between the two theories via the notion of variation of a function.

Chapter 12 uses the tools developed in Chapter 11 to construct the Black-Scholes-Merton PDE, the solution of which leads to the celebrated Black-Scholes formula for the price of a call option. A detailed analysis of the analytical properties of the formula is given in the last section of the chapter. The more technical proofs are relegated to appendices so as not to interrupt the main flow of ideas.

Chapter 13 gives a brief overview of those aspects of continuous-time martingales needed for risk-neutral pricing. The primary result is Girsanov's Theorem, which guarantees the existence of risk-neutral probability measures.

Chapters 14 and 15 provide a martingale approach to option pricing, using risk-neutral probability measures to find the value of a variety of derivatives. Rather than being encyclopedic, the material here is intended to convey the essential ideas of derivative pricing and to demonstrate the utility and elegance of martingale techniques in this regard.

The text contains numerous examples and over 250 exercises designed to help the reader gain expertise in the methods of financial calculus and, not incidentally, to increase his or her level of general mathematical sophistication. The exercises range from routine calculations to spreadsheet projects to the pricing of a variety of complex financial instruments. Solutions to the odd-numbered problems are given in Appendix D.

For greater clarity and ease of exposition (and to remain within the intended scope of the text), we have avoided stating results in their most general form. Thus interest rates are assumed to be constant, paths of stochastic processes are required to be continuous, and financial markets trade in a single risky asset. While these assumptions may not be wholly realistic, it is our belief that the reader who has obtained a solid understanding of the theory in this simplified setting will have little difficulty in making the transition to more general contexts.

While the text contains numerous examples and problems involving the use of spreadsheets, we have not included any discussion of general numerical techniques, as there are several excellent texts devoted to this subject. Indeed, such a text could be used to good effect in conjunction with the present one.

It is inevitable that any serious development of option pricing methods at the intended level of this book must occasionally resort to invoking a result whose proof falls outside the scope of the text. For the few times that this has occurred, we have tried either to give a sketch of the proof or, failing that, to give references where the reader may find a reasonably accessible proof.

The text is organized to allow as flexible use as possible. The first edition of the book has been successfully used in the classroom as a single semester course in discrete-time theory (Chapters 1–9), as a one-semester course giving an overview of both discrete-time and continuous-time models (Chapters 1–7, 11, and 12), and, with additional chapters, as a two-semester course. Sections with asterisks may be omitted on first reading.

A few words as to the major changes from the first edition: A significant part of the book was revised in an attempt to improve the exposition. In particular, details were added to many of the theorems. Also, about 50 new exercises have been added. Additionally, some minor restructuring of the chapters has occurred, again in the hopes of improving exposition. The major change is the inclusion of many new tables and graphs generated by over thirty VBA Excel programs available on the author's webpage. It is hoped that the reader will use these as models to write additional programs.

To those whose sharp eye caught typos, inconsistencies, and downright errors in the notes leading up to the book as well as in the first edition: thank you. Special thanks go to Farshad Foroozan, who made several valuable suggestions which found their way into the second edition, and also to Steve East, who tried to teach me the stock market from the bottom up.

Hugo D. Junghenn
Washington, D.C., USA

Chapter 1

Basic Finance

In this chapter we consider assets whose future values are completely determined by a fixed interest rate. If the asset is guaranteed, as in the case of an insured savings account or a government bond (which, typically, has only a small likelihood of default), then the asset is said to be *risk-free*. By contrast, a *risky asset*, such as a stock or commodity, is one whose future values cannot be determined with certainty. As we shall see in later chapters, mathematical models that describe the values of a risky asset typically include a risk-free component. Our first goal then is to describe how risk-free assets are valued, which is the content of this chapter.

1.1 Interest

Interest is a fee paid by one party for the use of assets of another. The amount of interest is generally time dependent: the longer the outstanding balance, the more interest is accrued. A familiar example is the interest generated by a money market account. The bank pays the depositor an amount that is a predetermined fraction of the balance in the account, that fraction derived from a prorated annual percentage called the *nominal rate*, denoted typically by the symbol r. In the following subsections we describe various ways that interest determines the value of an account.

Simple Interest

Consider first an account that pays *simple interest* at an annual rate of $r \times 100\%$. If an initial deposit of A_0 is made at time zero, then after one year the account has value $A_1 = A_0 + rA_0 = A_0(1+r)$, after two years the account has value $A_2 = A_0 + 2rA_0 = A_0(1+2r)$, and so forth. In general, after t years the account has value

$$A_t = A_0(1+tr) \tag{1.1}$$

which is the so-called *simple interest formula*. Notice that interest is paid only on the initial deposit.

Discrete-Time Compound Interest

Suppose now that an account pays the same annual rate $r \times 100\%$ but with interest *compounded m times per year*, for example, monthly ($m = 12$) or daily ($m = 365$). The interest rate *per period* is then $i := r/m$. In this setting, if an initial deposit of A_0 is made at time zero, then after the first period the value of the account is $A_1 = A_0 + iA_0 = A_0(1+i)$, after the second period the value is $A_2 = A_1 + iA_1 = A_1(1+i) = A_0(1+i)^2$, and so forth. In general, the value of the account at time n is $A_0(1+i)^n$. The formula reflects the fact that interest is paid not just on the principle A_0 but on the accrued amounts in the account. Since there are m periods per year, the value of the account after t years is

$$A_t = A_0(1 + r/m)^{mt} \tag{1.2}$$

which is the *compound interest formula.* In this context, A_0 is called the *present value* or *discounted value* of the account and A_t is the *future value at time t.*

Continuous-Time Compound Interest

Now consider what happens when the number m of compounding periods per year increases without bound. Write (1.2) as

$$A_t = A_0\left[(1 + 1/x)^x\right]^{rt}, \quad \text{where } x = m/r.$$

As $m \to \infty$, the term in brackets tends to e, the base of the natural logarithm. This leads to the formula for the value of an account after t years under *continuously compounded interest*:

$$A_t = A_0 e^{rt}. \tag{1.3}$$

Comparison of the Methods

TABLE 1.1: Input/output for **FutureValue**.

	A	B	C
3	Initial Deposit	Years	Annual Rate
4	1000	5.4	3.5%
6	Interest Method	Future Value	
7	simple	1189.00	
8	12	1207.71	
9	365	1208.03	
10	continuous	1208.04	

Table 1.1 depicts the spreadsheet generated by the module **FutureValue**, which uses formulas (1.1), (1.2), and (1.3). The initial deposit, number of years, and annual rate are entered in cells A4, B4, and C4, respectively, and the interest methods are entered in column A. Corresponding future values are generated in column B. Note that compounding continuously instead of daily adds only $.01 to the future value.

Effective Rate

In view of the various ways one can compute interest, it is useful to have a method to compare investment strategies. One such device is the *effective interest rate* r_e, defined as the simple interest rate that produces the same yield in one year as compound interest. Thus if interest is compounded m times a year, then the effective rate must satisfy the equation $A_0(1+r/m)^m = A_0(1+r_e)$ and so

$$r_e = (1 + r/m)^m - 1.$$

Similarly, if interest is compounded continuously, then $A_0e^r = A_0(1 + r_e)$, so

$$r_e = e^r - 1.$$

Table 1.2 below summarizes data generated by the module **EffectiveRate**. The annual rates are entered in column A and the compounding methods in column B. Effective rates are generated in column C using the above formulas. The table gives the effective interest rates of three investment schemes

TABLE 1.2: Input/output for **EffectiveRate**.

	A	B	C
3	Annual Rate	Compounding Method	Effective Rate
4	.1100	2	.11302500
5	.1076	12	.11306832
6	.1072	continuous	.11315686

summarized in rows 1–3. The scheme in row 3 has the highest effective rate and is therefore the best choice.

---

*1.2 Inflation

Inflation is defined as an increase over time of the general level of prices of goods and services, resulting in a decrease of purchasing power. For a mathematical model, assume that inflation is running at an annual rate of r_f. If the price of an item now is A_0, then the price after one year is $A_0(1 + r_f)$, the price after two years is $A_0(1 + r_f)^2$, and so on. Thus inflation is governed by a compound interest formula. For example, if inflation is running at 5% per year, then prices will increase by 5% each year, so an item costing \$100 now will cost $100(1.05)^2 = \$110.25$ in 2 years.

To see the combined effects of inflation and interest, suppose we make an investment whose (nominal) rate of return is r, compounded yearly. What is the actual annual rate of return r_a if inflation is taken into account? A dollar investment now will have future value $(1+r)$ in one year. In purchasing power this is equivalent to the amount $(1+r)/(1+r_f)$ now. Thus, in terms of constant purchasing units, the investment transforms \$1 of purchasing units into the amount $(1+r)/(1+r_f)$. The annual interest rate r_a that produces the same return is called the *inflation-adjusted rate of return.* Thus $1+r_a = (1+r)/(1+r_f)$ and so

$$r_a = \frac{1+r}{1+r_f} - 1 = \frac{r-r_f}{1+r_f}.$$

For small r_f we have the approximation

$$r_a \approx r - r_f,$$

which is frequently taken as the definition of r_a.

Now suppose you are to receive a payment of Q dollars one month from now, i.e., at time 1. If the investment pays interest at an annual rate of r, then, taking inflation into account, the time-0 value of this payment is (approximately) $A_0 = (1+i-i_f)^{-1}Q$, where $i = r/12$ and $i_f = r_f/12$. Similarly a payment of Q two months from now would have time-1 value $(1+i-i_f)^{-1}Q$ and hence a time-0 value of $(1+i-i_f)^{-2}Q$. In general if you are to receive a payment of Q in n months, then the present value of this payment, taking monthly inflation into account, is

$$A_0 = (1+i-i_f)^{-n}Q.$$

This is the same as the present value formula for an annual rate of $r - r_f$ spread over 12 months.

1.3 Annuities

An *annuity* is a sequence of periodic payments of a fixed amount P. The payments may be deposits into an account such as a pension fund or a layaway plan, or withdrawals from an account, as in a trust fund or retirement account.[1] Assume the account pays interest at an annual rate r compounded m times per year and that a deposit (withdrawal) is made at the *end* of each compounding interval. We seek the value A_n of the account at time n, that is, the value immediately after the nth deposit (withdrawal).

Deposits

For deposits, the quantity A_n is the sum of the time-n values of payments 1 to n. Since payment j accrues interest over $n-j$ compounding periods,

[1] An annuity of deposits is also called a *sinking fund.*

its time-n value is $P(1+r/m)^{n-j}$. The scheme is illustrated in Figure 1.1. Summing from $j=1$ to n we obtain $A_n = P(1+x+x^2+\cdots+x^{n-1})$, where

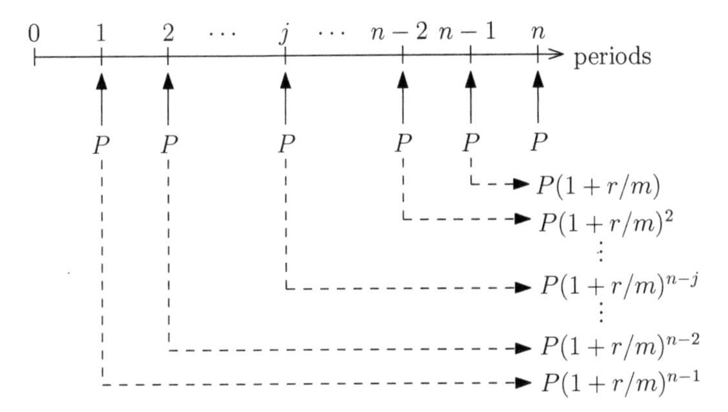

FIGURE 1.1: An annuity of deposits.

$x := 1+r/m$. The preceding geometric series sums to $(x^n-1)/(x-1)$, hence the value of the account at time n is

$$A_n = P\frac{(1+i)^n-1}{i}, \qquad i := \frac{r}{m}. \tag{1.4}$$

More generally, if a deposit of A_0 is made at time 0, then the time-n value of the account is

$$A_n = A_0(1+i)^n + P\frac{(1+i)^n-1}{i}.$$

Using this equation, one may easily calculate the number n of monthly deposits required to reach a goal of A. Indeed, setting $A = A_n$, solving for $(1+i)^n$, and then taking logarithms we see that the desired value n is the quantity

$$\frac{\ln(iA+P)-\ln(iA_0+P)}{\ln(1+i)},$$

rounded up to the nearest integer. This is the smallest integer n for which $A_n \geq A$.

TABLE 1.3: Input/output for **NumberOfMonthlyDeposits**.

	A	B
3	Initial deposit (optional)	1,410.00
4	Monthly deposit	312.00
5	Desired future value	26,340.00
6	Annual rate in %	4.0
7	Number of months	71

The spreadsheet depicted in Table 1.3 was generated using the VBA Excel module **NumberOfMonthlyDeposits**. The relevant data is entered in cells B3–B6 and the number of months then appears in cell B7. The reader may check that about three more months are needed if no initial deposit is made.

Withdrawals

For withdrawal annuities we argue as follows: Let A_0 be the initial value of the account and let A_n denote the value of the account immediately after the nth withdrawal. The value just *before* withdrawal is A_{n-1} plus the interest over that period. Making the withdrawal reduces that value by P and so

$$A_n = aA_{n-1} - P, \quad a := 1 + i.$$

Iterating, we obtain

$$A_n = a\big(aA_{n-2} - P\big) - P = a^2 A_{n-2} - (1 + a)P$$

and eventually

$$A_n = a^n A_0 - (1 + a + \cdots + a^{n-1})P.$$

Summing the geometric series yields

$$A_n = (1 + i)^n A_0 + P\,\frac{1 - (1 + i)^n}{i}. \tag{1.5}$$

Now assume that the account is drawn down to zero after N withdrawals. Setting $n = N$ and $A_N = 0$ in (1.5) and solving for A_0 we obtain, after simplification,

$$A_0 = P\,\frac{1 - (1 + i)^{-N}}{i}. \tag{1.6}$$

This is the initial deposit required to support exactly N withdrawals of size P from, say, a retirement account or trust fund. It may be seen as the sum of the present values of the N withdrawals. Solving for P in (1.6) we obtain the formula

$$P = A_0\,\frac{i}{1 - (1 + i)^{-N}}, \tag{1.7}$$

which may be used, for example, to calculate the mortgage payment for a mortgage of size A_0 (see below). Substituting (1.7) into (1.5) we obtain the following formula for the time-n value of an annuity supporting exactly N withdrawals:

$$A_n = A_0\,\frac{1 - (1 + i)^{n-N}}{1 - (1 + i)^{-N}}. \tag{1.8}$$

The following graph of A_n against time was generated by the module **WithdrawalAnnuityGraph** using the formula in (1.5). The module shows that an account with an annual rate of 7% and initial value \$4000 supports 465 withdrawals of \$25.00 each and a single payment of \$14.67.

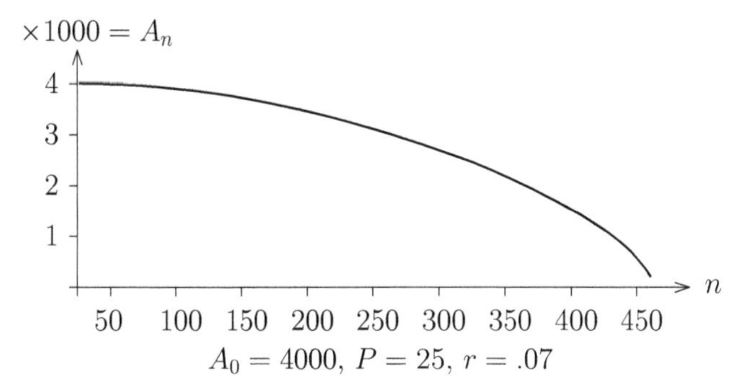

FIGURE 1.2: Withdrawal annuity graph.

We remind the reader that the annuity formulas (1.4), (1.5), and (1.8) assume that the compounding interval and the payment interval are the same, and that payment is made at the *end* of the compounding interval, thus describing what is called an *annuity-immediate*. If payments are made at the *beginning* of the period, as is the case, for example, with rents or insurance, one obtains an *annuity-due*, and the formulas change accordingly. (See Exercise 27.)

Application: Retirement

Suppose you make monthly deposits of size P into a retirement account with an annual rate r, compounded monthly. After t years you transfer the account to one that pays an annual rate of r' compounded monthly and you make monthly withdrawals of size Q from the new account for s years, drawing down the account to zero. This plan requires that the final value of the first

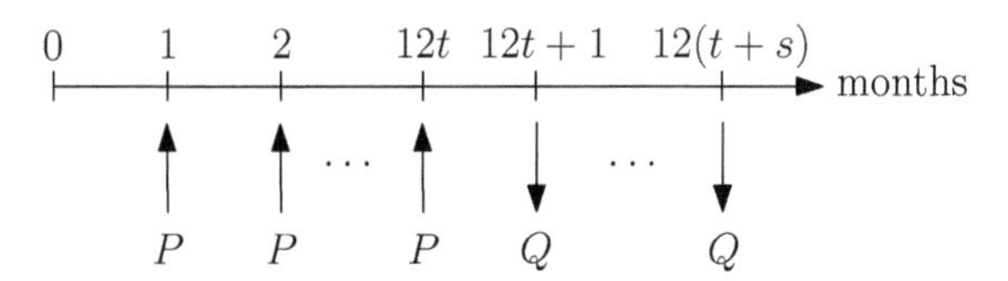

FIGURE 1.3: Retirement plan.

account is the initial value of the second. Thus, by (1.4) and (1.6),

$$P\,\frac{(1+i)^{12t}-1}{i} = Q\,\frac{1-(1+i')^{-12s}}{i'}, \qquad i := \frac{r}{12},\ i' := \frac{r'}{12}.$$

Solving for P we obtain the formula

$$P = \frac{i}{i'}\,\frac{1-(1+i')^{-12s}}{(1+i)^{12t}-1}\,Q. \tag{1.9}$$

A more realistic analysis takes into account the reduction of purchasing power due to inflation. Suppose that yearly inflation is running at $r_f \times 100\%$. In order for the nth withdrawal to have a value Q in *current* dollars, the amount of the withdrawal must be $Q(1+r_f/12)^{12t+n}$. We then have the modified scheme shown in Figure 1.4.

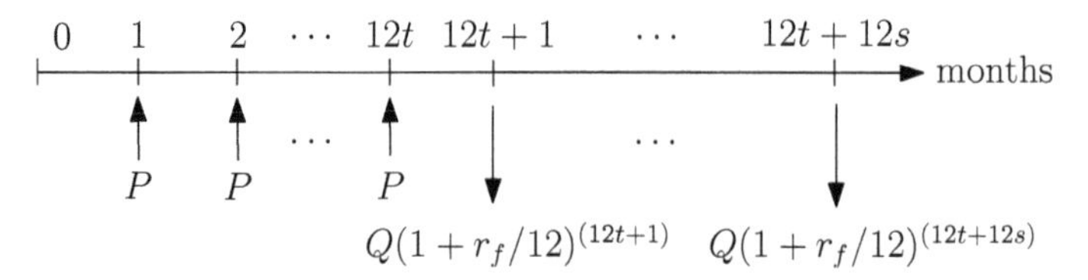

FIGURE 1.4: Retirement plan with inflation.

The final value of the first account must now satisfy

$$P\,\frac{(1+i)^{12t}-1}{i}=\sum_{n=1}^{12s} Q(1+i_f)^{12t+n}(1+i')^{-n}=Q(1+i_f)^{12t}\sum_{n=1}^{12s}\left(\frac{1+i_f}{1+i'}\right)^n.$$

Summing the geometric series and rearranging the resulting equation we obtain the formula

$$\frac{P}{Q}=\left(\frac{i(1+i_f)^{12t}}{(1+i)^{12t}-1}\right)\left(\frac{1-(1+i_f)^{12s}(1+i')^{-12s}}{i'-i_f}\right). \tag{1.10}$$

Note that if $r_i=0$, then (1.10) reduces to (1.9), as of course it should.

Table 1.4 below depicts a spreadsheet generated by the module **RetirementPlan**. Cells B4 and D4 contain, respectively, the number of payment years t and the number of withdrawal years s. Cell B3 contains the inflation rate and cells B5 and D5 contain the nominal percentage rates during the these years.

TABLE 1.4: Input/Output for **RetirementPlan**.

	A	B	C	D
3	Inflation rate	3		
4	Deposit years	40	Withdrawal years	30
5	Deposit annual rate	7	Withdrawal interest rate	5
6	Number of deposits	480	Number of withdrawals	360
7	Desired withdrawal Q	5000	Required deposit P	1705.41

The desired withdrawal Q in current dollars is entered in B7 and the required monthly deposit P is generated in D7. The table shows that with an inflation rate of 3%, monthly deposits of \$1705.41 are required to generate future withdrawals of \$5000. The reader may check by running the program that with no inflation the monthly deposits are only \$355, a reduction of over \$1300.

Application: Amortization

Suppose a home buyer takes out a mortgage in the amount A_0 for t years at an annual rate r, compounded monthly. From the point of view of the mortgage company, the monthly mortgage payments P constitute a withdrawal annuity with "initial deposit" A_0 and so are given by (1.7) with $i = r/12$ and $N = 12t$. The amount A_n still owed by the homeowner at the end of month n is given by (1.8).

Now let I_n and P_n denote, respectively, the portions of the nth payment that are interest and principle. Since A_{n-1} was owed at the end of month $n-1$, $I_n = iA_{n-1}$, hence by (1.8) and the formula $P_n = P - I_n$, we have

$$I_n = iA_0 \frac{1-(1+i)^{n-1-N}}{1-(1+i)^{-N}} \quad \text{and} \quad P_n = iA_0 \frac{(1+i)^{n-1-N}}{1-(1+i)^{-N}}. \tag{1.11}$$

The sequences (A_n), (P_n), and (I_n) form what is called an *amortization schedule* From (1.11) we see that

$$\frac{I_n}{P_n} = \frac{1-(1+i)^{n-1-N}}{(1+i)^{n-1-N}} = \frac{(1+i)^{N+1}}{(1+i)^n} - 1.$$

The ratio clearly decreases as n increases: as time passes more and more of the payment goes to paying off the principle. The maximum and minimum ratios are, respectively, $I_1/P_1 = (1+i)^N - 1$ and $I_N/P_N = i$.

For a concrete example, consider the following table, which depicts a spreadsheet generated by the module **AmortizationSchedule.**

TABLE 1.5: Input/output for **AmortizationSchedule.**

	A	B	C	D	E
3	Amount	Term	Annual Rate	Extra	
4	200,000	20	4.5%	100	
5	Month	Payment	Principle	Interest	Balance
6	1	1,365.30	615.30	750.00	199,384.70
7	2	1,365.30	617.61	747.69	198,767.10
⋮	⋮	⋮	⋮	⋮	⋮
217	212	1,365.30	1,355.44	9.86	1,274.78
218	213	1,279.56	1,274.78	4.78	0.00

The program is based on the equations

$$P = \frac{iA_0}{1-(1+i)^{-N}} + E, \quad A_n = (1+i)A_{n-1} - P, \quad I_n = iA_{n-1}, \quad P_n = P - I_n,$$

where E is an (optional) extra amount paid each month. In this example,

$$A_0 = 200,000, \quad t = 20, \quad r = .045, \quad \text{and} \quad E = 100,$$

which are entered in cells A4, B4, C4, and D4, respectively. As the reader may check by running the program, the extra monthly payment of $100 reduces the total number of payments by 27.

Continuous Income Stream

Suppose that an account that pays a nominal interest rate of r receives money at a rate of $f(t)$ dollars per year over T years. Thus, during a brief period from t to $t+\Delta t$, the account receives approximately $f(t)\Delta t$ dollars, assuming that f does not change much during the time interval (a valid assumption for continuous f). The present value of this amount is then $e^{rt}f(t)\Delta t$. Summing these amounts and taking limits as $\Delta t \to 0$ produces the integral $\int_0^T e^{-rt}f(t)\,dt$. Thus we have the following formula for the *present value of an income stream over T years*:

$$P = \int_0^T e^{-rt}f(t)\,dt. \tag{1.12}$$

1.4 Bonds

A bond is a financial contract issued by governments, corporations, or other institutions. It requires that the holder be reimbursed a specific amount at a prescribed time in the future. The simplest is the *zero coupon bond*, of which U.S. Treasury bills and U.S. savings bonds are common examples. Here, the purchaser of a bond pays an amount B_0 (which is frequently determined by bids) and receives a prescribed amount F, the *face value* or *par value* of the bond, at a prescribed time T, called the *maturity date*. The value B_t of the bond at time t may be expressed in terms of a continuously compounded interest rate r determined by the equation

$$B_0 = Fe^{-rT}.$$

Here, r is computed by the formula $r = [\ln(F) - \ln(B_0]/T$. The value B_t is then the face value of the bond discounted to time t:

$$B_t = Fe^{-r(T-t)} = B_0e^{rt}, \quad 0 \le t \le T.$$

Thus, during the time interval $[0,T]$, the bond acts like a savings account with continuously compounded interest.

The time restriction $0 \le t \le T$ may be theoretically removed as follows: At time T, reinvest the proceeds F from the bond by buying F/B_0 bonds, each for the amount B_0 and each with par value F and maturity date $2T$. At time $t \in [T, 2T]$ each bond has value $Fe^{-r(2T-t)} = Fe^{-rT}e^{-r(T-t)} = B_0e^{-rT}e^{rt}$, so the bond account has value

$$B_t = (F/B_0)B_0e^{-rT}e^{rt} = Fe^{-rT}e^{rt} = B_0e^{rt}, \quad T \le t \le 2T.$$

Continuing this process, we see that the formula $B_t = B_0 e^{rt}$ holds for all times $t \geq 0$, assuming par values are unchanged.

With a *coupon bond* one receives not only the amount F at time T but also a sequence of payments, called *coupons*, during the life of the bond. Thus, at prescribed times $t_1 < t_2 < \cdots < t_N$, the bond pays an amount C_n, and at maturity T one receives the face value F, as illustrated in the figure.

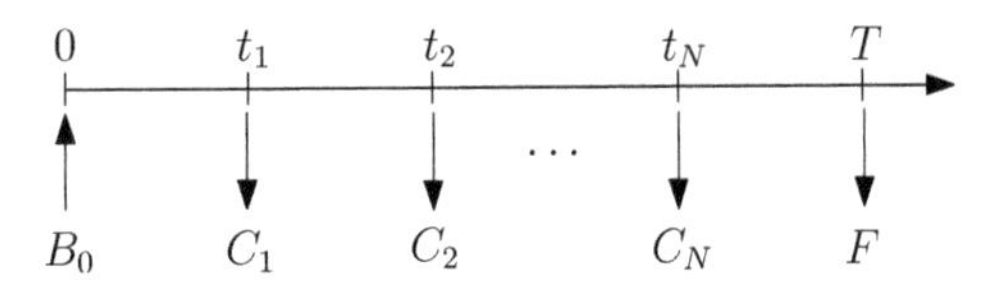

FIGURE 1.5: Coupons.

The price of the bond is the total present value

$$B_0 = \sum_{n=1}^{N} e^{-rt_n} C_n + F e^{-rT}, \tag{1.13}$$

which may be seen as the initial value of a portfolio consisting of $N + 1$ zero-coupon bonds maturing at times $t_1, t_2, \ldots, t_N$, and T.

---

*1.5 Internal Rate of Return

Internal rate of return is a measure used to evaluate a sequence of financial events. It is typically used to rank projects that involve future cash flows: the higher the rate of return the more desirable the project. For a mathematical model, consider a scenario that returns, for an initial investment of $P > 0$, an amount $A_n > 0$ at the end of periods $n = 1, 2, \ldots, N$, as illustrated in the diagram. Examples of such investments are annuities and coupon bonds. The

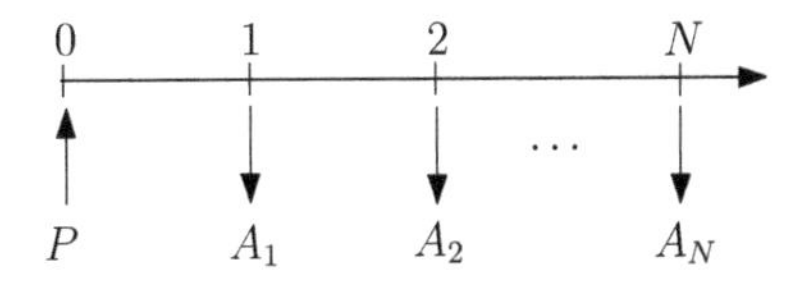

FIGURE 1.6: Investment and returns.

internal rate of return (IRR) of the investment P is defined as that periodic interest rate i for which the present value of the sequence of returns (under that rate) equals the initial payment P. Thus i satisfies the equation

$$P = \sum_{n=1}^{N} A_n (1 + i)^{-n}. \tag{1.14}$$

To see that Equation (1.14) has a unique solution $i > -1$, denote the right side by $R(i)$ and note that R is continuous on the interval $(-1, \infty)$ and satisfies

$$\lim_{i \to \infty} R(i) = 0 \quad \text{and} \quad \lim_{i \to -1^+} R(i) = \infty.$$

Since $P > 0$, the intermediate value theorem guarantees a solution $i > -1$ of the equation $R(i) = P$. Because R is strictly decreasing, the solution is unique.

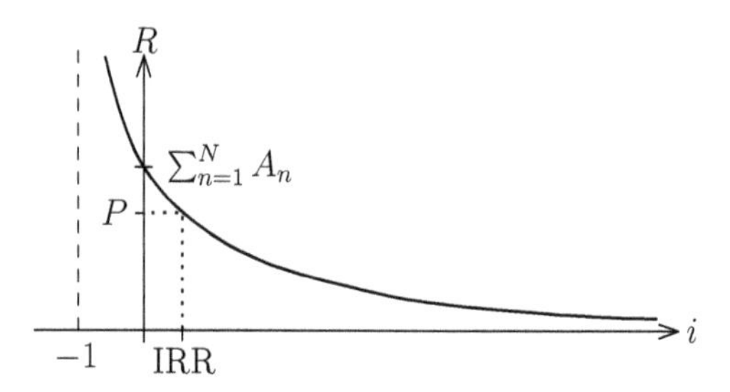

FIGURE 1.7: Rate of return.

Note that, because R is decreasing, $i > 0$ iff $R(i) < R(0)$, that is, iff $P < \sum_{n=1}^{N} A_n$. Thus for positive rates of return, the sum of the payoffs is greater than the initial investment and for negative rates of return, the sum of the payoffs is less than the initial investment.

To determine i, one can use Newton's method or an interval-halving procedure, or one can simply solve the equation by trial and error using a spreadsheet.

TABLE 1.6: Input/output for **RateOfReturn**.

	A	B	C
3	Initial Investment	Tolerance	Rate of return
4	100	.000001	7.66%
5	Returns		
6	18		
7	37		
8	39		
9	27		

Table 1.6 is based on a spreadsheet generated by the module **RateOfReturn**, which uses the interval-halving method. The initial investment and the sequence of returns are entered in column A; the rate of return appears in cell C4. The entry in cell B4 sets the accuracy of the output.

We remark that Excel has a function IRR that automatically calculates the internal rate of return (rounded). For this example, entering `= IRR(-100,18,37,39,27)` in a cell produces the desired output.

1.6 Exercises

1. What annual interest rate r would allow you to double your initial deposit in 6 years if interest is compounded quarterly? Continuously?

2. If you receive 6% interest compounded monthly, about how many years will it take for a deposit at time-0 to triple?

3. If you deposit \$400 at the end of each month into an account earning 8% interest compounded monthly, what is the value of the account at the end of 5 years? 10 years?

4. You deposit \$700 at the end of each month into an account earning interest at an annual rate of r compounded monthly. Use a spreadsheet to find the value of r that produces an account value of \$50,000 in 5 years.

5. You make an initial deposit of \$1000 at time-0 into an account with an annual rate of 5% compounded monthly and additionally you deposit \$400 at the end of each month. Use a spreadsheet to determine the minimum number of payments required for the account to have a value of at least \$30,000.

6. Suppose an account offers continuously compounded interest at an annual rate r and that a deposit of size P is made at the end of each month. Show that the value of the account after n deposits is

$$A_n = P\frac{e^{rn/12} - 1}{e^{r/12} - 1}.$$

7. Find a formula for the number N of monthly withdrawals needed to draw down to zero an account initially valued at A_0. Use the formula to determine how many withdrawals are required to draw down to zero an account with initial value \$200,000, if the account pays 6% compounded monthly.

8. An account pays an annual rate of 8% percent compounded monthly. What lump sum must you deposit into the account now so that in 10 years you can begin to withdraw \$4000 each month for the next 20 years, drawing down the account to zero?

9. A trust fund has an initial value of \$300,000 and earns interest at an annual rate of 6%, compounded monthly. If a withdrawal of \$5000 is made at the end of each month, use a spreadsheet to determine when the account will fall below \$150,000.

10. Referring to Equation (1.5), find the smallest value of A_0 in terms of P and i that will fund a *perpetual annuity*, that is, an annuity for which $A_n > 0$ for all n. What is the value of A_n in this case?

11. Suppose that an account offers continuously compounded interest at an annual rate r and that withdrawals of size P are made at the end of each month. If

the account is drawn down to zero after N withdrawals, show that the value of the account after n withdrawals is

$$A_n = P\,\frac{1 - e^{-r(N-n)/12}}{e^{r/12} - 1}.$$

12. In the retirement example, suppose that $t = 30$, $s = 20$, and $r = .12$. Find the payment amount P for withdrawals Q of \$3000 per month. If inflation is running at 2% per year, what value of P will give the first withdrawal the current purchasing power of \$3000? The last withdrawal?

13. For a 30-year, \$300,000 mortgage, use a spreadsheet to determine the annual rate you will have to lock in to have payments of \$1800 per month.

14. In the mortgage example, suppose that you must pay an inspection fee of \$1000, a loan initiation fee of \$1000, and 2 *points*, that is, 2% of the nominal loan of \$200,000. Effectively, then, you are receiving only \$194,000 from the lending institution. Use a spreadsheet to calculate the annual interest rate r' you will now be paying, given the agreed upon monthly payments of \$1667.85.

15. How large a loan can you take out at an annual rate of 15% if you can afford to pay back \$1000 at the end of each month and you want to retire the loan in 5 years?

16. Suppose you take out a 20-year, \$300,000 mortgage at 7% and decide after 15 years to pay off the mortgage. How much will you have to pay?

17. You can retire a loan either by paying off the entire amount \$8000 now, or by paying \$6000 now and \$6000 at the end of 10 years. Find a cutoff value r_0 such that if the nominal rate r is $< r_0$, then you should pay off the entire loan now, but if $r > r_0$, then it is preferable to wait. Assume that interest is compounded continuously.

18. You can retire a loan either by paying off the entire amount \$8000 now, or by paying \$6000 now, \$2000 at the end of 5 years, and an additional \$2000 at the end of 10 years. Find a cutoff value r_0 such that if the nominal rate r is $< r_0$, then you should pay off the entire loan now, but if $r > r_0$, then it is preferable to wait. Assume that interest is compounded continuously.

19. Referring the mortgage subsection, show that

$$P_n = (1+i)^{n-1}P_1, \quad \text{and} \quad I_n = \frac{1 - (1+i)^{n-N-1}}{1 - (1+i)^{-N}}I_1.$$

20. Suppose you take out a 30-year, \$100,000 mortgage at 6%. After 10 years, interest rates go down to 4%, so you decide to refinance the remainder of the loan by taking out a new 20-year mortgage. If the cost of refinancing is 3 points (3% of the new mortgage amount), what are the new payments? What threshold interest rate would make refinancing fiscally unwise? (Assume that the points are rolled in with the new mortgage.)

21. Referring to Section 1.4, find the time-t value B_t of a coupon bond for values of t satisfying $t_m \le t < t_{m+1}$ for $m = 0, 1, \ldots, N-1$, where $t_0 = 0$.

22. Use a spreadsheet to find the rate of return of a 4-year investment that, for an initial investment of \$1000, returns \$100, \$200, and \$300 at the end of years 1, 2, and 3, respectively, and, at the end of year 4, returns (a) \$350, (b) \$400, (c) \$550. Find the rates if the rate of return formula is based on continuously compounded interest.

23. Suppose for an initial investment of \$10,000, Plan A gives end-of-year returns \$3000, \$5000, \$7000, \$1000, and, for the same investment, Plan B gives end-of-year returns \$3500, \$4500, \$6500, \$1500. Determine which plan is best.

24. In Exercise 23, what is the smallest return in year 1 of Plan A that would make Plans A and B equally lucrative? Answer the same question for year 4.

25. You have the opportunity to invest \$10,000 in a scheme that would return \$3500 after the first year and \$2500 after each of years 2, 3, and 4. Suppose that your bank offers an annual rate of 6% compounded continuously. Should you invest in the scheme or deposit the money in your bank?

26. You can retire a loan either by paying off the entire amount \$10,000 now or by paying \$1400 per year for ten years, with the first payment made at the end of year one. Use a spreadsheet to find a cutoff value r_0 such that if the nominal rate r is $< r_0$, then you should pay off the entire loan now, but if $r > r_0$, then the second option is preferable. Assume that interest is compounded yearly.

27. Suppose that an account pays interest at an annual rate r compounded m times per year and that a withdrawal of size P is made at the *beginning* of each compounding interval (an *annuity-due*). Let $\widetilde{A}_n$ denote the value of the account at time n, that is, just before the nth withdrawal. (See figure.)

$$\begin{array}{cccc} 0 & 1 & \cdots\ n-1 & n \\ \widetilde{A}_0 & \widetilde{A}_1 & \widetilde{A}_{n-1} & \widetilde{A}_n \\ \downarrow & \downarrow & \cdots\ \downarrow & \downarrow \\ P & P & P & P \end{array}$$

Show that $\widetilde{A}_n = a^n \widetilde{A}_0 + \dfrac{aP}{i}(1 - a^n)$, where $a := 1 + i$.

28. An account pays compound interest at a rate of i per period. You decide to deposit an amount P at the end of each odd-numbered period and withdraw an amount $Q < P$ at the end of each even-numbered period. (See figure.)

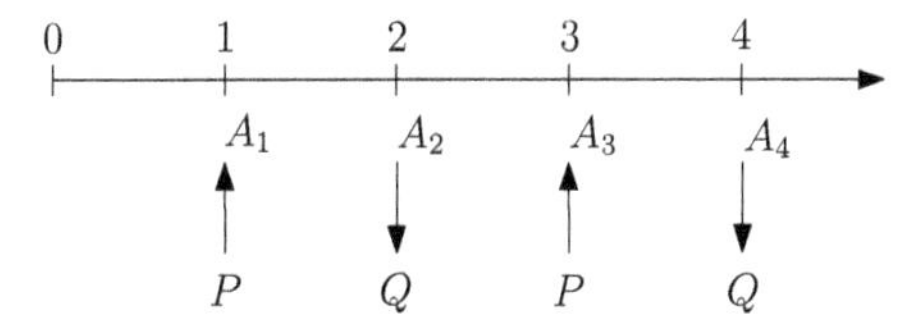

Find the value A_n of the account at time n immediately after the nth transaction. (You will have to consider cases $n = 2m$ and $n = 2m + 1$.)

29. Show that for a coupon bond that pays the amount C at each of the times $n = 1, 2, \ldots, N-1$ and pays the face value F at time N, the rate of return i satisfies
$$B_0 = C\frac{1-(1+i)^{-N}}{i} + F(1+i)^{-N}.$$

Chapter 2

Probability Spaces

Because financial markets are sensitive to a variety of unpredictable events, the value of a (risky) financial asset such as a stock or commodity is often interpreted as random, that is, subject to the laws of probability. Thus, while the future value of the asset cannot be predicted, measures of likelihood may be assigned to its possible values and from this one may deduce useful information. In this chapter we develop the probability theory needed to model the dynamic behavior of asset prices. The material here will require some familiarity with elementary set theory and basic combinatorics. A review of the relevant ideas may be found in Appendix A

2.1 Sample Spaces and Events

A *probability* is a number between 0 and 1 that expresses the likelihood of the occurrence of an event in an *experiment*. The experiment may be something as simple as the toss of a coin or as complex as the observation of stock prices over time. For our purposes, we shall consider an experiment to be any repeatable activity that produces observable outcomes. For example, tossing a die and noting the number of dots appearing on the top face is an experiment whose outcomes are the integers 1 through 6. Observing the average value of a stock over the previous week or noting the first time the stock dips below a prescribed level are experiments whose outcomes are nonnegative real numbers. Throwing a dart is an experiment whose outcomes may be viewed as the coordinates of the dart's landing position.

The set of all outcomes of an experiment is called the *sample space* of the experiment and is typically denoted by Ω. In probability theory, one starts with a given assignment of probabilities to certain subsets of Ω called *events*. This assignment must satisfy certain axioms and can be quite technical, depending on the sample space and the nature of the events. We begin with the simplest setting, that of a *discrete probability space*.

2.2 Discrete Probability Spaces

Definition and Examples

Consider an experiment whose outcomes may be represented by a finite or infinite sequence, say, ω_1, ω_2, ... Let p_n denote the probability of outcome ω_n. In practice, the determination of p_n may be based on relative frequency considerations, logical deduction, or analytical methods, and may be approximate or theoretical. For example, suppose that in a poll of 1000 people in a certain locality, 200 people prefer candidate A and 800 people prefer candidate B. If a person is chosen at random from the sample, then it is natural to assign a theoretical probability of .2 to the outcome that the person chosen prefers candidate A. If, however, the person is chosen from the general population, then pollsters take the probability of that outcome to be only approximately .2 and offer a margin of error. Similarly, if we flip a coin 10,000 times and find that exactly 5134 heads appear, we might assign the (approximate) statistical probability of $.5134 = 5134/10,000$ to the outcome that a single toss produces a head. On the other hand, for the "idealized coin" we would assign that same outcome a theoretical probability of .5. In this connection the reader may wish to use the module **CoinFlip**, which uses a random number generator to "flip a coin" with a user-specified theoretical probability. The outcome of one run is shown in Table 2.1. Note how well the statistical probability of a head approximates the theoretical probability if the number of flips is large. (This is an illustration of the Law of Large Numbers, discussed in Chapter 6.)

TABLE 2.1: Input/output for **CoinFlip**.

Theoretical probability of a single head	.6
Number of flips	1,000,000
Statistical probability of a single head	.599871

Whatever method is used to determine the probabilities p_n of an experiment, the resulting assignment is required to satisfy the following conditions:

$$0 \leq p_n \leq 1 \quad \text{and} \quad \sum_n p_n = 1.$$

The finite or infinite sequence $(p_1, p_2, \ldots)$ is called the *probability vector* or *probability distribution* for the experiment. The *probability* $\mathbb{P}(A)$ of a subset A of the sample space $\Omega = \{\omega_1, \omega_2, \ldots\}$ is then defined as

$$\mathbb{P}(A) = \sum_{\omega_n \in A} p_n, \tag{2.1}$$

where the notation indicates that the sum is taken over all indices n for which the outcome ω_n lies in A. The sum is either finite or a convergent infinite series.

(If $A = \emptyset$, the sum is interpreted as having the value zero.) The function $\mathbb{P}$ is called a *probability measure* for the experiment and the pair $(\Omega, \mathbb{P})$ is called a *discrete probability space*. The following proposition summarizes the basic properties of $\mathbb{P}$. We omit the proof.

2.2.1 Proposition. *For a discrete probability space,*

(a) $0 \le \mathbb{P}(A) \le 1$.

(b) $\mathbb{P}(\emptyset) = 0$ *and* $\mathbb{P}(\Omega) = 1$.

(c) *If* $A_1, A_2, \ldots$ *is a finite or infinite sequence of pairwise disjoint subsets of* Ω*, then*

$$\mathbb{P}\Big(\bigcup_n A_n\Big) = \sum_n \mathbb{P}(A_n).$$

Part (c) of the proposition is called the *additivity property* of $\mathbb{P}$. As we shall see, it is precisely the property needed to justify taking limits of random quantities.

2.2.2 Example. The experiment of tossing a pair of distinguishable dice and observing the sum of the dots on the top faces has sample space $\Omega = \{2, \ldots, 12\}$. The figure below gives the probability vector of the experiment in the form of

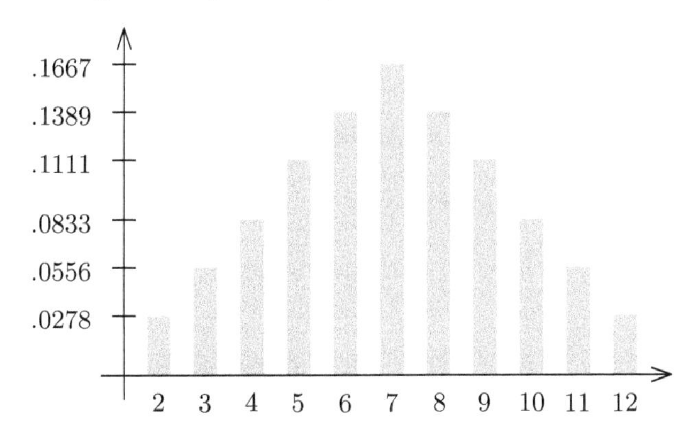

FIGURE 2.1: Dice sum distribution.

a column chart. The distribution is calculated by considering the number of ways each outcome $2, 3, \ldots, 12$ can occur and then dividing that number by 36. The chart was generated by the module **DiceToss**. The module also generates a statistical distribution (histogram) by simulating dice tosses using a random number generator. ◊

2.2.3 Example. There are 10 slips of paper in a hat, two of which are labeled with the number 1, three with the number 2, and five with the number 3. A slip is drawn at random from the hat, the label is noted, the slip is returned, and the process is repeated a second time. The sample space of the experiment is the set of all ordered pairs (j,k), where j is the number on the first slip and k is the number on the second. The event A that the sum of the numbers on the slips equals 4 consists of the outcomes $(1,3)$, $(3,1)$, and $(2,2)$. By relative frequency arguments, the probabilities of these outcomes are, respectively, .1, .1, and .09, hence, by the additivity property, $\mathbb{P}(A) = .29$. ◊

2.2.4 Example. Toss a fair coin infinitely often, producing an infinite sequence of heads H and tails T, and observe the first time an H occurs. The sample space is then $\Omega = \{0,1,2,3,\ldots\}$, where, for example, the outcome 2 means that the first toss comes up T and the second toss H, while the outcome 0 means that H never appears. To find the probability vector $(p_0, p_1, \ldots)$ for the experiment, we argue as follows: Since on the first toss the outcomes H or T are equally likely, we should set $p_1 = 1/2$. Similarly, the outcomes HH, HT, TH, TT of the first two tosses are equally likely, hence p_2, the probability that TH occurs, should be $1/4$. In general, we see that we should set $p_n = 2^{-n}$, $n \geq 1$. By additivity, the probability that a head eventually appears is

$$\sum_{n=1}^{\infty} p_n = \sum_{n=1}^{\infty} 2^{-n} = 1,$$

from which it follows that $p_0 = 0$. The probability vector for the experiment is therefore $\left(0, 2^{-1}, 2^{-2}, \ldots\right)$. Now let A be the event that the first H occurs on an even-numbered toss. Then A is the union of the pairwise disjoint events of the form

$$A_n = \text{"the first } H \text{ occurs on toss } 2n\text{"}, \quad n = 1, 2, \ldots$$

which may be described pictorially as

$$\overbrace{T \cdots T\, H}^{2n} * \; * \cdots$$

Since $\mathbb{P}(A_n) = 2^{-2n}$, we see from the additivity property that

$$\mathbb{P}(A) = \sum_{n=1}^{\infty} 2^{-2n} = 1/3. \qquad ◊$$

Equal Likelihood Probability Spaces

In the important special case where the sample space Ω is finite and each outcome is equally likely, $p_n = 1/|\Omega|$, hence (2.1) reduces to

$$\mathbb{P}(A) = \frac{|A|}{|\Omega|}, \quad A \subseteq \Omega,$$

where $|A|$ denotes the number of elements in a finite set A. The determination of probabilities is then simply a combinatorial problem.

2.2.5 Example. A poker deck consists of 52 cards with 13 denominations labeled 2 – 10, Jack, Queen, King, and Ace. Each denomination comes in four suits: hearts, diamonds, clubs, and spades. A poker hand consists of five randomly chosen cards from the deck. The total number of poker hands is therefore

$$\binom{52}{5} = 2,598,960.$$

We use simple counting arguments to show that three of a kind beats two pairs. (For example, (2, 2, 2, 9, 7) beats (A, K, K, Q, Q).)

By the multiplication principle (Appendix A), the number of poker hands consisting of three of a kind is

$$13 \cdot 4 \cdot \left(\frac{48 \cdot 44}{2}\right) = 54,912,$$

corresponding to the process of choosing a denomination for the triple, selecting three cards from that denomination, and then choosing the remaining two cards, avoiding the selected denomination as well as pairs. (The divisor 2 is needed since, for example, the choice $5, 8$ is the same as the choice $8, 5$.) The probability of getting a hand with three of a kind is therefore

$$\frac{54,912}{2,598,960} \approx .02113.$$

Similarly, the number of hands with two (distinct) pairs is

$$\binom{13}{2} \cdot \binom{4}{2}^2 \cdot 44 = 123,552,$$

corresponding to the process of choosing denominations for the pairs, choosing two cards from each of the denominations, and then choosing the remaining card, avoiding the selected denominations. The probability of getting a hand with two pairs is therefore

$$\frac{123,552}{2,598,960} \approx .04754,$$

more than twice the probability of getting three of a kind. ◊

2.3 General Probability Spaces

As we have seen, for a discrete probability space it is possible to assign a probability to each set of outcomes, that is, to each subset of the sample space Ω. In more general circumstances, this sort of flexibility may not be possible, and the assignment of probabilities must be narrowed to a suitably restricted collection of subsets of Ω called an *event σ-field.*

The σ-field of Events

A collection $\mathcal{F}$ of subsets of a sample space Ω is called a *σ-field* if it has the following properties:

(a) $\emptyset, \Omega \in \mathcal{F}$.

(b) $A \in \mathcal{F} \Rightarrow A' \in \mathcal{F}$.

(c) For any finite or infinite sequence of members A_n of $\mathcal{F}$, $\bigcup_n A_n \in \mathcal{F}$.

The members of $\mathcal{F}$ are called *events*. They may be viewed as the "observable outcomes" of the experiment.

The above properties are necessary for a robust and widely applicable mathematical theory. Property (a) asserts that the "sure event" Ω and the "impossible event" $\emptyset$ are always members of $\mathcal{F}$. Property (c) asserts that $\mathcal{F}$ is closed under countable unions. By (b), (c), and De Morgan's law,

$$\bigcap_n A_n = \left(\bigcup_n A_n'\right)',$$

hence $\mathcal{F}$ is also closed under countable intersections.

The trivial collection $\{\emptyset, \Omega\}$ and the collection of all subsets of Ω are examples of σ-fields. Here are more interesting examples.

2.3.1 Example. Let Ω be a finite, nonempty set and let $\mathcal{P}$ be a *partition* of Ω, that is, a collection of pairwise disjoint, nonempty sets with union Ω. The collection consisting of $\emptyset$ and all possible unions of members of $\mathcal{P}$ is a σ-field. For example, to illustrate property (b) of the above definition, suppose that $\mathcal{P} = \{A_1, A_2, A_3, A_4\}$. The complement of $A_1 \cup A_3$ is then $A_2 \cup A_4$. ◇

2.3.2 Example. Let $\mathcal{A}$ be a collection of subsets of Ω. In general, there are many σ-fields $\mathcal{F}$ that contain $\mathcal{A}$, an obvious one being $\mathcal{P}(\Omega)$. It is easy to see that the intersection of all σ-fields $\mathcal{F}$ containing $\mathcal{A}$ is again a σ-field containing $\mathcal{A}$. It is called the *σ-field generated by* $\mathcal{A}$ and is denoted by $\sigma(\mathcal{A})$. If Ω is finite and $\mathcal{A}$ is a partition of Ω, then $\sigma(\mathcal{A})$ is the σ-field of Example 2.3.1. If Ω is an interval of real numbers and $\mathcal{A}$ is the collection of all subintervals of Ω, then $\sigma(\mathcal{A})$ is called the *Borel σ-field* of Ω and its members the *Borel sets* of Ω. ◇

A σ-field $\mathcal{F}$ of events may be thought of as representing the available information in an experiment, information that is known only after an outcome of the experiment has been observed. For example, if we are contemplating buying a stock at time t, then the essential information available to us (barring insider information) is the price history of the stock up to time t. We show later that this information may be conveniently described by a time-dependent σ-field $\mathcal{F}_t$.

Once a sample space Ω and a σ-field of events have been specified, the next step is to assign probabilities. This is done in accordance with the following set of axioms, which may be seen as motivated by Proposition 2.2.1.

The Probability Measure

Let Ω be a sample space and $\mathcal{F}$ a σ-field of events. A *probability measure* for $(\Omega, \mathcal{F})$, or a *probability law* for the experiment, is a function $\mathbb{P}$ which assigns to each event $A \in \mathcal{F}$ a number $\mathbb{P}(A)$, called the *probability of* A, such that the following properties hold:

(a) $0 \leq \mathbb{P}(A) \leq 1$.

(b) $\mathbb{P}(\emptyset) = 0$ and $\mathbb{P}(\Omega) = 1$.

(c) If $A_1, A_2, \ldots$ is a finite or infinite sequence of pairwise disjoint events, then

$$\mathbb{P}\Big(\bigcup_n A_n\Big) = \sum_n \mathbb{P}(A_n).$$

The triple $(\Omega, \mathcal{F}, \mathbb{P})$ is then called a *probability space.*

A collection of events is said to be *mutually exclusive* if $\mathbb{P}(AB) = 0$ [1] for each pair of distinct members A and B in the collection. Pairwise disjoint sets are obviously mutually exclusive, but not conversely. It may be shown that the additivity axiom (c) holds more generally for mutually exclusive events A_n.

2.3.3 Proposition. *A probability measure* $\mathbb{P}$ *has the following properties:*

(a) $\mathbb{P}(A \cup B) = \mathbb{P}(A) + \mathbb{P}(B) - \mathbb{P}(AB)$.

(b) *If* $B \subseteq A$, *then* $\mathbb{P}(A - B) = \mathbb{P}(A) - \mathbb{P}(B)$; *in particular* $\mathbb{P}(B) \leq \mathbb{P}(A)$.

(c) $\mathbb{P}(A') = 1 - \mathbb{P}(A)$.

Proof. For (a) we note that $A \cup B$ is the union of the pairwise disjoint events AB', AB, and $A'B$. Therefore, by additivity,

$$\mathbb{P}(A \cup B) = \mathbb{P}(AB') + \mathbb{P}(AB) + \mathbb{P}(A'B). \tag{†}$$

Similarly,

$$\mathbb{P}(A) = \mathbb{P}(AB') + \mathbb{P}(AB) \quad \text{and} \quad \mathbb{P}(B) = \mathbb{P}(A'B) + \mathbb{P}(AB).$$

Adding the last two equations we have

$$\mathbb{P}(A) + \mathbb{P}(B) = \mathbb{P}(AB') + \mathbb{P}(A'B) + 2\mathbb{P}(AB). \tag{‡}$$

Subtracting equations (†) and (‡) yields property (a). Property (b) follows easily from additivity, as does (c) (using $\mathbb{P}(\Omega) = 1$). $\square$

From (a) of Proposition 2.3.3 we may deduce the important general *inclusion-exclusion rule*:

[1] For brevity we write AB for the intersection $A \cap B$. (See Appendix A.)

2.3.4 Proposition. *Let* $A = A_1 \cup A_2 \cup \cdots \cup A_n$*, where* $A_j \in \mathcal{F}$*. Then*

$$\mathbb{P}(A) = \sum_{i=1}^{n} \mathbb{P}(A_i) - \sum_{1 \le i < j \le n}^{n} \mathbb{P}(A_i A_j) + \sum_{1 \le i < j < k \le n}^{n} \mathbb{P}(A_i A_j A_k) - \cdots + (-1)^{n-1} \mathbb{P}(A_1 \cdots A_n).$$

Proof. (By induction on n). By (a) of Proposition 2.3.3,

$$\mathbb{P}(A_1 \cup \cdots \cup A_{n+1}) = \mathbb{P}(A \cup A_{n+1}) = \mathbb{P}(A) + \mathbb{P}(A_{n+1}) - \mathbb{P}(AA_{n+1}). \qquad (\dagger)$$

If we assume the proposition holds for n, then

$$\mathbb{P}(A) + \mathbb{P}(A_{n+1}) = \sum_{i=1}^{n+1} \mathbb{P}(A_i) - \sum_{1 \le i < j \le n}^{n} \mathbb{P}(A_i A_j) + \cdots + (-1)^{n-1} \mathbb{P}(A_1 \cdots A_n)$$

and

$$\mathbb{P}(AA_{n+1}) = \sum_{i=1}^{n} \mathbb{P}(A_i A_{n+1}) - \sum_{1 \le i < j \le n}^{n} \mathbb{P}(A_i A_j A_{n+1}) + \cdots + (-1)^{n-1} \mathbb{P}(A_1 \cdots A_n A_{n+1}).$$

Subtracting these equations and using (†) produces the desired formula for $\mathbb{P}(A_1 \cup \cdots \cup A_{n+1})$, completing the induction argument. □

Examples

We have seen several examples of discrete probability spaces. In this subsection we consider examples of non-discrete probability spaces. In each of these, the underlying experiment has a continuum of outcomes. As a consequence, the determination of the appropriate σ-field of events and the assignment of suitable probabilities requires some attention to technical detail.

2.3.5 Example. Consider the experiment of randomly choosing a real number from the interval $[0, 1]$. If we try to assign probabilities as in the discrete case, then we should assume that the outcomes x are equally likely and therefore set $\mathbb{P}(x) = p$ for some $p \in [0, 1]$. However, consider the event J that the number chosen is less than $1/2$. Following the discrete case, the probability of J should be $\sum_{x \in [0,1/2)} p$, which is either 0 or $+\infty$, if it has meaning at all.

Here is a more natural approach: Since we expect that half the time the number chosen will lie in the left half of the interval $[0, 1]$, we define $\mathbb{P}(J) = .5$. More generally, for any subinterval I, the probability that the selected number x lies in I should be the length of I, which is the theoretical proportion of times one would expect a random choice to lie in I. In this way every interval may be given a probability. Using this assignment of probabilities for intervals, one may show that every Borel subset of $[0, 1]$ (Example 2.3.2) may be assigned a probability consistent with the axioms of a probability space. Therefore, it is natural to take the event σ-field in this experiment to be the collection of all Borel sets.

As a concrete example, consider the event A that the selected number x has a decimal expansion $.d_1 d_2 d_3 \ldots$ with no digit d_j equal to 3. Set $A_0 = [0,1]$. Since $d_1 \neq 3$, A must be contained in the set A_1 obtained by removing from A_0 the interval $[.3, .4)$. Similarly, since $d_2 \neq 3$, A is contained in the set A_2 obtained by removing from A_1 the nine intervals of the form $[.d_1 3, .d_1 4)$, $d_1 \neq 3$. Having obtained A_{n-1} in this way, we see that A must be contained in the set A_n obtained by removing from A_{n-1} the 9^{n-1} intervals of the form $[.d_1 d_2 \ldots d_{n-1} 3, .d_1 d_2 \ldots d_{n-1} 4)$, $d_j \neq 3$. Since each of these intervals has length 10^{-n}, the additivity axiom implies that

$$\mathbb{P}(A_n) = \mathbb{P}(A_{n-1}) - 9^{n-1} 10^{-n} = \mathbb{P}(A_{n-1}) - (.1)(.9)^{n-1},$$

or

$$\mathbb{P}(A_n) - \mathbb{P}(A_{n-1}) = -(.1)(.9)^{n-1}.$$

Summing from 1 to N, acknowledging cancelations, we obtain

$$\mathbb{P}(A_N) - 1 = -(.1) \sum_{n=1}^{N} (.9)^{n-1} = (-.1) \frac{(.9)^N - 1}{.9 - 1} = (.9)^N - 1.$$

Therefore, $\mathbb{P}(A) \leq (.9)^N$ for all N, which implies that $\mathbb{P}(A) = 0$. Thus, with probability one, a number chosen randomly from the interval $[0,1]$ has the digit 3 somewhere in its decimal expansion (in fact, in infinitely many places). ◊

Here is a two–dimensional version of the preceding example.

2.3.6 Example. Consider a dartboard in the shape of a square with the origin of a coordinate system at the lower left corner and the point $(1,1)$ in the upper right corner. We throw a dart and observe the coordinates (x, y) of the landing spot. (If the dart lands off the board, we ignore the outcome.) The sample space of this experiment is $\Omega = [0,1] \times [0,1]$. Consider the region A below the curve $y = x^2$. The area of A is $1/3$, so we would expect that $1/3$ of the time the dart will land in A.[2] This suggests that we define the probability of the event A to be $1/3$. More generally, the probability of any "reasonable" region is defined as the area of that region. It turns out that probabilities may be assigned to all "two–dimensional" Borel subsets of Ω in a manner consistent with the axioms of a probability space. ◊

2.3.7 Example. In the coin tossing experiment of Example 2.2.4, we noted the first time a head appears. This gave us a discrete sample space consisting of the nonnegative integers. Suppose now we consider the entire sequence of

[2]This is borne out by Monte Carlo techniques, that is, computer simulations of an experiment repeated many times (see Chapter 3).

outcomes, giving us a sample space consisting of all sequences of H's and T's. That this sample space is not discrete may be seen as follows: Replace H and T by the digits 1 and 0, respectively, so that an outcome may be identified with the binary expansion of a number in the interval $[0,1]$. (For example, the outcome $THTHTHTH\ldots$ is identified with the number $.01010101\ldots = 1/3$.) The sample space of the experiment may therefore be identified with the interval $[0,1]$, which is uncountable.

To assign probabilities in this experiment, we begin by giving a probability of 2^{-n} to events that are described in terms of outcomes at n specified tosses. For example, the event A that H appears on the first and third tosses would have probability $1/4$. Note that under the above identification, the event A corresponds to the subset of $[0,1]$ consisting of all numbers with binary expansion beginning .101 or .111, which is the union of the intervals $[5/8, 3/4)$ and $[7/8, 1)$. The total length of these intervals is $1/4$, suggesting that the natural assignment of probabilities in this example is precisely that of Example 2.3.5 (which is indeed the case). ◊

2.4 Conditional Probability

Suppose we assign probabilities to the events A of an experiment and then learn that an event B has occurred. One would expect that this new information could have an effect on the original probabilities $\mathbb{P}(A)$. The (possibly) altered probability of an event A is called the *conditional probability of A given B* and is denoted by $\mathbb{P}(A \mid B)$. A precise mathematical definition of $\mathbb{P}(A \mid B)$ is suggested by the following example.

2.4.1 Example. Suppose that in a group of 100 people, exactly 40 people smoke, and that 15 of the smokers and 5 of the nonsmokers have lung cancer. A person is chosen at random from the group. Let A be the event that the person has lung cancer and let B denote the event that a person chosen is a nonsmoker. From the data we have $\mathbb{P}(A) = |A|/100 = .2$. Now suppose we are told that the person chosen is a nonsmoker, which is the event B has occurred. Then, in computing the new probability $\mathbb{P}(A \mid B)$ of A, we should replace Ω by the sample space B consisting of nonsmokers. This gives

$$\mathbb{P}(A \mid B) = |AB|/|B| = 5/60 = .083,$$

considerably smaller than the original probability of .2. ◊

Note that in the preceding example

$$\mathbb{P}(A \mid B) = \frac{|AB|}{|B|} = \frac{|AB|/|\Omega|}{|B|/|\Omega|} = \frac{\mathbb{P}(AB)}{\mathbb{P}(B)}.$$

This suggests the following definition: If A and B are events in a general probability space and $\mathbb{P}(B) > 0$, then the *conditional probability of A given B* is defined by

$$\mathbb{P}(A \mid B) = \frac{\mathbb{P}(AB)}{\mathbb{P}(B)}.$$

Note that $\mathbb{P}(A \mid B)$ is undefined if $\mathbb{P}(B) = 0$.

2.4.2 Example. In the dartboard experiment of Example 2.3.6 we assigned a probability of $1/3$ to the event A that the dart lands below the graph of $y = x^2$. Let B be the event that the dart lands in the left half of the board and C the event that the dart lands in the bottom half.

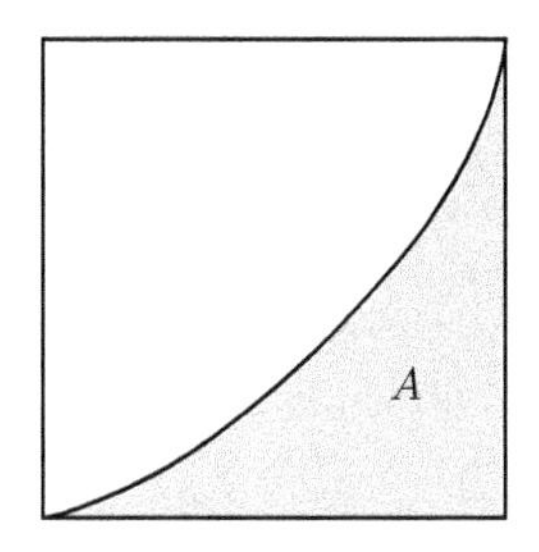

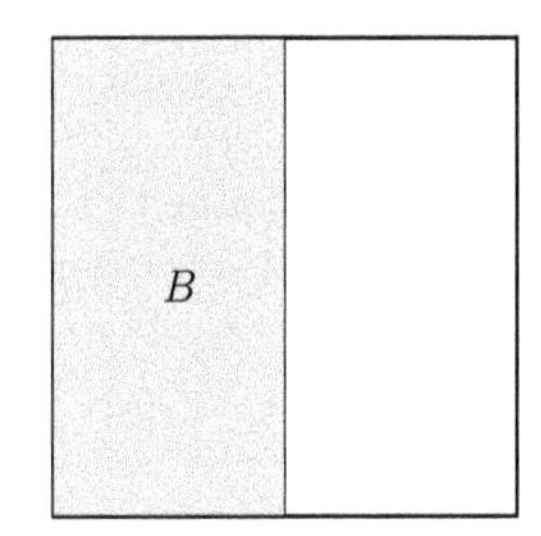

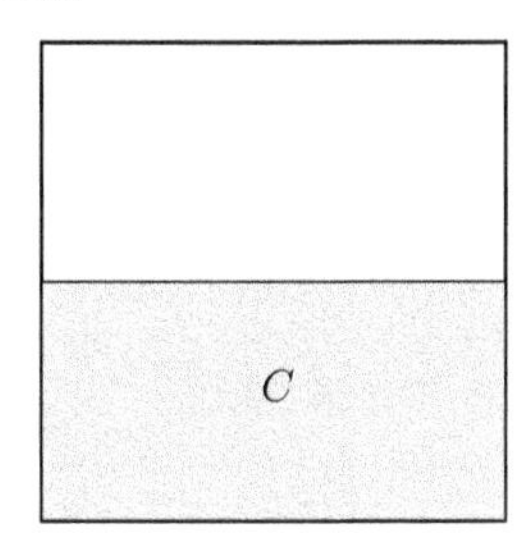

FIGURE 2.2: The events A, B, and C.

Recalling that probability in this experiment is defined as area, we see that

$$\mathbb{P}(B) = \mathbb{P}(C) = 1/2, \quad \mathbb{P}(AB) = 1/24, \quad \text{and} \quad \mathbb{P}(BC) = 1/4.$$

Therefore, $\mathbb{P}(A \mid B) = 1/12 \neq \mathbb{P}(A)$ and $\mathbb{P}(C \mid B) = 1/2 = \mathbb{P}(C)$. Thus knowledge of the event B changes the probability of A but not of C. ◊

2.4.3 Theorem (Multiplication Rule for Conditional Probabilities). *Suppose that $A_1, A_2, \ldots, A_n$ are events with $\mathbb{P}(A_1 A_2 \cdots A_{n-1}) > 0$. Then*

$$\mathbb{P}(A_1 A_2 \cdots A_n) = \mathbb{P}(A_1)\mathbb{P}(A_2 \mid A_1)\mathbb{P}(A_3 \mid A_1 A_2) \cdots \mathbb{P}(A_n \mid A_1 A_2 \cdots A_{n-1}). \tag{2.2}$$

Proof. (By induction on n.) The condition $\mathbb{P}(A_1 A_2 \cdots A_{n-1}) > 0$ ensures that the right side of (2.2) is defined. For $n = 2$, the equation follows from the definition of conditional probability. Suppose the equation holds for $n = k \geq 2$. If $A = A_1 A_2 \cdots A_k$, then, by definition of conditional probability,

$$\mathbb{P}(A_1 A_2 \cdots A_{k+1}) = \mathbb{P}(A A_{k+1}) = \mathbb{P}(A)\mathbb{P}(A_{k+1} \mid A).$$

Moreover, by the induction hypothesis,

$$\mathbb{P}(A) = \mathbb{P}(A_1)\mathbb{P}(A_2 \mid A_1)\mathbb{P}(A_3 \mid A_1 A_2) \cdots \mathbb{P}(A_k \mid A_1 A_2 \cdots A_{k-1}).$$

Combining these results we see that (2.2) holds for $n = k + 1$. □

2.4.4 Example. A jar contains 5 red and 6 green marbles. We randomly draw 3 marbles in succession without replacement. Let R_1 denote the event that the first marble is red, R_2 the event that the second marble is red, and G_3 the event that the third marble is green. The probability that the first two marbles are red and the third is green is

$$\mathbb{P}(R_1R_2G_3) = \mathbb{P}(R_1)\mathbb{P}(R_2 \mid R_1)\mathbb{P}(G_3 \mid R_1R_2) = (5/11)(4/10)(6/9) \approx .12.$$

The reader may wish to use the module **PolyaUrn** to generate more complex examples. ◊

2.4.5 Example. (Probability of a match). A deck of cards labeled with the numbers 1 to n is shuffled so that the cards are in random positions. What is the probability that, for some k, the card labeled k is in the kth position from the top of the deck?

Solution: Let A_k denote the event that the card labeled k is in the correct position. We seek the probability of the union $A_1 \cup A_2 \cup \ldots \cup A_n$. Now, for integers $1 \le i_1 < i_2 < \ldots < i_k \le n$, $A_{i_1}A_{i_2}\cdots A_{i_k}$ is the event that each of the cards labeled i_j is in the correct position. By the multiplication rule,

$$\begin{aligned}\mathbb{P}(A_{i_1}A_{i_2}\cdots A_{i_k}) &= \mathbb{P}(A_{i_1})\mathbb{P}(A_{i_2} \mid A_{i_1})\cdots\mathbb{P}(A_{i_k} \mid A_{i_1}A_{i_2}\cdots A_{i_{k-1}})\\ &= \frac{1}{n}\frac{1}{n-1}\cdots\frac{1}{n-k+1}\\ &= \frac{(n-k)!}{n!}.\end{aligned}$$

Since there are $\binom{n}{k}$ such sequences,

$$\sum_{1\le i_1<\ldots<i_k\le n} \mathbb{P}(A_{i_1}A_{i_2}\cdots A_{i_k}) = \binom{n}{k}\frac{(n-k)!}{n!} = \frac{1}{k!}.$$

Therefore, by Proposition 2.3.4,

$$\mathbb{P}\big(A_1 \cup \cdots \cup A_n\big) = 1 - \frac{1}{2} + \frac{1}{3!} - \cdots \frac{(-1)^{n-1}}{n!}.$$

Writing the series expansion $e^{-x} = 1 - x + x^2/2! - x^3/3! + \cdots$ in the form $1 - e^{-x} = x - x^2/2! + x^3/3! - \cdots$, we see that for large n the probability of a match is approximately $1 - 1/e = .6312\ldots$, a better than even chance. ◊

2.4.6 Theorem (Total Probability Law). *Let $B_1, B_2, \ldots$ be a finite or infinite sequence of mutually exclusive events whose union is Ω. If $\mathbb{P}(B_n) > 0$ for every n, then for any event A*

$$\mathbb{P}(A) = \sum_n \mathbb{P}(A \mid B_n)\mathbb{P}(B_n).$$

Proof. The events $AB_1, AB_2, \ldots$ are mutually exclusive with union A, hence

$$\mathbb{P}(A) = \sum_n \mathbb{P}(AB_n) = \sum_n \mathbb{P}(A \mid B_n)\mathbb{P}(B_n).$$ □

2.4.7 Example. (Investor's Ruin). Suppose you own a stock that each day increases by \$1 with probability p or decreases by \$1 with probability $q = 1-p$. Assume that the stock is initially worth x and that you intend to sell the stock as soon as its value is either a or b, whichever comes first, where $0 < a \le x \le b$. What is the probability that you will sell low?

Solution: Let $f(x)$ denote the probability of selling low, that is, of the stock reaching a before b, given that the stock starts out at x. Note that in particular $f(a) = 1$ and $f(b) = 0$. Let S_+ (S_-) be the event that the stock goes up (down) the next day and A the event of your selling low. Then $\mathbb{P}(A \mid S_+) = f(x+1)$, since if the stock goes up its value the next day is $x+1$. Similarly, $\mathbb{P}(A \mid S_-) = f(x-1)$. By the total probability law,

$$\mathbb{P}(A) = \mathbb{P}(A|S_+)\mathbb{P}(S_+) + \mathbb{P}(A|S_-)\mathbb{P}(S_-),$$

or, in terms of f,

$$f(x) = f(x+1)p + f(x-1)q.$$

Since $p + q = 1$, the last equation may be written

$$\Delta f(x) = r\Delta f(x-1), \quad r := \frac{q}{p},$$

where $\Delta f(z) := f(z+1) - f(z)$. Replacing x by $a + y$ and iterating, we have

$$\Delta f(a+y) = r\Delta f(a+y-1) = r^2\Delta f(a+y-2) = \cdots = r^y \Delta f(a),$$

hence

$$f(x) - f(a) = \sum_{y=0}^{x-a-1} \Delta f(a+y) = \Delta f(a) \sum_{y=0}^{x-a-1} r^y.$$

Since $f(a) = 1$, we obtain

$$f(x) = 1 + \Delta f(a) \sum_{y=0}^{x-a-1} r^y. \tag{2.3}$$

If $p = q$, then $r = 1$ and (2.3) reduces to $f(x) = 1 + (x-a)\Delta f(a)$. Setting $x = b$ and recalling that $f(b) = 0$, we have $\Delta f(a) = -1/(b-a)$ and hence

$$f(x) = 1 - \frac{x-a}{b-a} = \frac{b-x}{b-a}. \tag{2.4}$$

If $p \ne q$, then, by summing the geometric series in (2.3), we see that $f(x) = 1 + \Delta f(a)(r^{x-a} - 1)/(r-1)$. Setting $x = b$ and solving for $\Delta f(a)$ we obtain $\Delta f(a) = -(r-1)/\left(r^{b-a} - 1\right)$. Substituting this into the last equation gives

$$f(x) = 1 - \frac{r-1}{r^{b-a}-1}\,\frac{r^{x-a}-1}{r-1} = \frac{r^{b-a} - r^{x-a}}{r^{b-a}-1}. \tag{2.5}$$

Note that, by L'Hôpital's rule, the limit of the expression in (2.5) as $r \to 1$ (that is, as $p \to 1/2$) is precisely the expression in (2.4). ◊

Example 2.4.7 is a stock market version of what is called "gambler's ruin." The name comes from the standard formulation of the example, where the stock's value is replaced by winnings of a gambler. Selling low is then interpreted as "ruin."

The stock movement in this example is known as a *random walk.* We return to this notion later.

2.5 Independence

Events A and B in a probability space are said to be *independent* if

$$\mathbb{P}(AB) = \mathbb{P}(A)\mathbb{P}(B).$$

Note that if $\mathbb{P}(B) \neq 0$, then, dividing by $\mathbb{P}(B)$, we obtain the equivalent statement $\mathbb{P}(A \mid B) = \mathbb{P}(A)$, which asserts that the additional information provided by B is irrelevant in calculating the probability of A. Thus in Example 2.4.2 the events B and C are independent while A and B are not. Here are additional examples.

2.5.1 Example. Suppose in Example 2.4.4 that we draw two marbles in succession without replacement. Then, by the total probability law, the probability of getting a red marble on the second try is

$$\mathbb{P}(R_2) = \mathbb{P}(R_1)\mathbb{P}(R_2 \mid R_1) + \mathbb{P}(G_1)\mathbb{P}(R_2 \mid G_1) = \frac{5}{11}\frac{4}{10} + \frac{6}{11}\frac{5}{10} = \frac{5}{11}.$$

Since $\mathbb{P}(R_2 \mid R_1) = 4/10$, R_1 and R_2 are not independent. This agrees with our intuition, since drawing without replacement changes the configuration of the marbles in the jar. If instead we replace the first marble, then $\mathbb{P}(R_1R_2) = \mathbb{P}(R_1)\mathbb{P}(R_2)$: the events are independent. Note that in this experiment $\mathbb{P}(R_2) = \mathbb{P}(R_1)$ whether or not the marbles are replaced. This phenomenon holds generally (see Exercise 21). ◊

2.5.2 Example. Roll a fair die twice (or, equivalently, toss a pair of distinguishable dice once). A typical outcome can be described by the ordered pair (j, k), where j and k are, respectively, the number of dots on the upper face in the first and second rolls. Since the die is fair, each of the 36 outcomes has the same probability. Let A be the event that the sum of the dice is 7, B the event that the sum of the dice is 8, and C the event that the first die is even. Then $\mathbb{P}(AC) = 1/12 = \mathbb{P}(A)\mathbb{P}(C)$ but $\mathbb{P}(BC) = 1/12 \neq \mathbb{P}(B)\mathbb{P}(C)$. Thus the events A and C are independent, but B and C are not. ◊

Events in a collection $\mathcal{A}$ are said to be *independent* if for any n and any choice of $A_1, A_2, \ldots, A_n \in \mathcal{A}$,

$$\mathbb{P}(A_1A_2\cdots A_n) = \mathbb{P}(A_1)\mathbb{P}(A_2)\cdots\mathbb{P}(A_n).$$

This extends the definition of independence to arbitrarily many events.

2.5.3 Example. Toss a fair coin 3 times in succession and let A_j be the event that the jth coin comes up heads, $j = 1, 2, 3$. The events A_1, A_2, and A_3 are easily seen to be independent, which explains the use of the phrase "independent trials" in this and similar examples. ◊

2.6 Exercises

1. Show that $\mathbb{P}(A) + \mathbb{P}(B) - 1 \le \mathbb{P}(AB) \le \mathbb{P}(A \cup B) \le \mathbb{P}(A) + \mathbb{P}(B)$.

2. Let A and B be events with $\mathbb{P}(B) < 1$. Show that

$$\mathbb{P}(A' \mid B') = 1 + \frac{\mathbb{P}(AB) - \mathbb{P}(A)}{1 - \mathbb{P}(B)}.$$

3. Jack and Jill run up the hill. The probability of Jack reaching the top first is p, while that of Jill is q. They decide to have a tournament, the grand winner being the first one who wins 3 races. Find the probability that Jill wins the tournament. Assume that there are no ties ($p + q = 1$) and that the races are independent.

4. You have 6 history books, 5 English books, 4 mathematics books, and 3 chemistry books which you randomly place next to each other on a shelf. What is the probability that all books within the same subject will be adjacent?

5. An exam has parts A and B, both of which must be completed. In part A you can choose between answering 10 true-false questions or answering 8 multiple choice questions, each of which has 5 choices. In part B you can choose between answering 12 true-false questions or 6 multiple choice questions, 3 choices each. If one guesses on each question, what is the probability of getting a perfect score?

6. A group of women and men line up for a photograph. Find the probability that at least two people of the same gender are adjacent if there are (a) 5 men and 4 women; (b) 4 men and 4 women.

7. A "loaded" die has the property that the numbers 1 to 5 are equally likely but the number 6 is three times as likely to occur as any of the others. What is the probability that the die comes up even?

8. A *full house* is a poker hand with 3 cards of one denomination and 2 cards of another, for example, three kings and two jacks. Show that four of a kind beats a full house.

9. Balls are randomly thrown one at a time at a row of 30 open-topped jars numbered 1 to 30. Assuming that each ball lands in some jar, find the smallest number of throws so that there is a better than a 60% chance that at least two balls land in the same jar.

10. Toss a coin infinitely often and let p be the probability of a head appearing on any single toss ($0 < p < 1$). For $m \geq 2$, find the probability P_m that

    (a) a head appears on a toss that is a multiple of m.

    (b) the *first* head appears on a toss that is a multiple of m.

    (For example, in (a), P_2 is the probability that a head appears on an even toss, and in (b), P_2 is the probability that the *first* head appears on an even toss.)

    Show that in (b)

$$\lim_{m \to \infty} P_m = \lim_{p \to 1-} P_m = 0, \quad \text{and} \quad \lim_{p \to 0+} P_m = 1/m.$$

    Interpret probabilistically.

11. Toss a coin twice and let p be the probability of a head appearing on a single toss ($0 < p < 1$). Find the probability that

    (a) both tosses come up heads, given that at least one toss comes up heads.

    (b) both tosses come up heads, given that the first toss comes up heads.

    Can the probabilities in (a) and (b) ever be the same?

12. A jar contains $n - 1$ vanilla cookies and one chocolate cookie. You reach into the jar and choose a cookie at random. What is the probability that you will first get the chocolate cookie on the kth try if you (a) eat (b) replace each cookie you select. Show that if n is large compared to k then the ratio of the probabilities in (a) and (b) is approximately 1.

13. A hat contains six slips of paper numbered 1 through 6. A slip is drawn at random, the number is noted, the slip is replaced in the hat, and the procedure is repeated. What is the probability that after three draws the slip numbered 1 was drawn exactly twice, given that the sum of the numbers on the three draws is 8.

14. A jar contains 12 marbles: 3 reds, 4 greens, and 5 yellows. A handful of 6 marbles is drawn at random. Let A be the event that there are at least 3 green marbles and B the event that there is exactly 1 red. Find $\mathbb{P}(A|B)$. Are the events independent?

15. A number x is chosen randomly from the interval $[0, 1]$. Let A be the event that $x < .5$ and B the event that the second and third digits of the decimal expansion $.d_1d_2d_3\ldots$ of x are 0. Are the events independent? What if the inequality is changed to $x < .49$?

16. Roll a fair die twice. Let A be the event that the first roll comes up odd, B the event that the second roll is odd, and C the event that the sum of the dice is odd. Show that any two of the events A, B, and C are independent but the events A, B, and C are not independent.

17. Suppose that A and B are independent events. Show that in each case the given events are independent:

    (a) A and B'. (b) A' and B. (c) A' and B'.

18. The *odds for an event* E are said to be r *to* 1 if E is r times as likely to occur as E', that is, $\mathbb{P}(E) = r\mathbb{P}(E')$. *Odds* r *to* s means the same thing as odds r/s to 1, and *odds* r *to* s *against* means the same as odds s to r for. A bet of one dollar on an event E with odds r to s is *fair* if the bettor wins s/r dollars if E occurs and loses one dollar if E' occurs. (If E occurs, the dollar wager is returned to the bettor.)

    (a) If the odds for E are r to s, show that $\mathbb{P}(E) = \dfrac{r}{r+s}$.

    (b) Verify that a fair bet of one dollar on E returns $1/\mathbb{P}(E)$ dollars (including the wager) if E occurs.

    (c) Pockets of a roulette wheel are numbered 1 to 36, of which 18 are red and 18 are black. There are also green pockets numbered 0 and 00. If a \$1 bet is placed on black, the gambler wins \$2 (hence has a profit of \$1) if the ball lands on black, and loses \$1 otherwise (similarly for red). What should the winnings of the gambler be in order to make the wager fair?

19. Consider a race with only three horses, H_1, H_2, and H_3. Suppose that the odds *against* H_i winning are quoted as o_i to 1. If the odds are based solely on probabilities (determined by, say, statistics on previous races), then, by Exercise 18, the probability that horse H_i wins is $(1 + o_i)^{-1}$. Assuming there are no ties, the triple

    $$o := \left((1+o_1)^{-1}, (1+o_2)^{-1}, (1+o_3)^{-1}\right)$$

    is then a probability vector. However, it is typically the case that quoted odds are based on additional factors such as the distribution of wagers made before the race and profit margins for the bookmaker. In the following, the reader is asked to use elementary linear algebra to make a connection between quoted odds and betting strategies.

    (a) Suppose that for each i a bet of size b_i is made on H_i. The bets may be positive, negative, or 0. The vector $b = (b_1, b_2, b_3)$ is called a *betting strategy*. Show that if horse H_i wins, then the net winnings for the betting strategy b may be expressed as

    $$W_b(i) := (o_i + 1)b_i - (b_1 + b_2 + b_3).$$

    (b) A betting strategy b is said to be a *sure-win strategy*, or an *arbitrage*, if $W_b(i) > 0$ for each i. Show that there is a sure-win strategy iff there exist numbers $s_i < 0$ such that the system

    $$\begin{array}{rrrl} -o_1x_1 & +x_2 & +x_3 & = s_1 \\ x_1 & -o_2x_2 & +x_3 & = s_2 \\ x_1 & +x_2 & -o_3x_3 & = s_3 \end{array}$$

    has a solution $x = (x_1, x_2, x_3)$ (that solution being a sure-win betting strategy).

    (c) Let A be the coefficient matrix of the system in (b). Show that the determinant of A is $D := 2 + o_1 + o_2 + o_3 - o_1o_2o_3$.

(d) Suppose $D \neq 0$. Show that

$$A^{-1} = \frac{1}{D} \begin{bmatrix} o_2 o_3 - 1 & 1 + o_3 & 1 + o_2 \\ 1 + o_3 & o_1 o_3 - 1 & 1 + o_1 \\ 1 + o_2 & 1 + o_1 & o_1 o_2 - 1 \end{bmatrix}$$

and that, for any choice of negative numbers s_1, s_2, and s_3, the vector $A^{-1}s^T$ is a sure-win betting strategy, where s^T denotes the transpose of $s := (s_1, s_2, s_3)$.

(e) Show that if $D \neq 0$, then a sure-win betting strategy is

$$b = -\text{sgn}(D)(1 + o_2 + o_3 + o_2 o_3, 1 + o_1 + o_3 + o_1 o_3, 1 + o_1 + o_2 + o_1 o_2)$$

where $\text{sgn}(D)$ denotes the sign of D.

(f) Show that there is a sure-win betting strategy iff o is not a probability vector. (This assertion is a special case of the *Arbitrage Theorem*, a statement and proof of which may be found, for example, in [16].)

20. Verify the last assertion of Example 2.4.7.

21. An urn contains r red marbles and g green marbles. Suppose n marbles are drawn in succession without replacement, where $n < r + g$. In the notation of Example 2.4.4, show that $\mathbb{P}(R_n) = \mathbb{P}(R_1)$.

22. A jar contains g green and $2g$ red marbles. Three marbles are drawn in succession without replacement. Since there are twice as many reds as greens, it should be twice as likely that the last ball drawn is red rather than green. Verify this using the multiplication rule for conditional probabilities.

23. Referring to the dartboard experiment of Example 2.3.6, find the probability that dart lands above the graph of $y = 1 - x^2$ given that it lies below the graph of $y = x^2$. Are the events independent?

24. A jar contains 4 vanilla cookies, 6 strawberry cookies, and 3 chocolate cookies. A handful of six cookies is chosen at random. Let A be the event that there are at least 2 chocolate cookies and B the event that there are at least 3 vanilla cookies. Find $\mathbb{P}(A|B)$ and $P(B|A)$. Are the events A and B independent?

25. John and Mary order pizzas. The pizza shop offers only plain, anchovy, and sausage pizzas with no multiple toppings. The probability that John gets a plain (resp., anchovy) is .1 (resp., .2) and the probability that Mary gets a plain (resp., anchovy) is .3 (resp., .4). Assuming that John and Mary order independently, use Exercise 17 to find the probability that neither gets a plain but at least one gets an anchovy.

26. John and Mary order pizza from the same shop as in Exercise 25. The probability that John gets an anchovy is .1, the probability that Mary gets an anchovy is .2, and the probability that neither gets a plain but at least one gets an anchovy is .3. Assuming that John and Mary order independently, find the probability that at least one gets an anchovy but neither gets a sausage.

Chapter 3

Random Variables

3.1 Introduction

We have seen that outcomes of experiments are frequently real numbers. Such outcomes are called *random variables*. Before giving a formal definition, we introduce a convenient shorthand notation for describing sets involving real-valued functions X, Y etc. on Ω. The notation essentially leaves out the standard "$\omega \in \Omega$" part of the description, which is often redundant. Some typical examples are

$$\{X < a\} := \{\omega \in \Omega \mid X(\omega) < a\} \quad \text{and} \quad \{X \leq Y\} := \{\omega \in \Omega \mid X(\omega) \leq Y(\omega)\}.$$

We also use notation such as $\{X \in A,\ Y \in B\}$ for $\{X \in A\} \cap \{Y \in B\}$. For probabilities we write, for example, $\mathbb{P}(X < a)$ rather than $\mathbb{P}(\{X < a\})$. The following numerical example should illustrate the basic idea.

3.1.1 Example. The table below summarizes the distribution of grade-point averages for a group of 100 students. The first row gives the number of stu-

no. of students	7	13	19	16	12	10	8	6	5	4
grade pt. avg.	2.1	2.3	2.5	2.7	2.9	3.1	3.3	3.5	3.7	3.9

dents having the grade-point averages listed in the second row. If X denotes the grade-point average of a randomly chosen student, then, using the above notation, we see that the probability that a student has a grade point average greater than 2.5 but not greater than 3.3 is

$$\begin{aligned}\mathbb{P}(2.5 < X \leq 3.3) &= \mathbb{P}(X = 2.7) + \mathbb{P}(X = 2.9) + \mathbb{P}(X = 3.1) + \mathbb{P}(X = 3.3) \\ &= .16 + .12 + .10 + .08 = .46. \qquad \Diamond\end{aligned}$$

The preceding example suggests that is useful to be able to assign probabilities to sets of the form $\{X \in J\}$, where X is a numerical outcome of an experiment and J is an interval. Accordingly, we define a *random variable* on a probability space $(\Omega, \mathcal{F}, \mathbb{P})$ to be a function $X : \Omega \to \boldsymbol{R}$ such that for each interval J the set $\{X \in J\}$ is a member of $\mathcal{F}$. In case of ambiguity, we refer to X as an *$\mathcal{F}$-random variable* or say that X is *$\mathcal{F}$-measurable.*

An important example of a random variable, one that appears throughout the text, is the price of a stock at a prescribed time. The underlying probability space in this case must be carefully constructed and can take several forms. Particular concrete models are described in later chapters.

The following proposition can make the task of checking whether a function is a random variable considerably simpler.

3.1.2 Proposition. *A real-valued function X on a probability space $(\Omega, \mathcal{F}, \mathbb{P})$ is a random variable iff any one of the following conditions holds:*

(a) $\{X \le t\} \in \mathcal{F}$ *for all* $t \in \mathbf{R}$.

(b) $\{X < t\} \in \mathcal{F}$ *for all* $t \in \mathbf{R}$.

(c) $\{X \ge t\} \in \mathcal{F}$ *for all* $t \in \mathbf{R}$.

(d) $\{X > t\} \in \mathcal{F}$ *for all* $t \in \mathbf{R}$.

Proof. We illustrate by proving that if (a) holds, then $\{X \in (a, b)\} \in \mathcal{F}$ for any finite interval (a, b). This follows from the calculation

$$\{X \in (a,b)\} = \{X < b\} - \{X \le a\} = \bigcup_{n=1}^{\infty} \{X \le b - 1/n\} - \{X \le a\}$$

and the fact that $\mathcal{F}$ is closed under countable unions and set differences. Similar arguments show that each of the properties (a)–(d) implies that $\{X \in J\}$ is a member of $\mathcal{F}$ for all intervals J. □

Indicator Functions

The *indicator function* of a subset $A \subseteq \Omega$ is the function $\mathbf{1}_A$ on Ω defined by

$$\mathbf{1}_A(\omega) = \begin{cases} 1 & \text{if } \omega \in A, \\ 0 & \text{if } \omega \in A'. \end{cases}$$

3.1.3 Proposition. *Let $A \subseteq \Omega$. Then $\mathbf{1}_A$ is a random variable iff $A \in \mathcal{F}$.*

Proof. If $\mathbf{1}_A$ is a random variable, then $A = \{\mathbf{1}_A > 0\} \in \mathcal{F}$. Conversely, if $A \in \mathcal{F}$ and $a \in \mathbf{R}$, then

$$\{\mathbf{1}_A \le a\} = \begin{cases} \emptyset & \text{if } a < 0 \\ A' & \text{if } 0 \le a < 1, \text{ and} \\ \Omega & \text{if } a \ge 1. \end{cases}$$

Thus $\{\mathbf{1}_A \le a\} \in \mathcal{F}$ so, by Proposition 3.1.2, $\mathbf{1}_A$ is a random variable. □

The indicator function of an event provides a numerical way of expressing occurrence or non-occurrence of the event. It may be shown that random variables are, in a precise sense, completely generated by indicator functions.

A particularly important application of indicator functions occurs in the mathematical description of the payoff of a *barrier option*, where the function acts as an on-off switch, activating or deactivating the option depending on whether or not a certain barrier was breached by an asset. (See Chapter 4.)

3.2 General Properties of Random Variables

The following theorem describes an important technique for generating random variables.

3.2.1 Theorem. *Let $X_1,\dots, X_n$ be random variables. If $f(x_1,\dots,x_n)$ is a continuous function, then $f(X_1,\dots,X_n)$ is a random variable, where*

$$f(X_1,\dots,X_n)(\omega) := f(X_1(\omega),\dots,X_n(\omega)), \quad \omega\in\Omega.$$

Proof. We sketch the proof for the case $n = 1$, that is, for a single random variable X and a continuous function $f(x)$. For this, we use a standard result from real analysis which asserts that, because f is continuous, any set A of the form $\{x \mid f(x) < a\}$ is a union of a sequence of pairwise disjoint open intervals J_n. From this we conclude that

$$\{f(X) < a\} = \{X \in A\} = \bigcup_n \{X \in J_n\} \in \mathcal{F}.$$

Since a was arbitrary, $f(X)$ is a random variable. □

3.2.2 Corollary. *Let X and Y be random variables, $c \in \mathbf{R}$ and $p > 0$. Then $X+Y$, cX, XY, $|X|^p$, and X/Y (provided Y is never 0) are random variables.*

Proof. For example, to prove that X/Y is a random variable, apply the theorem to the continuous function $f(x,y) = x/y$, $y \neq 0$. □

The assertions of Corollary 3.2.2 regarding the sum and product of two random variables clearly extend to arbitrarily many random variables. Thus from Proposition 3.1.3 we have

3.2.3 Corollary. *A linear combination $\sum_{j=1}^n c_j \mathbf{1}_{A_j}$ of indicator functions $\mathbf{1}_{A_j}$ with $A_j \in \mathcal{F}$ is a random variable.*

3.2.4 Corollary. *If $\max(x,y)$ and $\min(x,y)$ denote, respectively, the larger and smaller of the real numbers x and y, and if X and Y are random variables, then $\max(X,Y)$ and $\min(X,Y)$ are random variables.*

Proof. The identity

$$\max(x,y) = y + (|x-y| + x - y)/2$$

shows that $\max(x,y)$ is continuous. Therefore, by Theorem 3.2.1, $\max(X,Y)$ is a random variable. A similar argument shows that $\min(X,Y)$ is a random variable (or one can use the identity $\min(x,y) = -\max(-x,-y)$). □

The Cumulative Distribution Function

The *cumulative distribution function* (cdf) of a random variable X is defined by

$$F_X(x) = \mathbb{P}(X \le x), \quad x \in \mathbf{R}.$$

A cdf can be useful in expressing certain probabilistic relations, as in the characterization of independent random variables described later.

3.2.5 Example. The cdf of the number X of heads that come up in three tosses of a fair coin is given by

$$F_X(x) = \begin{cases} 0 & \text{if } x < 0, \\ 1/8 & \text{if } 0 \le x < 1 \quad \text{(0 heads)}, \\ 1/2 & \text{if } 1 \le x < 2 \quad \text{(0 or 1 head)}, \\ 7/8 & \text{if } 2 \le x < 3 \quad \text{(0, 1, or 2 heads)}, \\ 1 & \text{if } x \ge 3 \quad \text{(0, 1, 2, or 3 heads)}. \end{cases}$$

The function may also be described by a linear combination of indicator functions:

$$F_X = \tfrac{1}{8}\mathbf{1}_{[0,1)} + \tfrac{1}{2}\mathbf{1}_{[1,2)} + \tfrac{7}{8}\mathbf{1}_{[2,3)} + \mathbf{1}_{[3,\infty)}. \qquad \diamond$$

3.3 Discrete Random Variables

A random variable X on a probability space $(\Omega, \mathcal{F}, \mathbb{P})$ is said to be *discrete* if the range of X is countable. The *probability mass function* (pmf) of a discrete random X is defined by

$$p_X(x) := \mathbb{P}(X = x), \quad x \in \mathbf{R}.$$

Since $p_X(x) > 0$ for at most countably many real numbers x, for an arbitrary subset A of $\mathbf{R}$, we may write

$$\mathbb{P}(X \in A) = \sum_{x \in A} p_X(x),$$

where the sum is either finite or a convergent infinite series (ignoring zero terms). In particular,

$$\sum_{x \in \mathbf{R}} p_X(x) = \mathbb{P}(X \in \mathbf{R}) = 1 \quad \text{and}$$

$$F_X(x) = \sum_{y \le x} p_X(x).$$

A linear combination of indicator functions of events is an example of a discrete random variable. In this regard, note that any random variable X with finite range may be written as a linear combination of indicator functions. Indeed, if X has the distinct values a_j, $j = 1, \ldots, n$, then

$$X = \sum_{j=1}^{n} a_j \mathbf{1}_{A_j} \quad \text{where} \quad A_j = \{X = a_j\}.$$

Here are some important concrete cases of discrete random variables X. In each case, the distribution of X is defined by a pmf.

Bernoulli Random Variable

A random variable X is said to be *Bernoulli with parameter* p $(0 < p < 1)$ if X has pmf defined by

$$p_X(1) = p \quad \text{and} \quad p_X(0) = 1 - p.$$

A Bernoulli random variable may be seen as describing an experiment with only two outcomes, frequently called "success" $(X = 1)$ and "failure" $(X = 0)$. Such an experiment is called a *Bernoulli trial.* For example, the single toss of a coin is a Bernoulli trial.

Binomial Random Variable

A random variable X is said to be *binomial with parameters* (N, p), where $0 < p < 1$ and N is a positive integer, if X has pmf defined by

$$p_X(n) = \binom{N}{n} p^n (1-p)^{N-n}, \quad n = 0, 1, \ldots, N.$$

In this case we write $X \sim B(N, p)$. Such a random variable appears natu-

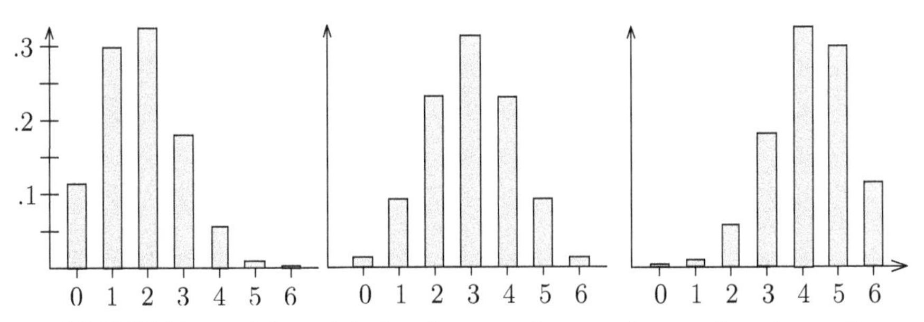

FIGURE 3.1: Binomial distributions for $p = .3$, $p = .5$ and $p = .7$.

rally as the number of successes in experiments consisting of N independent Bernoulli trials with parameter p. Indeed, in such a context the event $\{X = n\}$ can occur in $\binom{N}{n}$ ways, and since each of these has probability $p^n(1-p)^{N-n}$, the probability of exactly n successes is $p_X(n)$. Figure 3.3 shows the graph of the distribution of binomial random variables with parameters $(6, p)$, where $p = .3, .5, .7$. The distributions were generated by the module **BinomialDistribution**.

Binomial random variables are sometimes illustrated via the following experiment. Consider a jar that contains r red marbles and w white marbles. A random sample of size N is chosen from the jar by drawing N marbles in succession, each time noting the color and replacing the marble. A particular outcome of the experiment may be represented by a sequence of R's and W's. If X denotes the number of red marbles that were chosen, then the event $\{X = n\}$ consists of all possible sequences consisting of n R's and $N - n$ W's, each having probability

$$\left(\frac{r}{r+w}\right)^n \left(\frac{w}{r+w}\right)^{N-n}.$$

Since there are exactly $\binom{N}{n}$ such sequences,

$$\mathbb{P}(X = n) = \binom{N}{n} \left(\frac{r}{r+w}\right)^n \left(\frac{w}{r+w}\right)^{N-n}.$$

Setting $p = r/(r + w)$, which is the probability of getting a red marble on any particular draw, we see that $X \sim B(N, p)$. This experiment is sometimes called *sampling with replacement.*

Uniform Discrete Random Variable

A random variable X with finite range $R = \{x_1, \ldots, x_N\}$ is *uniformly distributed on* R if

$$p_X(x_n) = \tfrac{1}{N}, \quad n = 1, \ldots, N.$$

For example, if a jar contains n slips of paper labeled $1, \ldots, N$, then the number on a randomly chosen slip is uniformly distributed on $\{1, \ldots, N\}$.

Geometric Random Variable

A random variable X is said to be *geometric with parameter* $p \in (0, 1)$ if

$$p_X(n) = (1 - p)^n p, \quad n = 0, 1, \ldots. \tag{3.1}$$

For example, consider an experiment that consists of an infinite sequence of independent (see §3.6) Bernoulli trials with parameter p and let X denote the number of trials before the first success. For the event $\{X = n\}$ to occur, the first n trials must be failures and trial $n + 1$ a success. This happens with probability $(1 - p)^n p$. Note that

$$\sum_{k=0}^{\infty} p_X(k) = p \sum_{k=0}^{\infty} (1 - p)^k = \frac{p}{1 - (1 - p)} = 1,$$

so that p_X is indeed a pmf.

Hypergeometric Random Variable

Consider a jar that contains r red marbles and w white marbles. A random sample of size $z \le r+w$ is chosen from the jar by drawing the marbles in succession without replacement (or by simply drawing z marbles at once). The sample space Ω is the collection of all subsets of z marbles, hence $|\Omega| = \binom{r+w}{z}$. Let X denote the number of red marbles in the sample. The event $\{X = x\}$ is realized by choosing x red marbles from the r red marbles and $z - x$ white marbles from the w white marbles. Since the first task can be accomplished in $\binom{r}{x}$ ways and the second in $\binom{w}{z-x}$ ways, the probability of choosing exactly x red marbles is

$$\mathbb{P}(X = x) = \binom{r}{x}\binom{w}{z-x}\binom{r+w}{z}^{-1}. \tag{3.2}$$

Clearly, $0 \le x \le r$ and $0 \le z - x \le w$. Setting $M = r + w$, $p = r/M$, and $q = w/M$ (so that $p + q = 1$) we can write (3.2) as

$$p_X(x) = \binom{Mp}{x}\binom{Mq}{z-x}\binom{M}{z}^{-1}, \quad 0 \le x \le Mp, \; 0 \le z - x \le Mq. \tag{3.3}$$

A random variable X with this pmf is called a *hypergeometric random variable with parameters* (p, z, M). That p_X is indeed a pmf may be argued on probabilistic grounds or may be established analytically by using the identity

$$\sum_x \binom{Mp}{x}\binom{Mq}{z-x} = \binom{M}{z}. \tag{3.4}$$

The reader may wish to generate the distribution using the module **HypergeometricDistribution** and compare the result with the distribution generated by **BinomialDistribution**.

The experiment described in this example is called *sampling without replacement.* For an application, let the marbles represent individuals in a population of size M from which a sample of size z is randomly selected and then polled, red marbles representing individuals in the sample who favor candidate A for political office, white marbles representing those who favor candidate B. Pollsters use the sample to estimate the fraction of people in the general population who favor candidate A and also determine the margin of error in doing so. The hypergeometric pmf may be used for this purpose.[1] Marketing specialists apply similar techniques to determine product preferences of consumers.

[1] In practice, the more tractable normal distribution is used instead. This is justified by noting that for large M the hypergeometric distribution is very nearly binomial and that a (suitably adjusted) binomial random variable has an approximately normal distribution (by the central limit theorem).

3.4 Continuous Random Variables

A random variable X is said to be *continuous* if there exists a nonnegative (improperly) integrable function f_X such that

$$\mathbb{P}(X \in J) = \int_J f_X(t)\,dt \quad \text{for all intervals } J.$$

The function f_X is called the *probability density function* (pdf) of X. Note that the probability

$$\mathbb{P}(x \le X \le y) = \int_x^y f_X(t)\,dt$$

is simply the area under the graph of f_X between x and y. Setting $x = y$, we see that the probability that X takes on any particular value is zero. The cumulative distribution function of X takes the form

$$F_X(x) = \mathbb{P}(X \le x) = \int_{-\infty}^x f_X(t)\,dt.$$

Differentiating with respect to x, we have (at points of continuity of f_X),

$$F_X{}'(x) = f_X(x).$$

Here are some important examples of continuous random variables.

Uniform Continuous Random Variable

A random variable X is said to be *uniformly distributed on the interval* (a,b), written $X \sim U(a,b)$, if its pdf is of the form

$$f_X = \frac{1}{b-a}\mathbf{1}_{(a,b)},$$

where $\mathbf{1}_{(a,b)}$ is the indicator function of (a,b). For any subinterval (x,y) of

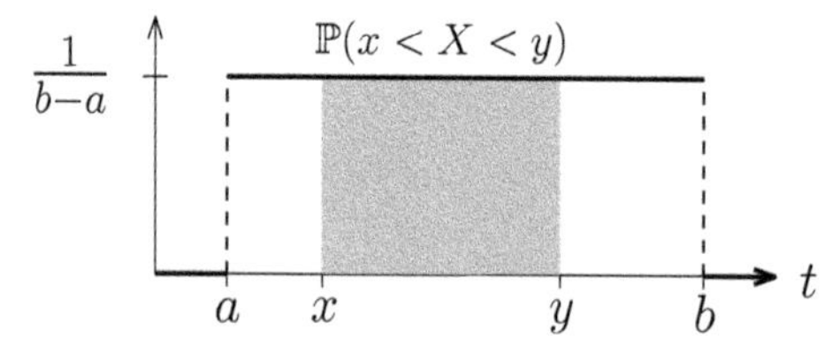

FIGURE 3.2: Uniform density function.

the interval (a,b), $\mathbb{P}(x < X < y) = \int_x^y f_X(t)\,dt = (y-x)/(b-a)$, which is the (theoretical) fraction of times a number chosen randomly from the interval (a,b) lies in the subinterval (x,y).

Normal Random Variable

Let σ and μ be real numbers with $\sigma > 0$. A random variable X is said to *normal with parameters μ and σ^2*, in symbols $X \sim N(\mu, \sigma^2)$, if it has pdf

$$f_X(x) = \frac{1}{\sigma\sqrt{2\pi}} e^{-(x-\mu)^2/2\sigma^2}, \quad -\infty < x < +\infty. \tag{3.5}$$

The normal density (3.5) is the familiar bell-shaped curve with maximum

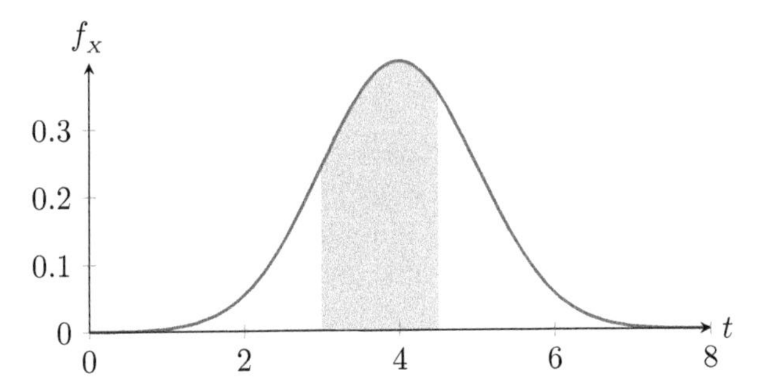

FIGURE 3.3: Normal density f_X.

occurring at μ. The parameter σ controls the spread of the bell: the larger the value of σ the flatter the bell. In Figure 3.3, the shaded region is the probability that a normal random variable takes values in the interval $(3, 4.5)$.

If $X \sim N(0,1)$, then X is said to have a *standard normal distribution*. In this case, we write φ for f_X and Φ for F_X. Thus,

$$\varphi(x) := \frac{1}{\sqrt{2\pi}} e^{-x^2/2} \quad \text{and} \quad \Phi(x) := \frac{1}{\sqrt{2\pi}} \int_{-\infty}^{x} e^{-t^2/2}\, dt. \tag{3.6}$$

Note that if $X \sim N(\mu, \sigma^2)$, then

$$F_X(x) = \Phi\left(\frac{x-\mu}{\sigma}\right),$$

as may be seen by making a simple substitution in the integral defining F_X.

Normal random variables arise in sampling from a large population of independent measurements such as test scores, heights, etc. They will figure prominently in option pricing formulas examined in later chapters.

Exponential Random Variable

A random variable X is said to have an *exponential distribution with parameter* $\lambda > 0$ if it has pdf

$$f_X(x) = \lambda e^{-\lambda x} \mathbf{1}_{[0,+\infty)}(x). \tag{3.7}$$

The cdf of X is then

$$F_X(x) = \int_{-\infty}^{x} f_X(y)\,dy = \begin{cases} 1 - e^{-\lambda x}, & \text{if } x \geq 0 \\ 0, & \text{if } x < 0. \end{cases}$$

In particular, for $x \geq 0$, $\mathbb{P}(X > x) = 1 - F_X(x) = e^{-\lambda x}$. It follows that for $s, t \geq 0$ we have the relation

$$\mathbb{P}(X > s + t, X > t) = \mathbb{P}(X > s + t) = e^{-\lambda(s+t)} = \mathbb{P}(X > t)\,\mathbb{P}(X > s),$$

which may be expressed in terms of conditional probabilities as

$$\mathbb{P}(X > s + t \mid X > t) = \mathbb{P}(X > s). \tag{3.8}$$

Equation (3.8) describes the so-called *memoryless* feature of an exponential random variable. If we take X to be the lifetime of some instrument, say, a light bulb, then (3.8) asserts (perhaps unrealistically) that the probability of the light bulb lasting at least $s + t$ hours, given that it has already lasted t hours, is the same as the initial probability that it will last at least s hours. Waiting times of buses, bank clerks, etc. are frequently assumed to be exponential random variables. It may be shown that the exponential distribution is the only continuous distribution for which Equation (3.8) holds.

3.5 Joint Distributions of Random Variables

In many situations one needs to consider several random variables simultaneously as, for example, in the case of a portfolio consisting of several stocks. In this section we consider the joint distribution of two random variables. The general case of finitely many random variables is entirely similar.

The *joint cumulative distribution function* $F_{X,Y}$ of a pair of random variables X and Y is defined by

$$F_{X,Y}(x, y) = \mathbb{P}(X \leq x,\ Y \leq y).$$

For properties of $F_{X,Y}$ we consider the discrete and continuous cases separately.

Discrete Joint Distribution

The *joint probability mass function* of discrete random variables X and Y is defined by

$$p_{X,Y}(x, y) = \mathbb{P}(X = x,\ Y = y).$$

In this case the joint cdf takes the form

$$F_{X,Y}(x, y) = \sum_{u \leq x,\ v \leq y} p_{X,Y}(u, v).$$

Note that the pmf's p_X and p_Y may be recovered from $p_{X,Y}$ by using the identities

$$p_X(x) = \mathbb{P}(X = x,\ Y \in \mathbf{R}) = \sum\nolimits_y p_{X,Y}(x,y),$$
$$p_Y(y) = \mathbb{P}(Y = y,\ X \in \mathbf{R}) = \sum\nolimits_x p_{X,Y}(x,y),$$

where the sums are taken, respectively, over all y in the range of Y and all x in the range of X. In this context, p_X and p_Y are called *marginal probability mass functions.*

Continuous Joint Distribution

Continuous random variables X and Y are said to be *jointly continuous* if there exists a nonnegative integrable function $f_{X,Y}$, called the *joint probability density function* of X and Y, such that

$$F_{X,Y}(x,y) = \int_{-\infty}^{x} \int_{-\infty}^{y} f_{X,Y}(s,t)\, dt\, ds \tag{3.9}$$

for all real numbers x and y. It follows from (3.9) that

$$\mathbb{P}(X \in J,\, Y \in K) = \int_K \int_J f_{X,Y}(x,y)\, dx\, dy = \iint\limits_{J \times K} f_{X,Y}(x,y)\, dx\, dy$$

for all intervals J and K. More generally,

$$\mathbb{P}\left((X,Y) \in A\right) = \iint\limits_{A} f(x,y)\, dx\, dy$$

for all sufficiently regular (e.g., Borel) sets $A \subseteq \mathbf{R}^2$. Differentiating (3.9) yields the following useful connection between $f_{X,Y}$ and $F_{X,Y}$:

$$f_{X,Y}(x,y) = \frac{\partial^2}{\partial x \partial y} F_{X,Y}(x,y).$$

The following proposition shows that, by analogy with the discrete case, the pdfs of jointly continuous random variables may be recovered from the joint pdf. In this context, f_X and f_Y are called *marginal density functions.*

3.5.1 Proposition. *If X and Y are jointly continuous random variables, then*

$$f_X(x) = \int_{-\infty}^{\infty} f_{X,Y}(x,y)\, dy \quad \textit{and} \quad f_Y(y) = \int_{-\infty}^{\infty} f_{X,Y}(x,y)\, dx.$$

Proof. For any interval J,

$$\mathbb{P}(X \in J) = \mathbb{P}(X \in J,\, Y \in \mathbf{R}) = \int_J \int_{-\infty}^{\infty} f_{X,Y}(x,y)\, dy\, dx.$$

The inner integral must then be the density f_X of X. □

3.6 Independent Random Variables

A collection $\mathcal{X}$ of random variables is said to be *independent* if each collection of events of the form $\{X \in J\}$, where $X \in \mathcal{X}$ and J is any interval, are independent. Here is a characterization of independence in terms of cdf's.

3.6.1 Proposition. *Let X and Y be random variables. Then X and Y are independent iff*

$$F_{X,Y}(x,y) = F_X(x)F_Y(y) \quad \textit{for all } x \textit{ and } y.$$

Proof. We give a partial proof of the sufficiency. If the equation holds, then

$$\begin{aligned}\mathbb{P}(a < X \le b, c < Y \le d) &= F_{X,Y}(b,d) - F_{X,Y}(a,d) - F_{X,Y}(b,c) + F_{X,Y}(a,c)\\ &= [F_X(b) - F_X(a)]\,[F_Y(d) - F_Y(c)]\\ &= \mathbb{P}(a < X \le b)\,\mathbb{P}(c < Y \le d).\end{aligned}$$

Thus the criterion for independence

$$\mathbb{P}(X \in J,\, Y \in K) = \mathbb{P}(X \in J)\,\mathbb{P}(Y \in K)$$

holds for all finite intervals J, K that are open on the left and closed on the right. This result may be extended to all intervals by using properties of probability. For example, for closed intervals we have

$$\begin{aligned}\mathbb{P}(a \le X \le b,\, c \le Y \le d) &= \lim_{n\to\infty} \mathbb{P}(a - 1/n < X \le b,\, c - 1/n < Y \le d)\\ &= \lim_{n\to\infty} \mathbb{P}(a - 1/n < X \le b)\,\mathbb{P}(c - 1/n < Y \le d)\\ &= \mathbb{P}(a \le X \le b)\,\mathbb{P}(c \le Y \le d).\end{aligned}$$

Here we have used the fact that if $A_1 \supseteq A_2 \cdots \supseteq A_n \cdots$, then

$$\mathbb{P}\Big(\bigcap_{n=1}^{\infty} A_n\Big) = \lim_{n\to\infty} \mathbb{P}(A_n),$$

a consequence of the additive property. □

3.6.2 Corollary. *Let X and Y be jointly continuous random variables. Then X and Y are independent iff*

$$f_{X,Y}(x,y) = f_X(x)f_Y(y) \quad \textit{for all } x \textit{ and } y.$$

Proof. If X and Y are independent, then, by Proposition 3.6.1,

$$F_{X,Y}(x,y) = \left(\int_{-\infty}^{x} f_X(s)\,ds\right)\left(\int_{-\infty}^{y} f_Y(t)\,dt\right) = \int_{-\infty}^{x}\int_{-\infty}^{y} f_X(s)f_Y(t)\,dt\,ds,$$

which shows that $f_X(x)f_Y(y)$ is the joint density function. Conversely, if the density condition in the statement of the corollary holds, then, reversing the argument, we see that $F_{X,Y}(x,y) = F_X(x)F_Y(y)$. □

Here is the discrete analog of Corollary 3.6.2:

3.6.3 Proposition. *Discrete random variables X and Y are independent iff*

$$p_{X,Y}(x,y) = p_X(x)p_Y(y) \quad \textit{for all } x \textit{ and } y.$$

Proof. The proof of the necessity is left to the reader. For the sufficiency, suppose that the above equation holds. Then, for any intervals J and K,

$$\begin{aligned}\mathbb{P}(X \in J, Y \in K) &= \sum_{x\in J,\ y\in K} p_{X,Y}(x,y) = \sum_{x\in J} p_X(x) \sum_{y\in K} p_Y(y)\\ &= \mathbb{P}(X \in J)\,\mathbb{P}(Y \in K),\end{aligned}$$

where the sums are taken over all $x \in J$ in the range of X and all $y \in K$ in the range of Y. Thus X and Y are independent. □

3.6.4 Proposition. *If X and Y are independent random variables and g and h are continuous functions, then $g(X)$ and $h(Y)$ are independent.*

Proof. Let F denote the joint cdf of $g(X)$ and $h(Y)$. As in the proof of Theorem 3.2.1, given real numbers a and b there exist sequences (J_m) and (K_n) of pairwise disjoint open intervals such that

$$\{g(X) < a\} = \bigcup_m \{X \in J_m\} \quad \text{and} \quad \{h(Y) < b\} = \bigcup_n \{Y \in K_n\}.$$

It follows that

$$\begin{aligned}\mathbb{P}(g(X) < a,\ h(Y) < b) &= \sum_{m,n} \mathbb{P}(X \in J_m,\ Y \in K_n) = \sum_{m,n} \mathbb{P}(X \in J_m)\mathbb{P}(Y \in K_n)\\ &= \mathbb{P}(g(X) < a)\,\mathbb{P}(h(Y) < b).\end{aligned}$$

Taking $a = x + 1/k$, $b = y + 1/k$, and letting $k \to \infty$ we have

$$\mathbb{P}(g(X) \le x,\ h(Y) \le y) = \mathbb{P}(g(X) \le x)\,\mathbb{P}(h(Y) \le y),$$

that is,

$$F(x,y) = F_{g(X)}(x)F_{h(Y)}(y).$$

By Proposition 3.6.1, $g(X)$ and $h(Y)$ are independent. □

The above results have natural extensions to finitely many random variables. We leave the formulations to the reader.

3.7 Identically Distributed Random Variables

A collection of random variables with the same cdf is said to be *identically distributed*. If the random variables are also independent, then the collection is said to be *iid*.

3.7.1 Example. Consider a sequence of independent Bernoulli trials with parameter p. Let X_1 be the number of trials before the first success, and for $k > 1$ let X_k be the number of trials between the $(k-1)$st and kth successes. The event $\{X_1 = m, X_2 = n\}$ $(m, n \geq 0)$ consists of sequences of the form

$$\underbrace{F\,F \cdots F}_{m}\; S\; \underbrace{F\,F \cdots F}_{n}\; S\; *\; * \cdots .$$

Since the event occurs with probability $q^m p q^n p = q^{m+n}p^2$ $(q := 1-p)$,

$$\mathbb{P}(X_2 = n) = \sum_{m=0}^{\infty} \mathbb{P}(X_1 = m, X_2 = n) = p^2 q^n \sum_{m=0}^{\infty} q^m = pq^n = \mathbb{P}(X_1 = n).$$

Therefore, X_1 and X_2 are identically distributed. Since

$$\mathbb{P}(X_1 = m)\,\mathbb{P}(X_2 = n) = pq^m pq^n = \mathbb{P}(X_1 = m, X_2 = n),$$

Proposition 3.6.3 shows that X_1 and X_2 are independent and hence iid. An induction argument using similar calculations shows that the entire sequence (X_n) is iid. Note that X_n is a geometric random variable with parameter p. ◊

3.8 Sums of Independent Random Variables

Discrete Case

If X and Y are independent discrete random variables then

$$p_{X+Y}(z) = \sum_x \mathbb{P}(X = x, Y = z - x) = \sum_x p_X(x) p_Y(z - x). \tag{3.10}$$

The sum on the right in (3.10) is called the *convolution* of the pmf's p_X and p_Y.

3.8.1 Example. Let X and Y be independent binomial random variables with $X \sim B(m, p)$ and $Y \sim B(n, p)$. We claim that $Z := X + Y \sim B(m+n, p)$. Indeed, by (3.10),

$$\begin{aligned} p_Z(z) &= \sum_x \binom{m}{x} p^x q^{m-x} \binom{n}{z-x} p^{z-x} q^{n-(z-x)} \\ &= p^z q^{m+n-z} \sum_x \binom{m}{x}\binom{n}{z-x}, \quad q := 1-p, \end{aligned}$$

where the sum is taken over all integers x satisfying the inequalities $0 \le x \le m$ and $0 \le z - x \le n$, that is, $\max(z-n, 0) \le x \le \min(z, m)$. By (3.4), the last sum is $\binom{m+n}{z}$, verifying the claim. More generally, one may show by induction that if $X_1, \ldots, X_k$ are independent and $X_j \sim B(n_j, p)$, then

$$X := X_1 + \cdots + X_k \sim B(n_1 + \cdots + n_k, p).$$

In particular, if X_j is Bernoulli with parameter p, then $X \sim B(k, p)$. ◊

Continuous Case

If X and Y are jointly continuous random variables, then

$$\begin{aligned} F_{X+Y}(z) &= \iint\limits_{x+y \le z} f_{X,Y}(x,y)\,dx\,dy = \int_{-\infty}^{\infty}\int_{-\infty}^{z-y} f_{X,Y}(x,y)\,dx\,dy \\ &= \int_{-\infty}^{\infty}\int_{-\infty}^{z} f_{X,Y}(x-y,y)\,dx\,dy = \int_{-\infty}^{z}\int_{-\infty}^{\infty} f_{X,Y}(x-y,y)\,dy\,dx, \end{aligned}$$

hence

$$f_{X+Y}(x) = \int_{-\infty}^{\infty} f_{X,Y}(x-y,y)\,dy. \tag{3.11}$$

If X and Y are independent, then, by Corollary 3.6.2, the joint density of X and Y is $f_X(x)f_Y(y)$, hence, by (3.11),

$$f_{X+Y}(x) = \int_{-\infty}^{\infty} f_X(x-y)f_Y(y)\,dy. \tag{3.12}$$

The integral in (3.12) is called the *convolution* of the densities f_X and f_Y.

3.8.2 Example. Let X and Y be independent normal random variables with $X \sim N(\mu, \sigma^2)$ and $Y \sim N(\nu, \tau^2)$. Then

$$X + Y \sim N(\mu + \nu, \sigma^2 + \tau^2).$$

To see this set $Z = X + Y$ and suppose first that $\mu = \nu = 0$. We need to show that $f_Z = g$, where

$$g(z) = \frac{1}{\varrho\sqrt{2\pi}} \exp\left(-z^2/2\varrho^2\right), \quad \varrho^2 := \sigma^2 + \tau^2.$$

Let $a := (2\pi\sigma\tau)^{-1}$. From (3.12),

$$\begin{aligned} f_Z(z) &= a\int_{-\infty}^{\infty} \exp\left\{-\frac{1}{2}\left[\frac{(z-y)^2}{\sigma^2} + \frac{y^2}{\tau^2}\right]\right\} dy \\ &= a\int_{-\infty}^{\infty} \exp\left\{\frac{\tau^2(z-y)^2 + \sigma^2 y^2}{-2\sigma^2\tau^2}\right\} dy. \end{aligned}$$

The expression $\tau^2(z-y)^2 + \sigma^2 y^2$ in the second integral may be written

$$\begin{aligned}
\tau^2 z^2 - 2\tau^2 yz + \varrho^2 y^2 &= \varrho^2\left(y^2 - \frac{2\tau^2 yz}{\varrho^2}\right) + \tau^2 z^2 \\
&= \varrho^2\left(y - \frac{\tau^2 z}{\varrho^2}\right)^2 + \tau^2 z^2 - \frac{\tau^4 z^2}{\varrho^2} \\
&= \varrho^2\left(y - \frac{\tau^2 z}{\varrho^2}\right)^2 + \frac{\tau^2\sigma^2 z^2}{\varrho^2}.
\end{aligned}$$

Thus, for suitable positive constants b and c,

$$\frac{\tau^2(z-y)^2 + \sigma^2 y^2}{-2\sigma^2\tau^2} = -b(y-cz)^2 - \frac{z^2}{2\varrho^2}.$$

It follows that

$$\begin{aligned}
f_Z(z) &= a\exp\left(-\frac{z^2}{2\varrho^2}\right)\int_{-\infty}^{\infty}\exp(-b(y-cz)^2)\,dy \\
&= a\exp\left(-\frac{z^2}{2\varrho^2}\right)\int_{-\infty}^{\infty}\exp(-bu^2)\,du \\
&= kg(z)
\end{aligned}$$

for some constant k and for all z. Since f_Z and g are densities, by integrating we see that $k = 1$, verifying the assertion for the case $\mu = \nu = 0$.

For the general case, observe that $X - \mu \sim N(0, \sigma^2)$ and $Y - \nu \sim N(0, \tau^2)$ so by the first part $Z - \mu - \nu \sim N(0, \varrho^2)$. Therefore, $Z \sim N(\mu + \nu, \rho^2)$. ◊

3.8.3 Example. Let S_n denote the price of a stock on day n. A common model assumes that the ratios $Z_n := S_n/S_{n-1}$, $n \ge 1$, are iid *lognormal random variables* with parameters μ and σ^2, that is, $\ln Z_n \sim N(\mu, \sigma^2)$. Note that $Z_n - 1$ is the fractional increase of the stock from day $n-1$ to day n. The probability that the stock price rises on each of the first n days is

$$\mathbb{P}(Z_1 > 1, \ldots, Z_n > 1) = [\mathbb{P}(\ln Z_1 > 0)]^n = \left[1 - \Phi\left(\frac{-\mu}{\sigma}\right)\right]^n = \left[\Phi\left(\frac{\mu}{\sigma}\right)\right]^n,$$

where the last equality follows from the identity $1 - \Phi(-x) = \Phi(x)$ (Exercise 9).

Now observe that, by Example 3.8.2, $\ln Z_1 + \ln Z_2 + \cdots + \ln Z_n \sim N(n\mu, n\sigma^2)$. Using this we see that the probability that on day n the price of the stock will be larger than its initial price S_0 is

$$\begin{aligned}
\mathbb{P}(S_n > S_0) &= \mathbb{P}(Z_1 Z_2 \cdots Z_n > 1) = \mathbb{P}(\ln Z_1 + \ln Z_2 + \cdots + \ln Z_n > 0) \\
&= 1 - \Phi\left(\frac{0 - n\mu}{\sigma\sqrt{n}}\right) \\
&= \Phi\left(\frac{\mu\sqrt{n}}{\sigma}\right).
\end{aligned}$$

◊

3.9 Exercises

1. Let A, B, and C be subsets of Ω. Prove that

   (a) $\mathbf{1}_{AB} = \mathbf{1}_A \mathbf{1}_B$. (b) $\mathbf{1}_{A\cup B} = \mathbf{1}_A + \mathbf{1}_B - \mathbf{1}_A \mathbf{1}_B$. (c) $\mathbf{1}_A \le \mathbf{1}_B$ iff $A \subseteq B$.

2. Let A_n, $B_n \subseteq \Omega$ such that $A_1 \subseteq A_2 \subseteq \cdots$ and $B_1 \supseteq B_2 \supseteq \cdots$. If $A = \bigcup_{n=1}^{\infty} A_n$ and $B = \bigcap_{n=1}^{\infty} B_n$, prove that for each ω

   (a) $\lim_{n\to\infty} \mathbf{1}_{A_n}(\omega) = \mathbf{1}_A(\omega)$. (b) $\lim_{n\to\infty} \mathbf{1}_{B_n}(\omega) = \mathbf{1}_B(\omega)$.

3. Use a spreadsheet to determine the smallest number of flips of a fair coin needed to be 99% sure that at least two heads will come up.

4. Use a spreadsheet to determine the smallest number of tosses of a pair of fair dice needed to be 99% sure that the event "the sum of the dice is 7" occurs at least three times.

5. Let X be a hypergeometric random variable with parameters (p, n, N). Show that

   $$\lim_{N\to\infty} p_X(k) = \binom{n}{k} p^k q^{n-k}.$$

   Thus for large populations the experiments of sampling with replacement and without replacement are probabilistically nearly identical.

6. Let X_1, X_2, ... be an infinite sequence of random variables such that the limit $X(\omega) := \lim_{n\to\infty} X_n(\omega)$ exists for each $\omega \in \Omega$. Verify that

   $$\{X < a\} = \bigcup_{m=1}^{\infty} \bigcup_{n=1}^{\infty} \bigcap_{k=n}^{\infty} \{X_k < a - 1/m\}.$$

   Conclude that X is a random variable.

7. Let $Y = aX + b$, where X is a continuous random variable and a and b are constants with $a \ne 0$. Show that

   $$f_Y(y) = |a|^{-1} f_X\left(\frac{y-b}{a}\right).$$

8. Let X be a random variable with density f_X and set $Y = X^2$. Show that

   $$f_Y(y) = \frac{f_X(\sqrt{y}) + f_X(-\sqrt{y})}{2\sqrt{y}} \mathbf{1}_{(0,+\infty)}(y).$$

   In particular, find f_Y if X is uniformly distributed over $(-1, 1)$.

9. Show that $1 - \Phi(x) = \Phi(-x)$. Conclude that $X \sim N(0,1)$ iff $-X \sim N(0,1)$.

10. Show that if $X \sim N(\mu, \sigma^2)$ and $a \ne 0$, then $aX + b \sim N(a\mu + b, a^2\sigma^2)$.

11. Show that $X \sim N(\mu, \sigma^2)$ iff $2\mu - X \sim N(\mu, \sigma^2)$.

12. Let X, Y be independent and $X, Y \sim N(0,1)$. Show that

$$\mathbb{P}(a < X - Y < b) = \frac{1}{\sqrt{2}}\Big[\Phi(b) - \Phi(a)\Big].$$

13. In the dartboard of Example 2.3.6, let Z be the distance from the origin to the landing position of the dart. Find F_Z and f_Z.

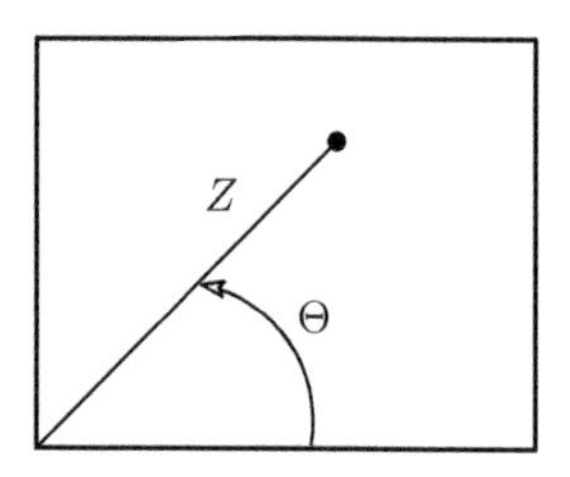

FIGURE 3.4: Illustrations for Exercises 13 and 14.

14. Referring to Example 2.3.6, let Θ be the angle from the x-axis to the landing position of the dart. Find F_Θ and f_Θ.

15. Let X and Y be independent random variables. Express the cdfs F_M of $M := \max(X, Y)$ and F_m of $m := \min(X, Y)$ in terms of F_X and F_Y. Conclude that

$$F_M + F_m = F_X + F_Y.$$

16. Let $Z = X + Y$, where X and Y are independent and uniformly distributed on $(0, 1)$. Show that

$$F_Z(z) = \tfrac{1}{2}z^2\mathbf{1}_{[0,1)}(z) + \left[1 - \tfrac{1}{2}(2 - z)^2\right]\mathbf{1}_{[1,2)}(z) + \mathbf{1}_{[2,\infty)}(z).$$

17. Refer to Example 3.8.3. Let $1 \le k \le n$. Find the probability that the stock

    (a) increases exactly k times in n days.

    (b) increases exactly k *consecutive* times in n days (decreasing on the other days).

    (c) has a run of exactly k consecutive increases (not necessarily decreasing on the other days), where $k \ge n/2$.

18. Let X_j, $j = 1, \ldots, n$, be independent Bernoulli random variables with parameter p. Set $Y_j = X_1 + \cdots + X_j$. Show that for $1 \le m < n$,

$$\mathbb{P}(Y_m = j,\, Y_n = k) = \binom{m}{j}\binom{n-m}{k-j}p^k q^{n-k},$$

where $1 \le k \le n$ and $\max(0, k - n + m) \le j \le \min(m, k)$. Conclude that the function

$$p_X(j) := \mathbb{P}(Y_m = j \mid Y_n = k)$$

is the pmf of a hypergeometric random variable X with parameters $(m/n, k, n)$.

19. Let $X_1, X_2, \ldots, X_n$ be integer-valued iid random variables with cdf F, and let

$$Y_{\min} = \min\{X_1, \ldots, X_n\} \text{ and } Y_{\max} = \max\{X_1, \ldots, X_n\}.$$

Show that

$$p_{Y_{\min}}(y) = F^n(y) - F^n(y-1) \text{ and } p_{Y_{\max}}(y) = 1 - \big(1 - F(y)\big)^n.$$

20. Let $X \sim B(4, 1/2)$. Find $\mathbb{P}(.8 \le X^2/(1+X^2) \le .9)$.

21. Let $X \sim N(1, 4)$. Find $\mathbb{P}(.1 < 1/(1+X^2) < .2)$.

22. Let X be uniformly distributed on the interval $[2, 7]$. Find the probability of the events

(a) $\{2/3 < X/(1+X) < 5/6\}$. (b) $\{1/9 < (X-3)^2 < 1/4\}$.

23. Let X and Y be uniformly distributed on the interval $[1, 3]$. If X and Y are independent, find the probability of the event $\{1/2 < Y/X < 2\}$.

24. Same as the preceding exercise, but now let X be uniformly distributed on the interval $[2, 5]$.

25. Let X and Y be independent geometric random variables with parameter p and let m and n be integers with $0 \le n \le m$. Show that

$$\mathbb{P}(X = n \mid X + Y = m) = \frac{1}{m+1},$$

which is the probability of randomly choosing an integer from 0 to m.

26. A discrete random variable X is said to be *Poisson with parameter* $\lambda > 0$ if

$$p_X(k) = \frac{\lambda^k}{k!} e^{-\lambda}, \quad k = 0, 1, 2, \ldots.$$

Let X_1 and X_2 be independent Poisson random variables with parameters λ_1 and λ_2, respectively, and let $0 \le n \le m$. Show that

$$\mathbb{P}(X_1 = n \mid X_1 + X_2 = m) = \binom{m}{n} \left(\frac{\lambda_1}{\lambda_1 + \lambda_2}\right)^n \left(\frac{\lambda_2}{\lambda_1 + \lambda_2}\right)^m.$$

Thus, given that $X_1 + X_2 = m$, $X_1 \sim B(m, p)$, where $p = \lambda_1(\lambda_1 + \lambda_2)^{-1}$.

27. Let X and Y be independent random variables with X, $Y > 0$. Show directly that $1/X$ and $1/Y$ are independent.

28. Let X_j, $j = 1, 2, \ldots, n$, be independent Bernoulli random variables with parameter p and set $Y_j = X_1 + \cdots + X_j$. Let $1 \le m < n$ and $1 \le j \le k \le n$. Show that

$$\mathbb{P}(Y_n = k \mid Y_m = j) = \mathbb{P}(Y_{n-m} = k - j).$$

29. Let $X_1, \ldots, X_n$ be independent Bernoulli random variables with parameter p and set $Y_m = X_1 + \cdots + X_m, \quad 1 \le m \le n$. Let p_X denote the pmf of a Bernoulli random variable X with parameter p.

    (a) Let $k_0, k_1, \ldots, k_n$ be integers with $k_0 = 0$. Show that

    $$\mathbb{P}(Y_1 = k_1,\, Y_2 = k_2, \ldots, Y_n = k_n) = \prod_{j=1}^{n} p_X(k_j - k_{j-1}) = p^{k_n} q^{n-k_n}.$$

    (b) Set $Z_n := Y_1 + \cdots + Y_n$. Show that the pmf of Z_n is given by

    $$p_{Z_n}(k) = \sum_{(k_1,\ldots,k_n)} p^{k_n} q^{n-k_n}, \quad 0 \le k \le n(n+1)/2,$$

    where the sum is taken over all nonnegative integers $k_1, \ldots, k_n$ such that $k_1 + \cdots + k_n = k$ and $0 \le k_j - k_{j-1} \le 1$.

    (c) Show that the joint pmf of (Z_n, Y_n) is given by

    $$\mathbb{P}(Z_n = k,\, Y_n = j) = \sum_{(k_1,\ldots,k_{n-1})} p^{k_{n-1}} q^{n-1-k_{n-1}} p_X(j - k_{n-1}),$$

    where the sum is taken over all nonnegative integers $k_1, \ldots, k_{n-1}$ such that $k_1 + \cdots + k_{n-1} = k - j$, $0 \le k_j - k_{j-1} \le 1$, $0 \le j - k_{n-1} \le 1$.

    (d) Use part (b) to find the probability distribution of $Y_1 + Y_2 + Y_3$.

30. Let X_j, $j = 1, \ldots, n$, be independent Bernoulli random variables with parameter p and set $Y_j = X_1 + \cdots + X_j$. Show that

    $$\mathbb{P}(Y_m < Y_n) = \binom{m}{j}\binom{n-m}{k-j} p^k (1-p)^{n-k} \quad (1 \le m < n).$$

31. Show that $Y \sim B(n,p)$ iff $n - Y \sim B(n, 1-p)$.

32. Let X and Y be independent random variables with X uniformly distributed on $(0,1)$ and Y uniformly distributed on $(1,2)$. Find the density of $Z := \ln(X+Y)$.

Chapter 4

Options and Arbitrage

Assets traded in financial markets fall into the following main categories:

- *stocks* (equity in a corporation or business).
- *bonds* (financial contracts issued by governments and corporations).
- *currencies* (traded on foreign exchanges).
- *commodities* (goods, such as oil, wheat, or electricity).
- *derivatives.*

A derivative is a financial instrument whose value is based on one or more underlying assets. Such instruments provide a way for investors to reduce the risks associated with investing by locking in a favorable price. The most common derivatives are forwards, futures, and options. In this chapter, we show how the simple assumption of no-arbitrage may be used to derive fundamental properties regarding the value of a derivative. In later chapters, we show how the no-arbitrage principle leads to the notion of replicating portfolios and ultimately to the Cox-Ross-Rubinstein and Black-Scholes option pricing formulas.

4.1 The Price Process of an Asset

A *financial market* is a system by which are traded finitely many securities $\mathcal{S}_1$, ..., $\mathcal{S}_d$. We assume the market allows unlimited trading in these securities as well as derivatives based on them. For reasons that will become clear later, we also assume there is available to investors a risk-free account that allows unlimited transactions in the form of deposits or withdrawals (including loans). As we saw in Section 1.4, this is equivalent to the availability of a risk-free bond $\mathcal{B}$ that may be purchased or *sold short* (that is, borrowed and then sold) in unlimited quantities. We follow the convention that borrowing an amount A is the same as lending (depositing) the amount $-A$, where A may be positive or negative. The array

$$(\mathcal{B}, \mathcal{S}_1, \ldots, \mathcal{S}_d)$$

will be called a *market.*

For ease of exposition, in this chapter we treat only the case $d = 1$. Thus we assume that the market allows unlimited trades of a single risky security $\mathcal{S}$. With the exception of Section 4.10 we also assume that $\mathcal{S}$ pays no dividends.

The value of a share of $\mathcal{S}$ at time t will be denoted by S_t. The term "risky" refers to the unpredictable nature of $\mathcal{S}$ and hence suggests a probabilistic setting for the model. Thus we assume that S_t is a random variable on some probability space $(\Omega, \mathcal{F}, \mathbb{P})$.[1] The set D of indices t is either a discrete set $\{0, 1, \ldots, N\}$ or an interval $[0, T]$. Since the initial value of the security is known at time 0, S_0 is assumed to be a constant. The collection $S = (S_t)_{t \in D}$ of random variables is called the *price process* of $\mathcal{S}$. The model is said to be *continuous* if $d = [0, T]$ and *discrete* if $D = \{0, 1, \ldots, N\}$. In the former case the risk-free asset is assumed to earn interest compounded continuously at a constant annual rate r. In the latter case, we assume compounding occurs at a rate i per time interval. Thus if B_0 is the initial value of the bond (or account), then the value of the bond at time t is given by

$$B_t = \begin{cases} B_0 e^{rt} & \text{in a continuous model } (0 \le t \le T, \text{ in years}), \\ B_0 (1+i)^t & \text{in a discrete model } (t = 0, 1, \ldots N, \text{ in periods}). \end{cases}$$

These conventions apply throughout the book.

4.2 Arbitrage

An *arbitrage* is an investment opportunity that arises from mismatched financial instruments. For an example, suppose Bank A is willing to lend money at an annual rate of 3%, compounded monthly, and Bank B is offering CDs at an annual rate of 4%, also compounded monthly. An investor recognizes this as an arbitrage, that is, a "sure win" opportunity. She simply borrows an amount from Bank A at 3% and immediately deposits it into Bank B at 4%. The transaction results in a risk-free positive profit.[2]

Clearly, a market cannot sustain such obvious instances of arbitrage. However, more subtle examples do occur, and while their existence may be fleeting they provide employment for market analysts whose job it is to discover them. (High-speed computers are commonly employed to ferret out and exploit arbitrage opportunities.) Absence of arbitrage, while an idealized condition, is necessary for market soundness and, indeed, for general economic stability.

[1] In this chapter the probability space is unspecified. Concrete probability models are developed in later chapters.

[2] In practice, to avoid fluctuations in asset prices, arbitrage transactions must occur simultaneously. For example, arbitrage can occur by buying securities on one exchange and immediately selling them on another. In general this is possible only if the transactions are carried out electronically.

Moreover, as we shall see, the assumption of no-arbitrage leads to a robust mathematical theory of option pricing.

A precise mathematical definition of arbitrage is given in Chapter 5. For now, we define an arbitrage as a trading strategy resulting in a portfolio with zero probability of loss and positive probability of gain.[3] A formal mathematical definition of portfolio is given in Chapter 5. For the purposes of the present chapter, it is sufficient to think of a portfolio informally as a collection of assets.

The assertion in the following example is the first of several in the book that show how the assumption of no-arbitrage has concrete mathematical consequences. The result will play a fundamental role in the pricing of derivatives in the binomial market model (Chapter 7).

4.2.1 Example. Suppose that the initial value of a single share of our security $\mathcal{S}$ is S_0 and that after one time period its value goes up by a factor of u with probability p or down by a factor of d with probability $q = 1 - p$, where $0 < d < u$ and $0 < p < 1$. In the absence of arbitrage, it must then be the case that

$$d < 1 + i < u, \tag{4.1}$$

where i is the interest rate during the time period. The verification of (4.1) uses the idea of selling a security *short*, that is, borrowing and then selling a security under the agreement that it will be returned at some later date. (You are *long* in the security if you actually own it.)

We prove (4.1) by contradiction. Suppose first that $u \le 1 + i$. We then employ the following trading strategy: At time 0, we sell the stock short for S_0 and invest the money in a risk-free account paying the rate i. This results in a portfolio consisting of -1 shares of $\mathcal{S}$ and an account worth S_0. No wealth was required to construct the portfolio. At time 1, the account is worth $S_0(1 + i)$, which we use to buy the stock, returning it to the lender. This costs us nothing extra since the stock is worth at most uS_0, which, by our assumption $u \le 1 + i$, is covered by the account. Furthermore, there is a positive probability q that our gain is the positive amount $(1 + i)S_0 - dS_0$. The strategy therefore constitutes an arbitrage.

Now assume that $1 + i \le d$. We then employ the reverse strategy: At time 0 we borrow S_0 dollars and use it to buy one share of $\mathcal{S}$. We now have a portfolio consisting of one share of the stock and an account worth $-S_0$, and, as before, no wealth was required to construct it. At time 1, we sell the stock and use the money to pay back the loan. This leaves us with at least $dS_0 - (1 + i)S_0 \ge 0$, and there is a positive probability p that our gain is the positive amount $uS_0 - (1 + i)S_0$, again, an arbitrage. Therefore, (4.1) must hold. ◊

[3] In games of chance, such as roulette, a casino has a *statistical* arbitrage (the so-called *house advantage*). Here, in contrast to financial arbitrage, the casino has a positive probability of loss. However, the casino has a positive expected gain, so in the long run the house wins.

The Law of One Price

The following consequence of no-arbitrage is frequently useful:

In an arbitrage-free market, two portfolios $\mathcal{X}$ and $\mathcal{Y}$ with the same value at time T must have the same value at all times $t < T$.

Indeed, if the time-t value of $\mathcal{X}$ were greater than that of $\mathcal{Y}$, one could obtain a positive profit at time t by taking a short position in $\mathcal{X}$ and a long position in the lower-priced portfolio $\mathcal{Y}$. The proceeds from selling $\mathcal{Y}$ at time T exactly cover the obligation assumed by shorting $\mathcal{X}$. (See Exercise 3 for a generalization.)

For the remainder of the chapter, we assume that the market admits no arbitrage opportunities. Also, for definiteness, we assume a continuous-time price process for an asset.

4.3 Classification of Derivatives

In the sections that follow, we examine various common types of derivatives. As mentioned above, a derivative is a financial contract whose value depends on that of another asset $\mathcal{S}$, called the *underlying*.[4] The expiration time of the contract is called the *maturity* or *expiration date* and is denoted by T. The price process of the underlying is given by $S = (S_t)_{0 \le t \le T}$.

Derivatives fall into four main categories:

- European
- American
- Path independent
- Path dependent

A *European derivative* is a contract that the holder may exercise only at maturity T. By contrast, an *American derivative* may be exercised at any time $\tau \le T$. A *path-independent derivative* has *payoff* (value) at exercise time τ that depends only on S_τ, while a *path-dependent* derivative has payoff at τ which depends on the entire history $S_{[0,\tau]}$ of the price process.

We begin our analysis with the simplest types of European path-independent derivatives: forwards and futures.

[4]The asset need not be a security, commodity, or currency but may be another derivative or even an intangible quantity. For example, in recent years derivatives based on weather have become available. This is not surprising when one considers that weather can affect crop yields, energy use and production, travel and tourism, etc.

4.4 Forwards

A *forward contract* between two parties places one party under the obligation to buy an asset at a future time T for a prescribed price K, the *forward price*, and requires the other party to sell the asset under those conditions. The party that agrees to buy the asset is said to assume the *long position*, while the party that will sell the asset has the *short position*. The payoff for the long position is $S_T - K$, while that for the short position is $K - S_T$. The forward price K is set so that there is no cost to either party to enter the contract.

The following examples illustrate how forwards may be used to hedge against unfavorable changes in commodity prices.

4.4.1 Example. A farmer expects to harvest and sell his wheat crop six months from now. The price of wheat is currently \$8.70 per bushel and he expects a crop of 10,000 bushels. He calculates that he would make an adequate profit even if the price dropped to \$8.50. To hedge against a larger drop in price, he takes the short position in a forward contract with $K = \$8.50 \times 10,000 = \$85,000$. A flour mill is willing to take the long position. At time T (six months from now) the farmer is under the obligation to sell his wheat to the mill for \$8.50 per bushel. His payoff is the difference between the forward price K and the price of wheat at maturity. Thus if wheat drops to \$8.25, his payoff is \$.25 per bushel; if it rises to \$8.80, his payoff is −\$.30 per bushel. ◊

4.4.2 Example. An airline company needs to buy 100,000 gallons of jet fuel in three months. It can buy the fuel now at \$4.80 per gallon and pay storage costs, or wait and buy the fuel three months from now. The company decides on the latter strategy, and to hedge against a large increase in the cost of jet fuel, it takes the long position in a forward contract with $K = \$4.90 \times 100,000 = \$490,000$. In three months, the airline is obligated to buy, and the fuel company to sell, 100,000 gallons of jet fuel for \$4.90 a gallon. The company's payoff is the difference between the price of fuel then and the strike price. If in three months the price rises to, say, \$4.96 per gallon, then the company's payoff is \$.06 per gallon. If it falls to \$4.83, its payoff is −\$.07 per gallon. ◊

Since there is no cost to enter a forward contract, the initial value of the forward is zero. However, as time passes, the forward may acquire a non-zero value. Indeed, the no-arbitrage assumption implies that the (long) value F_t of a forward at time t is given by

$$F_t = S_t - e^{-r(T-t)}K, \quad 0 \le t \le T. \tag{4.2}$$

This is clear for $t = T$. Suppose, however, that at some time $t < T$ the inequality $F_t < S_t - e^{-r(T-t)}K$ holds. We then take a short position on the security and a long position on the forward. This provides us with cash in the amount $S_t - F_t > 0$, which we deposit in a risk-free account at rate r. At maturity, we discharge our obligation to buy the security for the amount K, returning it to the lender. Since our account has grown to $(S_t - F_t)e^{r(T-t)}$, we now have cash in the amount $(S_t - F_t)e^{r(T-t)} - K$. As this amount is positive, our strategy constitutes an arbitrage. If $F_t > S_t - e^{-r(T-t)}K$, we employ the reverse strategy, taking a short position on the forward and a long position on the security. This requires cash in the amount $S_t - F_t$, which we borrow at the rate r. At time T, we discharge our obligation to sell the security for K and use this to settle our debt. This leaves us with the positive cash amount $K - (S_t - F_t)e^{r(T-t)}$, again implying an arbitrage. Therefore, (4.2) must hold.

One can also obtain (4.2) using the law of one price, one investment consisting of a long position in the forward, the other consisting of a long position in the stock and a short position in a bond with face value K. The investments have the same value at maturity ((4.2) with $t = T$) and therefore must have the same value for all $t \le T$.

Setting $t = 0$ in (4.2) and solving the resulting equation $F_0 = 0$ for K, we have

$$K = S_0 e^{rT}. \tag{4.3}$$

Substituting (4.3) into (4.2) yields the alternate formula

$$F_t = S_t - e^{rt} S_0, \quad 0 \le t \le T. \tag{4.4}$$

Thus the long value of a forward is that of a portfolio long in a share of the security and short in a bond with initial price S_0.

4.5 Currency Forwards

Currencies are traded over the counter on the *foreign exchange market* (FX), which determines the relative values of the currencies. The main purpose of the FX is to facilitate trade and investment so that business among international institutions may be conducted efficiently with minimal regard to currency issues. The FX also supports currency speculation, typically through hedge funds. In this case, currencies are bought and sold without the traders actually taking possession of the currency.

An *exchange rate* specifies the value of one currency relative to another, as expressed, for example, in the equation

$$1 \text{ euro} = 1.23 \text{ USD}.$$

Like stock prices, FX rates can be volatile. Indeed, rates may be influenced by a variety of factors, including government budget deficits or surpluses, balance of trade levels, political conditions, inflation levels, quantitative easing by central banks, and market perceptions. Because of this volatility there is a risk associated with currency trading and therefore a need for currency derivatives.

A forward contract whose underlying is a foreign currency is called a *currency forward*. For example, consider a currency forward that allows the purchase in US dollars of one euro at time T. Let K_t denote the forward price of the euro established at time t and let the time-t rate of exchange be 1 euro $= Q_t$ USD. Further, let r_d be the dollar interest rate and r_e the euro interest rate. To establish a formula relating K_t and Q_t, consider the following possible investment strategies made at time t:

- Enter into the forward contract and deposit $K_t e^{-r_d(T-t)}$ dollars in a US bank. At time T, the value of the account is K_t, which is used to purchase the euro.
- Buy $e^{-r_e(T-t)}$ euros for $e^{-r_e(T-t)}Q_t$ dollars and deposit the euro amount in a European bank. At time T, the account has exactly one euro.

As both strategies yield the same amount at maturity, the law of one price ensures that they must have the same value at time t, that is,

$$K_t e^{-r_d(T-t)} = Q_t e^{-r_e(T-t)} \text{ USD}.$$

Solving for K_t, we see that the proper time-t forward price of a euro is

$$K_t = Q_t e^{(r_d - r_e)(T-t)}.$$

In particular, the forward price at time zero is $K = K_0 = Q_0 e^{(r_d - r_e)T}$. This expression should be contrasted with the forward price $K = S_0 e^{r_d T}$ of a stock (4.3). In the latter case, only one instrument, the dollar account, makes payments. In the case of a currency, both accounts are active.

4.6 Futures

A *futures contract*, like a forward, is an agreement between two parties to buy or sell a specified asset for a certain price, the *futures price*, at a certain time T. However, there are two important differences: First, future contracts are traded at an exchange rather than *over-the-counter* (directly by the parties), the exchange acting as an intermediary between the buyer of the asset (long position) and seller (short position). And second, daily futures prices, denoted below by a sequence $F_0, F_1, \ldots, F_T$, are determined daily in a process called "daily settlement" or "marking to market."

Here are the details. The exchange requires both the buyer and the seller to put up an initial amount of cash, usually 10% to 20% of the price, forming what are called *margin accounts*. If the nth daily difference $F_n - F_{n-1}$ is positive, this amount is transferred from the seller's to the buyer's margin account. If it is negative, then the amount $F_{n-1} - F_n$ is transferred from the buyer's to the seller's account. Thus, throughout the life of the contract, the buyer receives

$$\sum_{n=1}^{T}(F_n - F_{n-1}) = F_T - F_0.$$

If an account dips below a certain pre-established level, called the *maintenance level*, then a *margin call* is made and the owner must restore the value of the account. The entire operation is implemented by brokers and has the effect of protecting the parties against default. Table 4.1 illustrates the process. The maintenance level is assumed to be \$700. A margin call is made on day 2, resulting in a \$400 increase in the long account. The table was generated by the VBA module **MarkingToMarket**.

TABLE 4.1: Daily settlements.

Day	Futures Prices	Long Margin Account	Short Margin Account
0	\$5500	\$1000	\$1000
1	\$5300	\$800	\$1200
2	\$5100	\$600 + \$400	\$1400
3	\$5200	\$1100	\$1300
4	\$5400	\$1300	\$1100
5	\$5500	\$1400	\$1000

4.6.1 Example. Suppose the farmer in Example 4.4.1 takes the short position on a futures contract on day 0. On each day j, a futures price F_j for wheat with delivery date $T = 180$ days is quoted. The price depends on the current prospects for a good crop, the expected demand for wheat, etc. The long holder of the contract (the wheat buyer) receives $F_1 - F_0$ on day one, $F_2 - F_1$ on day two, and so on, until delivery day T, when he receives $F_T - F_{T-1}$, where F_T is the spot price of wheat that day.[5] On day T, the buyer has cash in the amount of $F_T - F_0$, pays F_T, and receives his wheat. The net cost to him is F_0. Since this amount is known on day 0, F_0 acts like a forward price. The difference is that the payoff $F_T - F_0$ is paid gradually rather than all at once at maturity. ◊

[5]The *spot price* of a commodity is its price ready for delivery.

*4.7 Equality of Forward and Future Prices

Under the assumption of a constant continuously compounded interest rate r with daily rate $i := r/365$, the futures price and forward price in contracts with the same maturity T (in days) and underlying $\mathcal{S}$ are equal. To verify this we consider two investment strategies:

Strategy I.

- **Day 0**: Deposit F_0 in a risk-free account, so that at maturity T you will have F_0e^{Ti}. Additionally, enter into e^i futures contracts.
- **Day 1**: The profit from the futures transaction on day 0 is $e^i(F_1 - F_0)$, which may be negative. Deposit this into the account.[6] At maturity you will have
$$e^{(T-1)i}e^i(F_1 - F_0) = e^{Ti}(F_1 - F_0).$$
Additionally, increase your holdings to e^{2i} futures contracts.
- **Day 2**: The profit from the futures transaction on day 1 is $e^{2i}(F_2 - F_1)$. Deposit this into the account. At maturity you will have from this transaction
$$e^{(T-2)i}e^{2i}(F_2 - F_1) = e^{Ti}(F_2 - F_1).$$
Additionally, increase your holdings to e^{3i} futures contracts. Continue this process, eventually arriving at
- **Day T-1**: The profit from the futures transaction on the previous day is $e^{(T-1)i}(F_{T-1} - F_{T-2})$. Deposit this into the account. At maturity you will have
$$e^ie^{(T-1)i}(F_{T-1} - F_{T-2}) = e^{Ti}(F_{T-1} - F_{T-2}).$$
Additionally, increase your holdings to e^{Ti} futures contracts.
- **Day T**: The profit from the futures transaction on the previous day is $e^{Ti}(F_T - F_{T-1})$.

Adding the proceeds we see that Strategy I yields the payoff
$$e^{Ti}\big[F_0 + (F_1 - F_0) + (F_2 - F_1) + \cdots + (F_T - F_{T-1})\big] = F_Te^{Ti} = S_Te^{Ti}.$$

Strategy II.

- Obtain e^{iT} forward contracts on day 0, each with strike price K, and additionally deposit the amount K in the account. The payoff at time T for this transaction is
$$e^{Ti}(S_T - K) + Ke^{Ti} = S_Te^{Ti}.$$

The payoffs of strategies I and II are seen to be the same. Since the initial cost of strategy I was F_0 and that of strategy II was K, the law of one price dictates that $F_0 = K$, as claimed.

---

[6]Recall that depositing a negative amount A means borrowing an amount $-A$.

4.8 Call and Put Options

Options are derivatives similar to forwards but have the additional feature of limiting an investor's loss to the cost of the option. Specifically, an *option* is a contract between two parties, the *holder* and the *writer*, which gives the former the right, *but not the obligation*, to buy from or sell to the writer a particular security under terms specified in the contract. An option has value, since the holder is under no obligation to exercise the contract and could gain from the transaction if she does so. The holder of the option is said have the *long position*, while the writer of the option has the *short position*.

Each of the two basic types of options, the *call option* and the *put option*, comes in two styles, *American* and *European*. We begin with the latter.

European Call Option

A *European call option* is a contract with the following conditions: At a prescribed time T, called the *maturity*, the holder (buyer) of the option may purchase from the writer (seller) a prescribed security $\mathcal{S}$ for a prescribed amount K, the *strike price*, established at contract opening. For the holder, the contract is a right, not an obligation. On the other hand, the writer of the option *does* have a potential obligation: he *must* sell the asset if the holder chooses to exercise the option. Since the holder has a right with no obligation, the option has a value and therefore a price, called the *premium*. The premium must be paid by the holder at the time of opening of the contract. Conversely, the writer, having assumed a financial obligation, must be compensated.

To find the payoff of the option at maturity T we argue as follows: If $S_T > K$, the holder will exercise the option and receive the payoff $S_T - K$.[7] On the other hand, if $S_T \le K$, then the holder will decline to exercise the option, since otherwise he would be paying $K - S_T$ more than the security is worth. The payoff for the holder of the call option may therefore be expressed as $(S_T - K)^+$, where, for a real number x,

$$x^+ := \max(x, 0).$$

The option is said to be *in the money* at time t if $S_t > K$, *at the money* if $S_t = K$, and *out of the money* if $S_t < K$.

European Put Option

A *European put option* is a contract that allows the holder, at a prescribed time T, to sell to the writer an asset for a prescribed amount K. The holder is under no obligation to exercise the option, but if she does so the writer must

[7] This assumes that an investor will always act in his or her own self-interest—an assumption that we make throughout the book.

buy the asset. Whereas the holder of a call option is betting that the asset price will rise (the wager being the premium), the holder of a put option is counting on the asset price falling in the hopes of buying low and selling high.

An argument similar to that in the call option case shows that the payoff of a European put option at maturity is $(K - S_T)^+$. Here, the option is *in the money at time t* if $S_t < K$, *at the money* if $S_t = K$, and *out of the money* if $S_t > K$.

Figure 4.1 shows the payoffs, graphed against S_T, for the holder of a call and the holder of a put. The writer's payoff in each case is the negative of the holder's payoff. Thus the transaction is a zero-sum game.

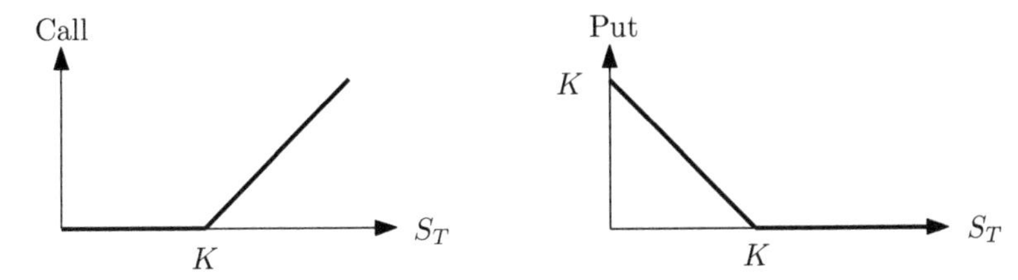

FIGURE 4.1: Call and put payoffs for the holder.

To distinguish among several options with varying strike prices, we shall sometimes refer to a call (put) with strike price K as a *K-call* (*K-put*).

Hedging and Speculation with Options

Options, like forwards and futures, may be used to hedge against price fluctuations. For instance, the farmer in Example 4.4.1 could buy a put option that guarantees a price K for his harvest in six months. If the price drops below K, he would exercise the option. Similarly, the airline company in Example 4.4.2 could take the long position in a call option that gives the company the right to buy fuel at a pre-established price K.

Options may also be used for speculation. A third party in the airline example might take the long position in a call option, hoping that the price of fuel will go up. Of course, in contrast to forwards and futures, option-based hedging and speculation strategies have a cost, namely, the price of the option. The determination of that price is the primary goal of this book.

Note that, while the holder of the option has only the price of the option to lose, the writer stands to take a significantly greater loss. To offset this loss, the writer could use the money received from selling the option to start a portfolio with a maturity value sufficiently large to settle the claim of the holder. Such a portfolio is called a *hedging strategy*. For example, the writer of a call option could take the long position in one share of the security, resulting in a portfolio called a *covered call*. This requires borrowing an amount c, the price of the security minus the income received from selling the call. At time T, the writer's net profit is $S_T - ce^{rT}$, the value of the security less the loan repayment. If the option is exercised, the writer can use the portfolio to settle his obligation of $S_T - K$. [8] The writer will have successfully hedged the short position of the call. In the next chapter, we consider the more complicated process of dynamic hedging, where the holdings in the assets are time-dependent.

Put and Call Combination Portfolios

Options may be used in various combinations to achieve a variety of payoffs and hedging effects. For example, consider a portfolio that consists of a long position in a call option with strike price K_1 and a short position in a call option with strike price $K_2 > K_1$, each with underlying $\mathcal{S}$ and maturity T. Such a portfolio is called a *bull call spread.* Its payoff $(S_T - K_1)^+ - (S_T - K_2)^+$ gains in value as the stock goes up (the investor's outlook is "bullish"). Explicitly

$$(S_T - K_1)^+ - (S_T - K_2)^+ = \begin{cases} 0 & \text{if } S_T \leq K_1, \\ S_T - K_1 & \text{if } K_1 < S_T \leq K_2, \\ K_2 - K_1 & \text{if } K_2 < S_T. \end{cases}$$

Figure 4.2 shows the payoff of the bull call spread graphed against S_T.

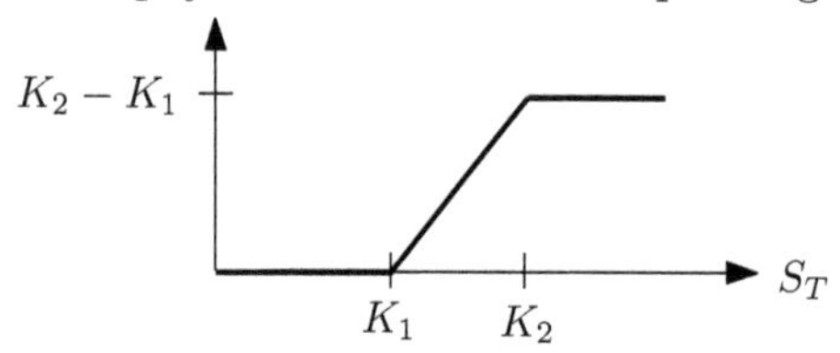

FIGURE 4.2: Bull call spread payoff.

Reversing positions in the calls produces a *bear call spread,* which benefits from a decrease in stock value. The payoff in this case is simply the negative of that of the bull call spread, hence the payoff diagram is the reflection in the S_T axis of that of the bull call spread.

Of more importance to the investor than the payoff is the *profit* of a portfolio, that is, the payoff minus the cost of starting the portfolio. Consider,

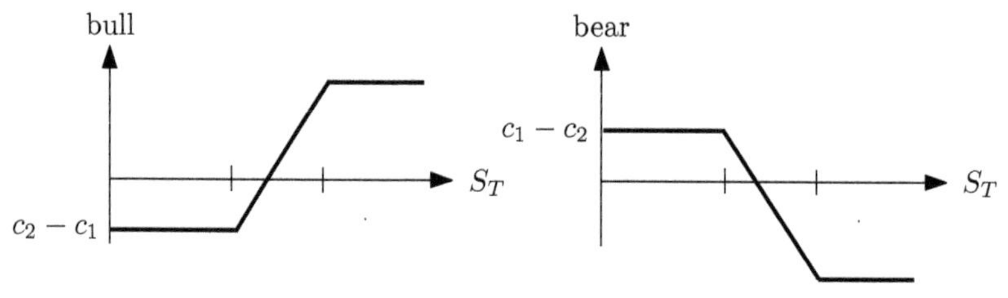

FIGURE 4.3: Bull call and bear call spread profits.

for example, the profit from a bull call spread. Let c_j denote the cost the cost

[8] That $ce^{rT} \leq K$, and consequently $S_T - ce^{rT} \geq S_T - K$, follows from the put-call parity formula $S_0 + P_0 - C_0 = Ke^{-rT}$, discussed in Section 4.9.

of the K_j-call. The cost of the portfolio is then $c_1 - c_2$, and because a lower strike price produces a greater payoff the cost is positive. The profit diagram is the bull call spread payoff graph shifted down by an amount $c_1 - c_2$. By contrast, the bear call spread starts off with a credit of $c_1 - c_2$, and the profit graph is the payoff graph shifted up by $c_1 - c_2$.

American Options

So far, we have considered only European options, these characterized by the restriction that the contracts may be exercised only at maturity. By contrast, American options may be exercised at any time up to and including the expiration date. American options are more difficult to analyze as there is the additional complexity introduced by the holder's desire to find the optimal time to exercise the option and the writer's need to construct a hedging portfolio that will cover the claim at any time $t \leq T$. Nevertheless, as we shall see in the next section, some properties of American options may be readily deduced from those of their European counterparts. We consider American options in more detail in Chapter 10.

4.9 Properties of Options

The options considered in this section are assumed to have maturity T, strike price K, and underlying one share of $\mathcal{S}$ with price process $(S_t)_{t\geq 0}$.

Equivalence of European and American Call Options

The following proposition asserts that an American call option, despite its greater flexibility, has the same value as that of a comparable European call option. (This is *not* the case for put options, however.)

4.9.1 Proposition. *There is no advantage in exercising an American call option early. Thus the cost of an American call option is the same as that of a comparable European call option.*

Proof. Because of its extra flexibility, the cost of the American call is at least that of the European call. Hence, the second statement follows from the first.

To verify the first statement, suppose that an investor holds an American option to buy one share of $\mathcal{S}$ at time T for the price K. If she exercises the option at some time $t < T$, she will immediately realize the payoff $S_t - K$, which, if invested in a risk-free account, will yield the amount $e^{r(T-t)}(S_t - K)$ at maturity. But suppose instead that she sells the stock short for the amount S_t, invests the proceeds, and then purchases the stock at time T (returning it to the lender). If $S_T \leq K$, she will pay the market price S_T; if $S_T > K$, she will exercise the option and pay the amount K. In each case she pays $\min(S_T, K)$.

Therefore, under this strategy, she will have cash $e^{r(T-t)}S_t - \min(S_T, K)$ at time T. Since this amount is at least as large as $e^{r(T-t)}(S_t - K)$, the second strategy is generally superior to the first. Since the second strategy did not require that the option be exercised early, there is no advantage in doing so. □

Henceforth, we shall use the following notation:

- C_t = time-t value of a European call option.
- P_t = time-t value of a European put option.

The Put-Call Parity Formula

The next result uses an arbitrage argument to obtain an important connection between P_t and C_t.

4.9.2 Proposition (Dividend-Free Put-Call Parity Formula). *Consider a call option (either European or American) and a European put option, each with strike price K, maturity T, and underlying one share of $\mathcal{S}$. Then*

$$S_t + P_t - C_t = Ke^{-r(T-t)}, \quad 0 \leq t \leq T. \tag{4.5}$$

Proof. The left side of the equation is the time-t value of a portfolio long in the security and the put and short in the call, and the right side is the time-t value of a bond with face value K. The first portfolio has payoff

$$S_T + (K - S_T)^+ - (S_T - K)^+.$$

By considering cases, we see that this expression reduces to K, the payoff of the second portfolio. By the law of one price, the two portfolios must have the same value at all times t. □

Proposition 4.9.2 shows that the price of a European put option may be expressed in terms of the price C_0 of a call. To find C_0, one uses the notion of a *self-financing portfolio*, described in the next chapter. We shall see later that C_0 depends on a number of factors, including

- the initial price of $\mathcal{S}$ (a large S_0 suggests that a large S_T is likely).
- the strike price K (the smaller the value the greater the payoff).
- the volatility of $\mathcal{S}$.
- the maturity T.
- the interest rate r (which affects the discounted value of the strike price).

The precise quantitative dependence of C_0 on these factors will be examined in detail in the context of the Black-Scholes-Merton model in Chapter 12.

4.10 Dividend-Paying Stocks

In this section, we illustrate how dividends can affect option properties by proving a version of the put-call parity formula for dividend-paying stocks.

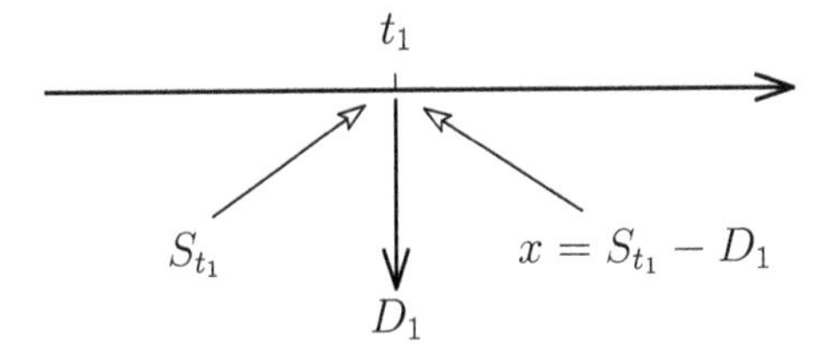

FIGURE 4.4: Dividend D_1.

We begin by observing that in the absence of arbitrage, when a stock pays a dividend, the stock's value is immediately reduced by the amount of the dividend. To see this, suppose that the stock pays a dividend D_1 at time t_1. If the stock's value x immediately after t_1 were greater than $S_{t_1} - D_1$, then at time t_1 an arbitrageur could buy the stock for S_{t_1}, get the dividend, and then sell the stock for x, realizing a positive profit of $D_1 + x - S_{t_1}$. On the other hand, if x were less than $S_{t_1} - D_1$, then at time t_1 she could sell the stock short for S_{t_1}, buy the stock immediately after t_1 for the amount x, and return it along with the dividend (as is required) for a profit of $S_{t_1} - x - D_1 > 0$. Thus, in the absence of arbitrage, $x = S_{t_1} - D_1$.

To derive a put-call parity formula for a dividend-paying stock, assume that a dividend D_j is paid at time t_j, where $0 < t_1 < t_2 < \cdots < t_n \leq T$. Let D denote the present value of the dividend stream:

$$D := e^{-rt_1}D_1 + e^{-rt_2}D_2 + \cdots + e^{-rt_n}D_n. \tag{4.6}$$

Consider two portfolios, one consisting of a long position on a European put with maturity T, strike price K, and underlying $\mathcal{S}$; the other consisting of a long position on the corresponding call, a bond with face value K and maturity T, bonds with face values D_k and maturity times t_k, $k = 1, 2, \ldots, n$, and a short position on one share of the stock. The initial values of the portfolios are P_0 and $C_0 + Ke^{-rT} + D - S_0$, respectively. At maturity, the value of the first portfolio is $(K - S_T)^+$. Assuming that the dividends are deposited into a risk-free account and recalling that in the short position the stock as well as the dividends with interest must be returned, we see that the value of the second portfolio at maturity is

$$(S_T - K)^+ + K + \sum_{k=1}^{n} D_k e^{r(T-t_k)} - \left(S_T + \sum_{k=1}^{n} D_k e^{r(T-t_k)} \right) = (K - S_T)^+.$$

Since the portfolios have the same value at maturity, the law of one price ensures that the portfolios have the same initial value. We have proved

4.10.1 Proposition (Dividend Put-Call Parity Formula). *Consider a call option and a European put option, each with strike price K, maturity T, and underlying one share of S, assumed to be dividend-paying. Then*

$$S_0 + P_0 - C_0 = Ke^{-rT} + D, \tag{4.7}$$

where D is given by (4.6).

4.11 Exotic Options

In recent years, an assortment of more complex derivatives has appeared. These are frequently called *exotic options* and are distinguished by the fact that their payoffs are no longer simple functions of the value of the underlying at maturity. Prominent among these are the path-dependent options, whose payoffs depend not only on the value of the underlying at maturity but on earlier values as well. The main types of path-dependent options are *Asian options*, *lookback options*, and *barrier options*. A brief description of each of these options follows. Valuation formulas are given in Chapter 7 in the context of the binomial model and in Chapter 15 in the Black-Scholes-Merton model.

Asian Options

An *Asian* or *average option* has a payoff that depends on an average $A(S)$ of the price process S of the underlying asset over the contract period. The most common types of Asian options are the

- *fixed strike Asian call* with payoff $(A(S) - K)^+$.
- *floating strike Asian call* with payoff $(S_T - A(S))^+$.
- *fixed strike Asian put* with payoff $(K - A(S))^+$.
- *floating strike Asian put* with payoff $(A(S) - S_T)^+$.

For example, a floating strike put option gives the holder the right to sell the security for an amount which is an average of a predetermined type of the values of S. The average is typically one of the following:

- *discrete arithmetic average*: $A(S) = \dfrac{1}{n}\sum_{j=1}^{n} S_{t_j}$.
- *continuous arithmetic average*: $A(S) = \dfrac{1}{T}\displaystyle\int_0^T S_t\,dt$.
- *discrete geometric average*: $A(S) = \left(\prod_{j=1}^{n} S_{t_j}\right)^{1/n}$.
- *continuous geometric average*: $A(S) = \exp\left(\dfrac{1}{T}\displaystyle\int_0^T \ln S_t\,dt\right)$.

The discrete times t_j satisfy $0 \le t_1 < t_2 < \cdots < t_n \le T$, and the integral $\int_0^T S_t\,dt$ is defined as the random variable whose value at ω is $\int_0^T S_t(\omega)\,dt$.

Asian options are usually less expensive than standard options and have the advantage of being less sensitive to brief manipulation of the underlying asset price, the effect dampened by the averaging process. They are also useful as hedges for an investment plan involving a series of purchases of a commodity with changing price.

Lookback Options

The payoff of a *lookback option* depends on the maximum or minimum value of the asset over the contract period. There are two main categories of lookback options, *floating strike* and *fixed strike*. The holder of a floating strike lookback call option has the right at maturity to buy the stock for its lowest value over the duration of the contract, while the holder of a floating strike lookback put option may sell the stock at its high. Fixed strike lookback options are similar, but involve a predetermined strike price.

To describe the payoffs of lookback options, define

$$M^S := \max\{S_t \mid 0 \le t \le T\} \quad \text{and} \quad m^S := \min\{S_t \mid 0 \le t \le T\}.$$

The payoffs are then

- floating strike lookback call: $S_T - m^S$.
- floating strike lookback put: $M^S - S_T$.
- fixed strike lookback call: $(M^S - K)^+$.
- fixed strike lookback put: $(K - m^S)^+$.

Barrier Options

The payoff for a *barrier option* depends on whether the value of the asset has crossed a predetermined level, called a *barrier*. Because of this added condition, barrier options are generally cheaper than standard options. They are important because they allow the holder to forego paying a premium for scenarios deemed unlikely, while still retaining the essential features of a standard option. For example, if an investor believes that a stock will never fall below \$20, he could buy a barrier call option on the stock whose payoff is $(S_T - K)^+$ if the stock remains above \$20, and zero otherwise.

The payoff for a barrier call option is $(S_T - K)^+\mathbf{1}_A$, while that of a put is $(K - S_T)^+\mathbf{1}_A$, where A is a *barrier event*. The indicator function acts as a switch, activating or deactivating the option if the barrier is breached. Barrier events are typically described in terms of the random variables M^S and m^S

defined above. The most common barrier events are [9]

$\{M^S \le c\}$:	*up-and-out option*	deactivated if asset rises above c.
$\{m^S \ge c\}$:	*down-and-out option*	deactivated if asset falls below c.
$\{M^S \ge c\}$:	*up-and-in option*	activated if asset rises above c.
$\{m^S \le c\}$:	*down-and-in option*	activated if asset falls below c.

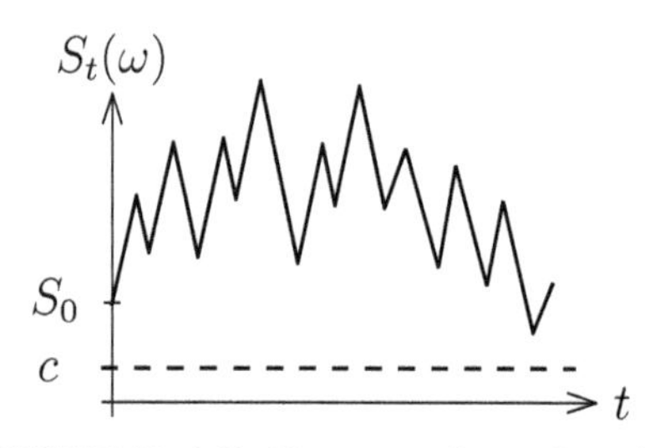

FIGURE 4.5: Down-and-out barrier.

In the first two cases, the so-called *knock-out* cases, the barrier is set so that the option is initially active. For the *knock-in* cases, the option is initially inactive. For example, in the case of an up-and-out option, $S_0 < c$, while for a down-and-in option $S_0 > c$.

Swaps

A *swap* is a contract between two parties to exchange financial assets or cash streams (payment sequences) at some specified time in the future. The most common of these are *currency swaps*, that is, exchanges of one currency for another, and *interest-rate swaps*, where a cash stream with a fixed interest rate is exchanged for one with a floating rate. A *swaption* is an option on a swap. It gives the holder the right to enter into a swap at some future time. A detailed analysis of swaps and swaptions may be found in [7]. A *credit default swap* (CDS) is a contract that allows one party to make a series of payments to another party in exchange for a future payoff if a specified loan or bond defaults. A CDS is, in effect, insurance against default but is also used for hedging purposes and by speculators, as was famously the case in the subprime mortgage crisis of 2007.[10]

[9] Strict inequalities may be used here, as well.

[10] The crisis was triggered by a sharp increase in subprime mortgage defaults. Speculators who realized early the extent of the crisis and purchased CDS's on mortgage-backed securities were compensated as a result of the defaults. See [13] for an interesting account of this and an overall view of the housing market collapse.

*4.12 Portfolios and Payoff Diagrams

The payoff diagram of a portfolio based on $\mathcal{S}$ is simply the graph of the payoff function $f(S_T)$. In this section, we consider the problem of finding the payoff diagram of a portfolio and, conversely, of constructing a portfolio from a given payoff diagram. We consider portfolios consisting of cash, shares of $\mathcal{S}$, and various call and put options on $\mathcal{S}$ with the same maturity T.

Generating a Payoff Diagram from a Portfolio

Portfolios may be conveniently described in tabular form. For example, Table 4.2 describes a portfolio that is long one \$5-call, one \$15-call, one \$10-put, one \$15-put, and one share of the stock, and is short two \$10-calls and a bond with par value \$15. The payoff function, obtained directly from the table, is

$$(S_T - 5)^+ + (S_T - 15)^+ - 2(S_T - 10)^+ + (10 - S_T)^+ + (15 - S_T)^+ + S_T - 15.$$

TABLE 4.2: Portfolio data.

	A	B	C	D	E
5	Strike prices	Calls	Puts	Shares	Cash
6	5	1		1	−\$15
7	10	−2	1		
8	15	1	1		

Figure 4.6 gives the payoff diagram for the portfolio. It was generated from the data in the above table by the VBA module **PortfolioDiagram**. The module may be used to generate payoff diagrams for portfolios of arbitrary complexity.

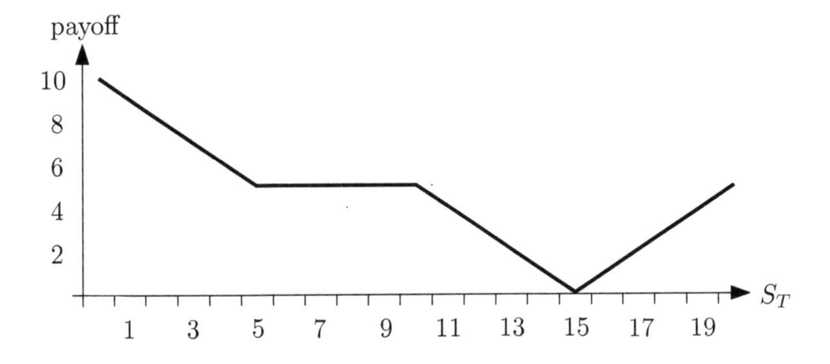

FIGURE 4.6: Portfolio payoff diagram.

Generating a Portfolio from a Payoff Diagram

Consider a portfolio consisting of cash, shares of $\mathcal{S}$, and various call and put options on $\mathcal{S}$ with maturity T. Suppose that the distinct strike prices of the options are $K_1 < K_2 < \cdots < K_n$. If $f(S_T)$ denotes the payoff of the portfolio, then the payoff diagram of the portfolio consists of a sequence of line segments joined at the *vertices* $(K_j, f(K_j))$, $j = 0, 1, \ldots, n$, where $K_0 := 0$ and the last segment has infinite extent. (The bull call spread and bear call spread are examples.) Such a graph is called *piecewise linear* and satisfies

$$f(S_T) = \begin{cases} f(K_j) + m_j(S_T - K_j) & \text{if } K_j \le S_T \le K_{j+1}, \\ f(K_n) + m_n(S_T - K_n) & \text{if } S_T \ge K_n, \end{cases} \tag{4.8}$$

where m_n is a given quantity and

$$m_j = \frac{f(K_{j+1}) - f(K_j)}{K_{j+1} - K_j}, \quad j = 0, \ldots, n-1. \tag{4.9}$$

(See Figure 4.7.)

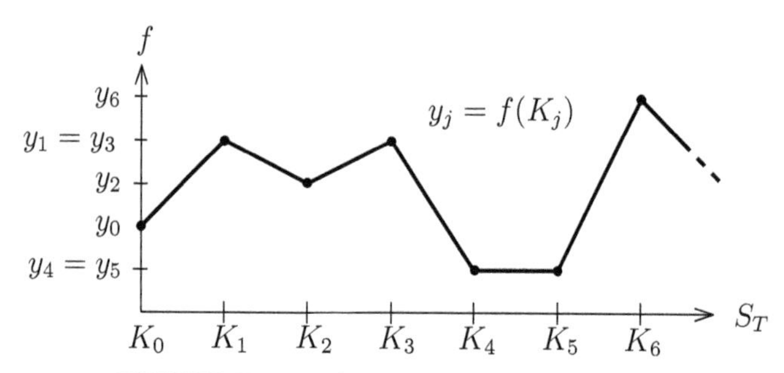

FIGURE 4.7: A piecewise linear function.

In view of (4.8) and (4.9), we see that the payoff of a portfolio may be completely described ("encoded") by the pairs $(K_j, f(K_j))$ and the slope m_n of the last line segment. Given such "payoff code" it is possible to construct a corresponding portfolio. The following theorem is the basis for the construction. It gives a formula for the payoff of a portfolio consisting of an amount $f(0)$ of cash, m_0 shares of a security with maturity T, and $m_i - m_{i-1}$ call options $(1 \le i \le n)$.

4.12.1 Theorem. *Given a function f with piecewise linear graph with last slope m_n and vertices $(K_j, f(K_j))$, where $K_0 = 0 < K_1 < K_2 < \cdots < K_n$, there exists a portfolio with payoff function*

$$f(S_T) = f(0) + m_0 S_T + \sum_{i=1}^{n} (m_i - m_{i-1})(S_T - K_i)^+, \tag{4.10}$$

where, for $0 \le j \le n-1$, m_j is given by (4.9).

Proof. Set $y_j = f(K_j)$. A value of S_T must satisfy either $K_j \le S_T \le K_{j+1}$ for some $0 \le j \le n-1$ or $S_T \ge K_n$. In each case the right side of (4.10) is

$$y_0 + m_0 S_T + \sum_{i=1}^{j}(m_i - m_{i-1})(S_T - K_i) = y_0 + m_j S_T - \sum_{i=1}^{j}(m_i - m_{i-1})K_i$$

for some $1 \le j \le n$. The last sum on the right of this equation may be written

$$m_j K_j - \sum_{i=0}^{j-1} m_i(K_{i+1} - K_i) = m_j K_j - \sum_{i=0}^{j-1}(y_{i+1} - y_i) = m_j K_j - y_j + y_0.$$

Thus the right side of (4.10) is $y_0 + m_j S_T - (m_j K_j - y_j + y_0) = m_j(S_T - K_j) + y_j$, which, by (4.8), is $f(S_T)$. □

4.12.2 Example. Suppose you want a portfolio with the payoff $f(S_T)$ given in the diagram below, where the slope of the last segment is -1. The module

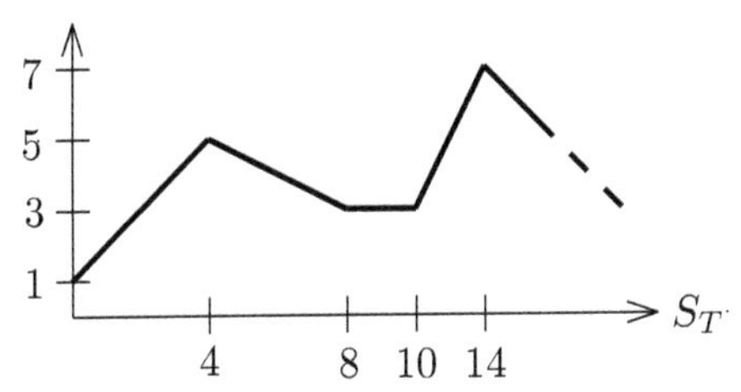

FIGURE 4.8: Payoff diagram for $f(S_T)$.

PortfolioFromDiagram was used to generate a corresponding portfolio of call options, which is summarized in Table 4.3. From this we see that the portfolio

TABLE 4.3: Input/output for **PortfolioFromDiagram**

K_j	$f(K_j)$	Last slope	Calls	Shares	Cash
0	1	-1		1	1.00
4	5		-1.5		
8	3		.5		
10	3		1		
14	7		-2		

should be long on a bond with face value \$1, long in one share of the stock, short in one and a half \$4-calls and two \$14-calls, and long in half an \$8-call and one \$10-call. The payoff is

$$1 + S_T - 1.5(S_T - 4)^+ + .5(S_T - 8)^+ + (S_T - 10)^+ - 2(S_T - 14)^+. \quad \Diamond$$

4.13 Exercises

1. Show that $S_0 - Ke^{-rT} \le C_0 \le S_0$ and $P_0 \le Ke^{-rT}$.

2. (Generalization of Exercise 1.) Show that

   (a) $C_0 + e^{-rT}\min(S_T, K) \le S_0$. (b) $e^{rT}P_0 + \min(S_T, K) \le K$.

3. Let two portfolios $\mathcal{X}$ and $\mathcal{Y}$ have time-t values X_t and Y_t, respectively. Show that in the absence of arbitrage

   (a) if $X_T \ge Y_T$ then $X_t \ge Y_t$ for all $0 \le t \le T$.

   (b) if $X_T > Y_T$ then $X_t > Y_t$ for all $0 \le t \le T$.

4. Consider two call options $\mathcal{C}$ and $\mathcal{C}'$ with the same underlying, the same strike price, and with prices C_0 and C_0', respectively. Suppose that $\mathcal{C}$ has maturity T and $\mathcal{C}'$ has maturity $T' < T$. Explain why $C_0' \le C_0$.

5. Consider two American put options $\mathcal{P}$ and $\mathcal{P}'$ with the same underlying, the same strike price K, and with prices P_0 and P_0', respectively. Suppose that $\mathcal{P}$ has maturity T and that $\mathcal{P}'$ has maturity $T' < T$. Explain why $P_0' \le P_0$.

6. Graph the payoffs against S_T of a portfolio that is

   (a) long in a stock $\mathcal{S}$ and short in a call on $\mathcal{S}$.

   (b) short in $\mathcal{S}$ and long in a call on $\mathcal{S}$.

   (c) long in $\mathcal{S}$ and short in a put on $\mathcal{S}$.

   (d) short in $\mathcal{S}$ and long in a put on $\mathcal{S}$.

   Assume the calls and puts are European with strike price K.

7. Graph the payoffs against S_T for an investor who

   (a) buys one call and two puts. (b) buys two calls and one put.

   Assume the calls and puts are European with strike price K, maturity T, and underlying $\mathcal{S}$. (The portfolios in (a) and (b) are called, respectively, a *strip* and a *strap*.) What can you infer about the strategy underlying each portfolio?

8. Let $0 < K_1 < K_2$. Graph the payoff against S_T for an investor who

   (a) holds one K_1-call and writes one K_2-put.

   (b) holds one K_1-put and writes one K_2-call.

   Assume that the options are European and have the same underlying and maturity T.

9. A *strangle* is a portfolio consisting of a K_1-put and a K_2-call with $K_2 > K_1$. Graph the payoff diagram. Assume that the options are European and have the same underlying asset and expiration date T.

10. A *straddle* is a portfolio with long positions in a European call and put with the same strike price, maturity, and underlying. The straddle is seen to benefit from a movement in either direction away from the strike price. Show that the payoff of a straddle is $|S_T - K|$ and construct the payoff diagram.

11. Consider two call options with the same underlying and same maturity T, one with price C_0 and strike price K, the other with price C_0' and strike price $K' > K$. Give a careful arbitrage argument to show that $C_0 \geq C_0'$.

12. Consider two European put options with the same underlying and same maturity T, one with price P_0 and strike price K, the other with price P_0' and strike price $K' > K$. Give a careful arbitrage argument to show that $P_0 \leq P_0'$.

13. Let C_0' and C_0'' denote the costs of call options with strike prices K' and K'', respectively, where $0 < K' < K''$, and let C_0 be the cost of a call option with price $K := (K' + K'')/2$, all with the same maturity and underlying.

    (a) Show that $C_0 \leq (C_0' + C_0'')/2$.

    Hint. Assume that $C_0 > (C_0' + C_0'')/2$. Write two options with strike price K, buy one with strike price K' and another with strike price K'', giving you cash in the amount $2C_0 - C_0' - C_0'' > 0$. Consider the cases obtained by comparing S_T with K, K', and K''.

    (b) The portfolio described in the hint is called a *butterfly spread.* Assuming the portfolio has a positive cost $c := C_0' + C_0'' - 2C_0$, graph the profit diagram of the portfolio.

14. Referring to Exercises 11 and 12, show that

$$\max(P_0' - P_0, C_0 - C_0') \leq (K' - K)e^{-rT}.$$

15. A *capped option* is a standard option with profit capped at a pre-established amount A. The payoff of a capped call option is $\min\big((S_T - K)^+, A\big)$. Which of the following portfolios has time-t value equal to that of a capped call option:

    (a) strip (b) strap (c) straddle (d) strangle (e) bull spread (f) bear spread.

16. Show that a bull spread can be created from a combination of positions in put options and a bond.

17. Find a portfolio that has the payoff diagram in Figure 4.9, where the slope of the last segment is 0. (See Section 4.12.)

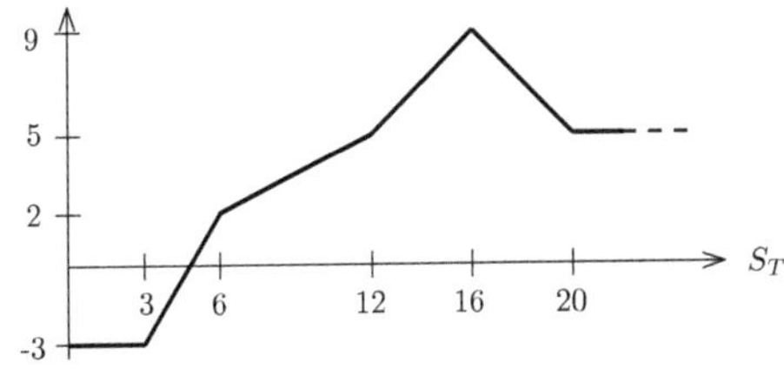

FIGURE 4.9: Payoff diagram for Exercise 17.

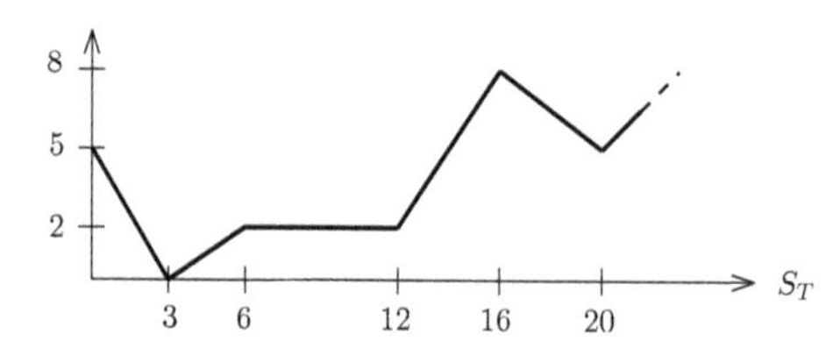

FIGURE 4.10: Payoff diagram for Exercise 18.

18. Find a portfolio that has the payoff diagram in Figure 4.10, where the slope of the last segment is 1. (See Section 4.12.)

19. Consider an investor who buys two European calls and sells a European put, all based on one share of $\mathcal{S}$ and with same strike price K and maturity T, and who sells short one share of $\mathcal{S}$. Assume all transaction cash is borrowed or invested and there is no arbitrage. Show that if at time T the calls are in the money, then the total profit of the investor is $S_T - K - e^{rT}C_0$.

20. Consider an investor who buys two European puts, sells a European call, all with underlying one share of $\mathcal{S}$, with same strike price K and maturity T, and who buys two shares of $\mathcal{S}$. Assume all transaction cash is borrowed or invested and there is no arbitrage. Show that if at time T the call is in the money, then the total profit of the investor is $3(S_T - K) - e^{rT}C_0$.

21. Show that if the market is arbitrage-free then $\mathbb{P}(S_T \geq S_0e^{rT}) = 1$.

22. Show that if the market is arbitrage-free and $\mathbb{P}(S_T > K) > 0$, then $C_0 > 0$ and $P_0 > Ke^{-rT} - S_0$.

23. Show that for an arbitrage-free market $\mathbb{P}(S_T \geq K + C_0e^{rT}) = 1$.

24. The *dual* of a portfolio is obtained by replacing each K-call by a K-put and vice versa. Prove that the butterfly spread of Exercise 13 has the same payoff as its dual.

Chapter 5

Discrete-Time Portfolio Processes

In this chapter we introduce the notion of a self-financing, replicating portfolio, a key component of option valuation models. Such a portfolio is based on underlying assets whose values are assumed to be discrete-time random processes, defined in the first section of this chapter. Portfolios based on assets with continuous-time value random processes are considered in later chapters.

5.1 Discrete Time Stochastic Processes

Many experiments are *dynamic*, that is, they continue over time. Repeated tosses of a coin or the periodic observation of price fluctuations of a stock are simple examples. If the outcome of the experiment at time n is described by a random variable X_n, then the sequence (X_n) may be taken as a mathematical model of the experiment. In this section we formalize these ideas.

Definition of a Process

A *discrete-time stochastic process* on a probability space $(\Omega, \mathcal{F}, \mathbb{P})$ is a finite or infinite sequence $X = (X_n)$ of random variables X_0, $X_1, \ldots$ on the sample space Ω. [1] We shall also refer to X synonymously as a *random process* or simply a *process*. If each X_n is a constant, then the process is said to be *deterministic*. If $X^1, X^2, \ldots, X^d$ are processes and $\boldsymbol{X}_n$ is the random vector $(X_n^1, X_n^2, \ldots, X_n^d)$, then the sequence $\boldsymbol{X} = (\boldsymbol{X}_n)$ is called a *d-dimensional stochastic process*. For example, a portfolio consisting of d stocks gives rise to a d-dimensional stochastic process $\boldsymbol{S} = (S^1, \ldots, S^d)$, where S_n^j denotes the price of stock j on day n. In finance, one usually considers discrete-time processes indexed by the finite set $\{0, 1, \ldots, N\}$, reflecting the fact that financial contracts typically have a time horizon, for example, the maturity of an option.

Discrete Time Filtrations

A *discrete-time filtration* on a probability space $(\Omega, \mathcal{F}, \mathbb{P})$ is a finite or infinite sequence $(\mathcal{F}_n)$ of σ-fields contained in $\mathcal{F}$ such that $\mathcal{F}_n \subseteq \mathcal{F}_{n+1}$ for

[1] The process could start with the index 1 as well.

all n. A filtration $(\mathcal{F}_n)$ may be viewed as a mathematical description of the information gradually revealed by an experiment evolving in time, for example, one consisting of repeated trials: at the end of trial n, the information revealed by the outcomes of this and previous trials is encapsulated in a σ-field $\mathcal{F}_n$.

5.1.1 Example. A jar contains r red marbles and w white marbles. A marble is drawn at random from the jar, its color noted, and the marble is replaced. The procedure is repeated until N marbles have been drawn and replaced. The sample space consists of sequences $\omega_1\omega_2\ldots\omega_N$, where $\omega_n = R$ or W. To construct a filtration modeling the evolution of information revealed by the experiment, for each fixed sequence $\omega = \omega_1\omega_2\ldots\omega_n$ of R's and W's appearing in the first n draws, let A_ω denote the set of all possible sequences of the form $\omega\eta$, where η is any sequence of R's and W's of length $N-n$, these representing outcomes of the last $N-n$ draws. For example, if $N = 4$, then

$$A_{RW} = \{\underline{\mathrm{RW}}\mathrm{RR}, \underline{\mathrm{RW}}\mathrm{RW}, \underline{\mathrm{RW}}\mathrm{WR}, \underline{\mathrm{RW}}\mathrm{WW}\}.$$

We show that the sets A_ω generate a filtration that describes the flow of information during the experiment.

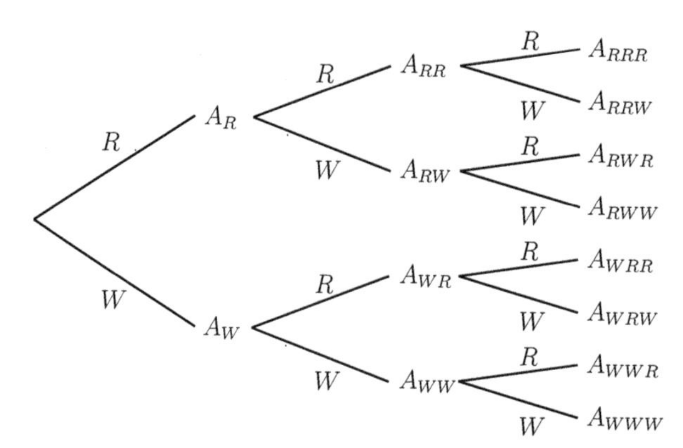

FIGURE 5.1: Filtration generators for three draws.

Before the first draw, we have no foreknowledge of the outcomes of the experiment. The σ-field $\mathcal{F}_0$ corresponding to this void of information is $\mathcal{F}_0 = \{\emptyset, \Omega\}$. After the first draw, we know whether the outcome was R or W but we have no information regarding subsequent draws. The σ-field $\mathcal{F}_1$ describing the information gained on the first toss is therefore $\{\emptyset, \Omega, A_R, A_W\}$. After the second toss, we know which of the outcomes $\omega_1\omega_2 = RR$, RW, WR, WW has occurred, but we have no knowledge regarding the impending third draw. The σ-field $\mathcal{F}_2$ describing the information gained on the first two draws consists of $\emptyset$, Ω, and all unions of the sets $A_{\omega_1\omega_2}$. It follows from $A_R = A_{RR} \cup A_{RW}$ and $A_W = A_{WR} \cup A_{WW}$ that $\mathcal{F}_1 \subseteq \mathcal{F}_2$. Continuing in this way, we obtain a filtration $(\mathcal{F}_n)_{n=0}^N$, where $\mathcal{F}_n$ $(n \geq 1)$ is the σ-field consisting of all unions of (that is, generated by) the pairwise disjoint sets A_ω (see Example 2.3.1). Note that after N tosses we have complete information, as is seen by the fact that $\mathcal{F}_N$ consists of all subsets of Ω. $\Diamond$

In the preceding example, the filtration consisted of σ-fields generated by partitions. It turns out that *any* filtration on a finite sample space is of this form. This follows from

5.1.2 Proposition. *Let $\mathcal{G}$ be a σ-field of subsets of a finite sample space Ω. Then $\mathcal{G}$ is generated by a partition of Ω.*

Proof. Let $B_1, B_2, \ldots, B_m$ be the distinct members of $\mathcal{G}$. For each m-tuple $\epsilon = (\epsilon_1, \epsilon_2, \ldots, \epsilon_m)$ with $\epsilon_j = \pm 1$, define $B^\epsilon = B_1^{\epsilon_1} B_2^{\epsilon_2} \cdots B_m^{\epsilon_m}$, where $B_j^{\epsilon_j} = B_j$ if $\epsilon_j = 1$, and $B_j^{\epsilon_j} = B_j'$ if $\epsilon_j = -1$. The sets B^ϵ are pairwise disjoint since two distinct ϵ's will differ in some coordinate j, and the intersection of the corresponding sets B^ϵ will be contained in $B_j B_j' = \emptyset$. Some of the sets B^ϵ are empty, but every B_j is a union of those B^ϵ for which $\epsilon_j = 1$. Denoting the nonempty sets B^ϵ by $A_1, \ldots, A_n$, we see that $\mathcal{G}$ is generated by the partition $\mathcal{P} = \{A_1, \ldots, A_n\}$. $\square$

5.1.3 Remark. Suppose $(\mathcal{F}_n)_{n=1}^N$ is a filtration on Ω such that $\mathcal{F}_n$ is generated by the partition $\{A_{n,1}, A_{n,2}, \ldots, A_{n,m_n}\}$. Since $\mathcal{F}_n \subseteq \mathcal{F}_{n+1}$, each $A_{n,j}$ is a union of some of the sets $A_{n+1,1}$, $A_{n+1,2}$, $\ldots$, $A_{n+1,m_{n+1}}$. We can assign to

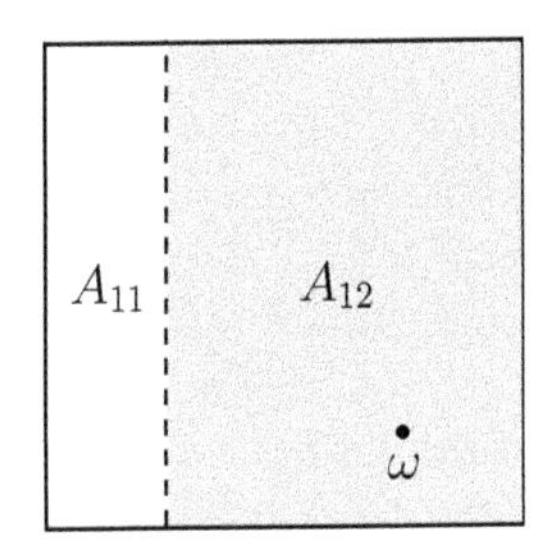

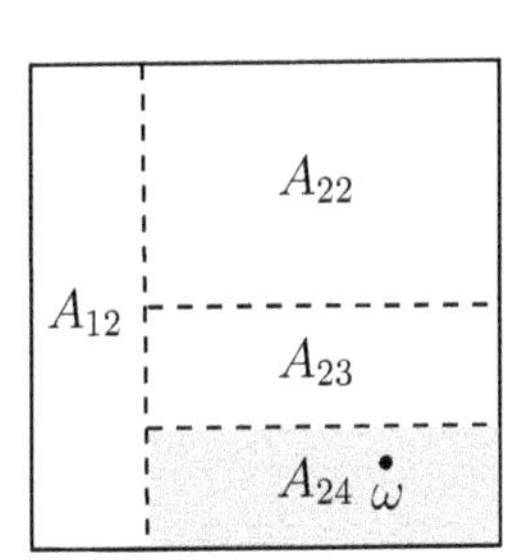

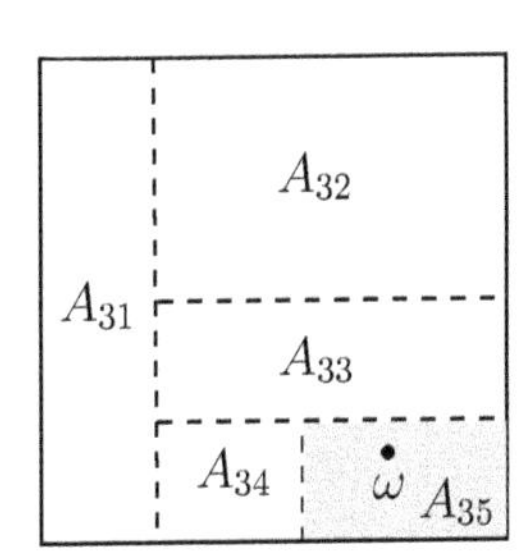

FIGURE 5.2: ω is described by the sequence $(2, 4, 5)$.

each outcome $\omega \in \Omega$ a unique sequence $(j_1, j_2, \ldots, j_N)$ with the property that $\omega \in A_{n,j_n}$ for each n. This provides a dynamic interpretation of an abstract experiment. Figure 5.2 illustrates the case $N = 3$. $\Diamond$

Adapted Processes

A stochastic process (X_n) is said to be *adapted to a filtration* $(\mathcal{F}_n)$ if for each n the random variable X_n is $\mathcal{F}_n$-measurable. A d-dimensional stochastic process is *adapted to a filtration* if each component process is adapted to $(\mathcal{F}_n)$.

An important special case of adaptability is the following. Let (X_n) be a stochastic process. For each n, let

$$\mathcal{F}_n^X = \sigma(X_j : j \leq n) = \sigma(X_1, \ldots, X_n)$$

denote the σ-field generated by all events $\{X_j \in J\}$, where $j \le n$ and J is an arbitrary interval of real numbers. The filtration $(\mathcal{F}_n^X)$ is called the *natural filtration for* (X_n) or the *filtration generated by the process* (X_n). It is the smallest filtration (in the sense of inclusion) to which (X_n) is adapted.

As noted earlier, a filtration is a mathematical description of the accumulation of information over time. If (X_n) is adapted to a filtration $(\mathcal{F}_n)$, then the σ-field $\mathcal{F}_n$ includes all relevant information about the process up to time n, this information being fundamentally of the form $\{a < X_n < b\}$. The σ-field $\mathcal{F}_n^X$ includes all relevant information about the process up to time n but nothing more.

5.1.4 Example. Referring to Example 5.1.1, denote by X_n the number of red marbles appearing in the first n draws. The filtration $(\mathcal{F}_n)$ in that example is the filtration generated by (X_n). This means that for each n, every set of the form A_ω, where ω has length n, may be completely described by events of the form $\{X_j = x_j\}$, $j \le n$; and, conversely, each event $\{X_j = x_j\}$ may be completely described by the sets A_ω. For example,

$$\{X_4 = 2\} = A_{\text{RRWW}} \cup A_{\text{RWRW}} \cup A_{\text{RWWR}} \cup A_{\text{WWRR}} \cup A_{\text{WRWR}} \cup A_{\text{WRRW}}$$

and

$$A_{\text{RWRW}} = \{X_1 = 1\} \cap \{X_2 = 1\} \cap \{X_3 = 2\} \cap \{X_4 = 2\}. \qquad \Diamond$$

Predictable Processes

A stochastic process $X = (X_n)$ is said to be *predictable with respect to a filtration* $(\mathcal{F}_n)$ if X_n is $\mathcal{F}_{n-1}$ measurable for each $n \ge 1$. A d-dimensional stochastic process is *predictable* if each component process is predictable.

Every predictable process is adapted, but not conversely. The difference may be understood as follows: For an adapted process, X_n is determined by events in $\mathcal{F}_n$. For a predictable process, it is possible to determine X_n by events in $\mathcal{F}_{n-1}$, that is, by the events occurring at the earlier time $n-1$.

5.1.5 Example. Suppose that the balls in Example 5.1.1 are not replaced after the draws. A gambler places a wager before each draw. If the original number of red and white balls is known, then, by keeping track of the number of marbles of each color in the jar after each draw, the gambler may make an informed decision regarding the size of the wager. For example, if the first $n-1$ draws result in more red marbles than white, the gambler should bet on white for the nth draw, the size of the bet determined by the configuration of white and red marbles left in the jar. As in Example 5.1.1, the flow of information may be modeled by a filtration $(\mathcal{F}_n)$. Because the gambler is not prescient, the nth wager Y_n must be determined by events in $\mathcal{F}_{n-1}$. Thus the wager process (Y_n) is predictable. On the other hand, if X_n denotes the number of red balls in n draws, then, as before, (X_n) is adapted to the filtration, but it is not predictable. For example, the event $\{X_4 = 2\}$, two reds in four draws, cannot be described solely by events in $\mathcal{F}_3$. $\Diamond$

5.2 Portfolio Processes and the Value Process

Consider a market $(\mathcal{B}, \mathcal{S}^1, \ldots, \mathcal{S}^d)$ consisting of d risky assets $\mathcal{S}^j$ and a risk-free bond $\mathcal{B}$ earning compound interest at a rate of i per period. We assume that the price of the jth security $\mathcal{S}^j$ is given by a stochastic process $S^j = (S^j_n)_{n=0}^N$ on some probability space $(\Omega, \mathcal{F}, \mathbb{P})$, where S^j_0 is constant (the price at time 0 is known). Let

$$\boldsymbol{S} := (S^1, S^2, \ldots, S^d)$$

denote the vector price process of the securities and let $(\mathcal{F}^{\boldsymbol{S}}_n)_{n=0}^N$ denote the natural filtration for $\boldsymbol{S}$. Thus, for each n, $\mathcal{F}^{\boldsymbol{S}}_n$ is generated by all events of the form

$$\{S^j_k \in J_{jk}\}, \quad \text{where } J_{jk} \text{ is an interval, } 0 \le k \le n, \text{ and } 1 \le j \le d.$$

Note that because S^j_0 is constant, $\mathcal{F}^{\boldsymbol{S}}_0$ is the trivial σ-field $\{\emptyset, \Omega\}$. The filtration $(\mathcal{F}^{\boldsymbol{S}}_n)$ models the accumulation of stock price information in the discrete time interval $[0, N]$. Because the focus of interest is only on this information, we may take $\mathcal{F} = \mathcal{F}^{\boldsymbol{S}}_N$.

Assume that the market allows unlimited transactions in the assets $\mathcal{S}^1, \ldots, \mathcal{S}^d$ and the bond $\mathcal{B}$. For simplicity, the value of $\mathcal{B}$ at time 0 is taken to be \$1.00. Note that the price process $B = (B_n)_{n=0}^N$ of $\mathcal{B}$ is deterministic; indeed, $B_n = (1+i)^n$. A *portfolio* or *trading strategy* for $(\mathcal{B}, \mathcal{S}^1, \ldots, \mathcal{S}^d)$ is a $(d+1)$-dimensional predictable stochastic process $(\phi, \theta^1, \ldots \theta^d)$ on $(\Omega, \mathcal{F}, \mathbb{P})$, where ϕ_n and θ^j_n denote, respectively, the number of bonds $\mathcal{B}$ and the number of shares of $\mathcal{S}^j$ held at time n. We set

$$\boldsymbol{\theta} := (\theta^1, \ldots \theta^d).$$

The *initial investment* or *initial wealth of the portfolio* is defined as

$$V_0 := \phi_1 + \boldsymbol{\theta}_1 \cdot \boldsymbol{S}_0 = \phi_1 + \sum\nolimits_{j=1}^d \theta^j_1 S^j_0. \tag{5.1}$$

The *value* of the portfolio at time n is

$$V_n := \phi_n B_n + \boldsymbol{\theta}_n \cdot \boldsymbol{S}_n = \phi_n B_n + \sum\nolimits_{j=1}^d \theta^j_n S^j_n, \quad n = 1, \ldots, N.$$

Here, we have used the dot product notation from vector analysis. The stochastic process $V = (V_n)_{n=0}^N$ is called the *value process* or *wealth process* of the portfolio.

5.3 Self-Financing Trading Strategies

The idea behind a trading strategy is this: At time 0, the number ϕ_1 of bonds $\mathcal{B}$ and the number of shares θ_1^j of $\mathcal{S}^j$ are chosen to satisfy the initial wealth equation (5.1). These are constants, as the values S_0^j are known. At time $n \geq 1$, the value of the portfolio before the prices S_n^j are known (and before the bond's new value is noted) is

$$\phi_n B_{n-1} + \boldsymbol{\theta}_n \cdot \boldsymbol{S}_{n-1}.$$

When S_n becomes known, the portfolio has value

$$\phi_n B_n + \boldsymbol{\theta}_n \cdot \boldsymbol{S}_n. \tag{5.2}$$

At this time, the number of units ϕ_n of $\mathcal{B}$ and shares θ_n^j of $\mathcal{S}^j$ may be adjusted,

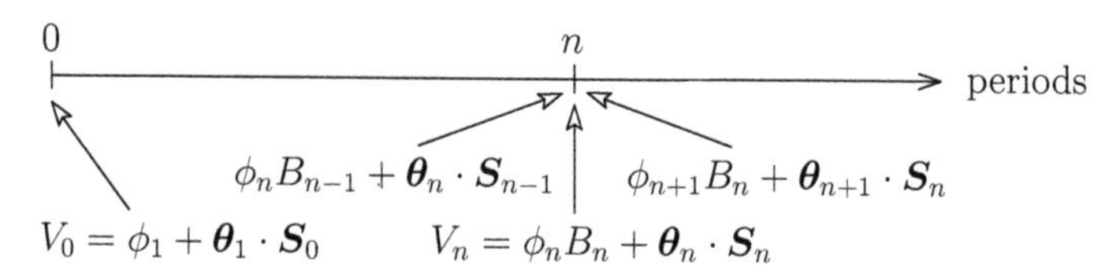

FIGURE 5.3: Portfolio readjustment.

the strategy based on the price history $S_0, S_1, \ldots, S_n$. The process is illustrated in Figure 5.3. Predictability of the portfolio process underlies this procedure: the new values ϕ_{n+1} and θ_{n+1} are determined using only information provided by $\mathcal{F}^{\boldsymbol{S}_n}$. After readjustment, the value of the portfolio during the time interval $(n, n+1)$ is

$$\phi_{n+1} B_n + \boldsymbol{\theta}_{n+1} \cdot \boldsymbol{S}_n. \tag{5.3}$$

At time $n+1$, the procedure is repeated.

Now suppose that each readjustment of the portfolio is carried out without changing its current value, that is, without the extraction or infusion of wealth. Shares of $\mathcal{S}^j$ and units of $\mathcal{B}$ may be bought or sold, but the net value of the transactions is zero. Mathematically, this simply means that the quantities in (5.2) and (5.3) are equal, that is,

$$(\phi_{n+1} - \phi_n) B_n + (\boldsymbol{\theta}_{n+1} - \boldsymbol{\theta}_n) \cdot \boldsymbol{S}_n = 0.$$

This is the notion of *self-financing portfolio.* Using the notation

$$\Delta x_n := x_{n+1} - x_n, \quad \text{and} \quad \Delta(x_n^1, x_n^2, \ldots, x_n^d) := (\Delta x_n^1, \Delta x_n^2, \ldots, \Delta x_n^d),$$

the above equation may be written

$$B_n \Delta\phi_n + \boldsymbol{S}_n \cdot \Delta\boldsymbol{\theta}_n = 0, \quad n = 1, 2, \ldots, N-1. \tag{5.4}$$

This is the *self-financing condition.* As we shall see in the next section, there are several ways of expressing this fundamental notion.

5.4 Equivalent Characterizations of Self-Financing

In Theorem 5.4.1 below, we give several alternate ways of characterizing a self-financing portfolio. One of these uses the following important notion: Let $X = (X_n)_{n=0}^N$ be a price process. The *discounted process* $\widetilde{X} = (\widetilde{X}_n)_{n=0}^N$ is defined by

$$\widetilde{X}_n = (1+i)^{-n} X_n, \quad n = 0, 1, \ldots, N.$$

The process $\widetilde{X}$ measures the current value of X in terms of the current value of the bond $\mathcal{B}$. In this context, the bond process is said to be a *numeraire.* Converting from one measure of value to another is referred to as a *change of numeraire.* For example, consider the case $d = 1$. If $i = .01$ and $S_{10}(\omega) = \$5000$, where ω is a particular market scenario, then

$$\widetilde{S}_{10}(\omega) = \frac{5000}{(1.01)^{10}} \approx 4526 \text{ current bond units,}$$

as the bond is now worth $\$(1.01)^{10}$ (the value of one current bond unit). This provides a dynamic way of measuring the value of the process (S_n) against the bond.

For a vector process $\boldsymbol{X} = (X^1, \ldots, X^d)$, the *discounted vector process* $\widetilde{\boldsymbol{X}}$ is defined by

$$\widetilde{\boldsymbol{X}}_n = \frac{1}{(1+i)^n}(X_n^1, \ldots, X_n^d) = (\widetilde{X}_n^1, \ldots, \widetilde{X}_n^d) \quad n = 0, 1, \ldots, N.$$

We may now prove the following theorem.

5.4.1 Theorem. *For a trading strategy* $(\phi, \boldsymbol{\theta}) = (\phi, \theta^1, \ldots, \theta^d)$ *with value process* V*, the following statements are equivalent:*

(a) $(\phi, \boldsymbol{\theta})$ *is self-financing, that is,* (5.4) *holds.*

(b) $\Delta V_n = \phi_{n+1} \Delta B_n + \boldsymbol{\theta}_{n+1} \cdot \Delta \boldsymbol{S}_n, \quad n = 0, 1, \ldots N-1.$

(c) *For* $n = 0, 1, \ldots, N-1$ *and* $a := 1 + i$*,*

$$V_{n+1} = \boldsymbol{\theta}_{n+1} \cdot \boldsymbol{S}_{n+1} + a\big(V_n - \boldsymbol{\theta}_{n+1} \cdot \boldsymbol{S}_n\big) = aV_n + \boldsymbol{\theta}_{n+1} \cdot (\boldsymbol{S}_{n+1} - a\boldsymbol{S}_n).$$

(d) $\Delta \widetilde{V}_n = \boldsymbol{\theta}_{n+1} \cdot \Delta \widetilde{\boldsymbol{S}}_n, \quad n = 0, 1, \ldots, N-1.$

(e) $\phi_{n+1} = V_0 - \sum_{k=0}^n \widetilde{\boldsymbol{S}}_k \cdot \Delta \boldsymbol{\theta}_k, \quad n = 0, 1, \ldots, N-1,$ *where* $\boldsymbol{\theta}_0 := (0, \ldots, 0)$.

Proof. For $n = 0, 1, \ldots, N-1$, set

$$\begin{aligned} Y_n &= B_n \Delta \phi_n + \boldsymbol{S}_n \cdot \Delta \boldsymbol{\theta}_n = (\phi_{n+1} - \phi_n) B_n + (\boldsymbol{\theta}_{n+1} - \boldsymbol{\theta}_n) \cdot \boldsymbol{S}_n \\ &= \phi_{n+1} B_n + \boldsymbol{\theta}_{n+1} S_n - \phi_n B_n + \boldsymbol{\theta}_n \cdot \boldsymbol{S}_n. \end{aligned}$$

Recalling that $V_n = \phi_n B_n + \boldsymbol{\theta}_n \cdot \boldsymbol{S}_n$ we see that

$$Y_n = \phi_{n+1} B_n + \boldsymbol{\theta}_{n+1} \cdot \boldsymbol{S}_n - V_n.$$

and so

$$V_n = \phi_{n+1} B_n + \boldsymbol{\theta}_{n+1} \cdot \boldsymbol{S}_n - Y_n. \tag{5.5}$$

Thus

$$\begin{aligned}\Delta V_n &= \phi_{n+1} B_{n+1} + \boldsymbol{\theta}_{n+1} \cdot \boldsymbol{S}_{n+1} - (\phi_{n+1} B_n + \boldsymbol{\theta}_{n+1} \cdot \boldsymbol{S}_n - Y_n) \\ &= \phi_{n+1} \Delta B_n + \boldsymbol{\theta}_{n+1} \cdot \Delta \boldsymbol{S}_n + Y_n.\end{aligned}$$

Since the portfolio is self-financing iff $Y_n = 0$ for all n, we see that (a) and (b) are equivalent. Further, by (5.5),

$$\phi_{n+1} B_{n+1} = (1+i)\phi_{n+1} B_n = (1+i)(Y_n + V_n - \boldsymbol{\theta}_{n+1} \cdot \boldsymbol{S}_n),$$

hence

$$V_{n+1} = \boldsymbol{\theta}_{n+1} \cdot \boldsymbol{S}_{n+1} + \phi_{n+1} B_{n+1} = \boldsymbol{\theta}_{n+1} \cdot \boldsymbol{S}_{n+1} + (1+i)(Y_n + V_n - \boldsymbol{\theta}_{n+1} \cdot \boldsymbol{S}_n).$$

This shows that (a) and (c) are equivalent.

That (c) and (d) are equivalent follows by dividing the equation in (c) by a^{n+1} to obtain the equivalent equation

$$\widetilde{V}_{n+1} = \widetilde{V}_n + \boldsymbol{\theta}_{n+1} \cdot (\widetilde{\boldsymbol{S}}_{n+1} - \widetilde{\boldsymbol{S}}_n).$$

To show that (a) and (e) are equivalent, assume first that (a) holds. Then, by (5.4),

$$B_k \Delta \phi_k = -\boldsymbol{S}_k \cdot \Delta \boldsymbol{\theta}_k, \quad k = 1, 2, \ldots, N-1.$$

Multiplying by $B_k^{-1} = (1+i)^{-k}$ and summing, we have

$$\sum_{k=1}^{n} \Delta \phi_k = -\sum_{k=1}^{n} \widetilde{\boldsymbol{S}}_k \cdot \Delta \boldsymbol{\theta}_k, \quad n = 1, \ldots, N-1.$$

The left side of this equation collapses to $\phi_{n+1} - \phi_1$, hence

$$\phi_{n+1} = \phi_1 - \sum_{k=1}^{n} \widetilde{\boldsymbol{S}}_k \cdot \Delta \boldsymbol{\theta}_k, \quad n = 1, 2, \ldots, N-1.$$

Finally, noting that $\phi_1 = V_0 - \theta_1 S_0 = V_0 - \boldsymbol{S}_0 \cdot \Delta \boldsymbol{\theta}_0$ (recalling that $\boldsymbol{\theta}_0 = (0, \ldots, 0)$), we have

$$\phi_{n+1} = V_0 - \sum_{k=0}^{n} \widetilde{\boldsymbol{S}}_k \cdot \Delta \boldsymbol{\theta}_k, \quad n = 0, 1, \ldots, N-1,$$

which is (e). Since the steps in this argument may be reversed, (a) and (e) are equivalent, completing the proof of the theorem. □

5.4.2 Corollary. *Given a predictable process $\boldsymbol{\theta} = (\theta^1, \ldots, \theta^d)$ and initial wealth V_0, there exists a unique predictable process ϕ such that the trading strategy $(\phi, \boldsymbol{\theta})$ is self-financing.*

Proof. Define a predictable process (ϕ_n) by part (e) of the theorem. The resulting trading strategy $(\phi, \boldsymbol{\theta})$ is then self-financing. □

5.4.3 Remarks. The equation in part (c) of the theorem is an example of a *recursion equation*: it expresses V_{n+1} in terms of the earlier value V_n. The quantity $V_n - \boldsymbol{\theta}_{n+1} \cdot \boldsymbol{S}_n$ in that equation is the cash left over from the transaction of buying $\boldsymbol{\theta}_{n+1}$ shares of the stocks at time n and therefore represents the value of the bond account. Thus, the new value V_{n+1} of the portfolio results precisely from the change in the value of the stocks and the growth of the bond account over the time interval $(n, n+1)$, as expected. ◊

5.5 Option Valuation by Portfolios

Recall that an arbitrage was defined as a trading strategy with zero probability of loss and positive probability of net gain. We may now give a precise definition in terms of the value process of a portfolio: An *arbitrage* is a self-financing portfolio $(\phi, \boldsymbol{\theta})$ whose value process V satisfies

$$\mathbb{P}(V_N \geq 0) = 1 \text{ and } \mathbb{P}(V_N > 0) > 0, \text{ where } V_0 = 0.$$

That is, for a portfolio with no startup cost, there is zero probability of loss and a positive probability of gain. An equivalent definition is

$$\mathbb{P}(V_N \geq V_0) = 1 \text{ and } \mathbb{P}(V_N > V_0) > 0$$

for portfolios with startup cost V_0. Self-financing portfolios may be used to establish the fair market (arbitrage-free) value of a derivative. To describe the method with sufficient generality, we introduce some terminology:

A *contingent claim* is an $\mathcal{F}_N^S$-random variable H. A *hedging strategy* or *hedge* for H is a self-financing trading strategy with value process V satisfying $V_N = H$. If a hedge for H exists, then H is said to be *attainable* and the hedge is called a *replicating portfolio.* A market is *complete* if every contingent claim is attainable. European options are the most common examples of contingent claims. For example, the holder of a call option based on a single asset has a claim against the writer, namely, the value (payoff) $(S_T - K)^+$ of the option at maturity, that value dependent on whether or not S_T exceeds K.

To see the implications of completeness, suppose that we write a European contract with payoff H in a complete, arbitrage-free market. At maturity we are obligated to cover the claim, which, by assumption, is attainable by a self-financing portfolio with value process V. Our strategy is to sell the contract for V_0 and use this amount to start the portfolio. At time N, our portfolio has value V_N, which we use to cover our obligation. The entire transaction costs us nothing, since the portfolio is self-financing; we have hedged the short position of the contract. It is natural then to define the time-n value of the contract to be V_n; any other value would result in an arbitrage. (This is another instance of the law of one price.) We summarize the discussion in the following theorem.

5.5.1 Theorem. *In a complete, arbitrage-free market, the time-n value of a European contingent claim H with maturity N is V_n, where V is the value process of a self-financing portfolio with final value $V_N = H$. In particular, the fair price of the claim is V_0.*

In Chapter 7, we apply Theorem 5.5.1 to the special case of a single security $\mathcal{S}$ that follows the binomial model.

5.6 Exercises

1. A hat contains three slips of paper numbered 1, 2, and 3. The slips are randomly drawn from the hat one at a time without replacement. Let X_n denote the number on the nth slip drawn. Describe the sample space Ω of the experiment and the natural filtration $(\mathcal{F}_n)_{n=1}^3$ associated with the process $(X_n)_{n=1}^3$.

2. Rework Exercise 1 if the second slip is replaced.

3. Consider the following variation of Example 5.1.1. Marbles are randomly drawn one at a time from a jar initially containing a known equal number of red and white marbles. At each draw, the marble drawn is replaced by one of the opposite color. For example, if there are initially two reds and two whites and the first draw is red, then the red marble is replaced by a white marble so the jar now contains one red and three whites. Assume that there are initially one red marble and one white marble and the game stops after five draws. Find the filtration of the experiment.

4. Same as Exercise 3 but the urn initially contains two red marbles and two white marbles and the game stops after four draws.

5. Complete the proof of Theorem 5.4.1 by showing that (e) implies (a).

6. Show that for $d = 1$ and $n = 0, 1, \ldots, N-1$,

$$\theta_{n+1} = \frac{\Delta \widetilde{V}_n}{\Delta \widetilde{S}_n}, \quad \text{and} \quad \phi_{n+1} = \frac{\Delta V_n \Delta \widetilde{S}_n - \Delta \widetilde{V}_n \Delta S_n}{i(1+i)^{-1} S_{n+1} - i S_n}.$$

   These equations express the trading strategy in terms of the stock price and value processes.

7. The *gain* of a portfolio $(\phi, \boldsymbol{\theta})$ in the time interval $(n-1, n]$ is defined as

$$(\phi_n B_n + \boldsymbol{\theta}_n \cdot \boldsymbol{S}_n) - (\phi_n B_{n-1} + \boldsymbol{\theta}_n \cdot \boldsymbol{S}_{n-1}) = \phi_n \Delta B_{n-1} + \boldsymbol{\theta}_n \cdot \Delta \boldsymbol{S}_{n-1},$$

where $n = 1, 2, \ldots, N$. The *gain up to time* n is the sum G_n of the gains over the time intervals $(j-1, j]$, $1 \le j \le n$:

$$G_n = \sum_{j=1}^{n} (\phi_j \Delta B_{j-1} + \boldsymbol{\theta}_j \cdot \Delta \boldsymbol{S}_{j-1}), \quad n = 1, 2, \ldots N, \quad G_0 = 0$$

The process $(G_n)_{n=0}^N$ is called the *gains process* of the portfolio. Show that the portfolio is self-financing iff for each n the time-n value of the portfolio is its initial value plus the gain G_n, that is,

$$V_n = V_0 + G_n, \quad n = 1, 2, \ldots, N.$$

8. Show that $\widetilde{V}_{n+1} = V_0 + \sum_{j=0}^{n} \boldsymbol{\theta}_{j+1} \cdot \Delta \widetilde{\boldsymbol{S}}_j$. Thus the discounted value of the portfolio is the initial wealth plus the changes in the discounted stock holdings.

Chapter 6

Expectation

The expectation of a random variable is a probabilistic generalization of the notion of average. For example, as we shall see later, the arbitrage-free cost of a financial derivative may be expressed in terms of an expectation of the future value of the derivative. Expectation may be defined for a very general class of random variables. However, for our purposes it is sufficient to treat two important special cases: discrete and continuous random variables.

6.1 Expectation of a Discrete Random Variable

To motivate the definition, consider the following grade distribution of a class of 100 students:

grade	A	B	C	D	F
no. of students	15	25	40	12	8

The class average may be expressed in weighted form as

$$4(.15) + 3(.25) + 2(.4) + 1(.12) + 0(.8) = 2.27,$$

where A is a 4, B a 3, etc. To give this a probabilistic interpretation, let X be the grade of a student chosen at random and $\mathbb{P}(X = n)$ the probability that the grade of the student is n. The average may then be written as

$$4 \cdot \mathbb{P}(X = 4) + 3 \cdot \mathbb{P}(X = 3) + 2 \cdot \mathbb{P}(X = 2) + 1 \cdot \mathbb{P}(X = 1) + 0 \cdot \mathbb{P}(X = 0),$$

or simply as $\sum_x x p_X(x)$, where p_X is the pmf of X and the sum is taken over all the values of X. In this form, the average is called the *expectation* of the random variable X.

Here is the formal definition: Let X be an arbitrary discrete random variable with pmf p_X. The *expectation* (*expected value, mean*) of X is defined as

$$\mathbb{E}\,X = \sum_x x p_X(x), \tag{6.1}$$

where the sum is taken over all values of X. The expression on the right is either a finite sum or an infinite series (ignoring zero terms). If the series diverges, then $\mathbb{E}\,X$ is undefined. In this chapter we consider only random variables with finite expectation. Note that since $\mathbb{P}(X = x)$ is the sum of terms $\mathbb{P}(\omega)$, where $X(\omega) = x$, the term $xp_X(x)$ in (6.1) may be expanded as

$$x \sum_{\omega:X(\omega)=x} \mathbb{P}(\omega) = \sum_{\omega:X(\omega)=x} X(\omega)\mathbb{P}(\omega).$$

Summing over x and noting that the pairwise disjoint sets $\{X = x\}$ partition Ω, we obtain the following characterization of expected value of a discrete random variable:

$$\mathbb{E}\,X = \sum\nolimits_{\omega\in\Omega} X(\omega)\mathbb{P}(\omega). \tag{6.2}$$

6.1.1 Examples.

(a) Let $A \in \mathcal{F}$ and $X = \mathbf{1}_A$. Then

$$\mathbb{E}\,\mathbf{1}_A = 1 \cdot p_X(1) + 0 \cdot p_X(1) = 1 \cdot \mathbb{P}(A) + 0 \cdot \mathbb{P}(A') = \mathbb{P}(A).$$

A special case is the Bernoulli random variable $X = \mathbf{1}_{\{1\}}$ with parameter p, giving us $\mathbb{E}\,X = p$.

(b) Let $X \sim B(n,p)$. Since $p_X(k) = \binom{n}{k} np^k q^{n-k}$ $(q := 1-p)$, we have

$$\mathbb{E}\,X = \sum_{k=1}^{n} k\binom{n}{k}p^k q^{n-k} = np\sum_{k=0}^{n-1}\binom{n-1}{k}p^k q^{n-1-k} = np(p+q)^{n-1} = np.$$

(c) Let X be a geometric random variable with parameter p. Since $p_X(n) = pq^n$ $(q := 1-p)$, we have

$$\mathbb{E}\,X = pq\sum_{n=1}^{\infty} nq^{n-1} = pq\sum_{n=1}^{\infty}\frac{d}{dq}q^n = pq\frac{d}{dq}\sum_{n=1}^{\infty}q^n = pq\frac{d}{dq}\frac{q}{p} = \frac{q}{p}. \qquad \diamond$$

The following theorem gives a formula for the expectation of a function $h(X)$ of a random variable X in terms of p_X. Note that by taking $h(x) = x$ one recovers the definition of $\mathbb{E}\,X$.

6.1.2 Theorem. *If X is a discrete random variable and $h(x)$ is any function, then*

$$\mathbb{E}\,h(X) = \sum_x h(x)p_X(x),$$

where the sum is taken over all x in the range of X.

Proof. For y in the range of $h(X)$, we have

$$\mathbb{P}(h(X) = y) = \sum_x \mathbb{P}(X = x, h(x) = y) = \sum_{x:h(x)=y} p_X(x),$$

hence

$$\mathbb{E}\, h(X) = \sum_y y \sum_{x:h(x)=y} p_X(x) = \sum_y \sum_{x:h(x)=y} h(x)p_X(x) = \sum_x h(x)p_X(x). \quad \square$$

6.1.3 Example. Let $X \sim B(N, p)$ and $q = 1 - p$. By Theorem 6.1.2,

$$\mathbb{E}\, X^2 = \sum_{k=0}^{n} k^2 \binom{n}{k} p^k q^{n-k} = \sum_{k=2}^{n} k(k-1) \binom{n}{k} p^k q^{n-k} + \mathbb{E}\, X. \qquad (\dagger)$$

The summation on the extreme right may be written, after cancellations, as

$$n(n-1)p^2 \sum_{k=2}^{n} \binom{n-2}{k-2} p^{k-2} q^{n-2-(k-2)} = n(n-1)p^2(p+q)^{n-2} = n(n-1)p^2.$$

Since $\mathbb{E}\, X = np$ (Example 6.1.1(b), we see from (†) that

$$\mathbb{E}\, X^2 = n(n-1)p^2 + np = np(np+q). \qquad \lozenge$$

Here is a version of Theorem 6.1.2 for jointly distributed discrete random variables. The proof is similar.

6.1.4 Theorem. *If X and Y are discrete random variables and $h(x, y)$ is any function, then*

$$\mathbb{E}\, h(X, Y) = \sum_{x,y} h(x, y)p_{X,Y}(x, y),$$

where the sum is taken over all x in the range of X and y in the range of Y.

6.2 Expectation of a Continuous Random Variable

Let X be a continuous random variable with probability density function f. The *expectation* (*expected value, mean*) of X is defined as

$$\mathbb{E}\, X = \int_{-\infty}^{\infty} x f_X(x)\, dx,$$

provided the integral converges. We shall consider only random variables for which this is the case.

6.2.1 Example. Let X be uniformly distributed on the interval (a, b). Recalling that $f_X = (b-a)^{-1}\mathbf{1}_{(a,b)}$, we have

$$\mathbb{E}\, X = \frac{1}{b-a} \int_a^b x\, dx = \frac{a+b}{2}.$$

In particular, the average value of a number selected randomly from $(0, 1)$ is $1/2$. $\lozenge$

6.2.2 Example. Let $X \sim N(0,1)$. Then

$$\mathbb{E}\, X = \frac{1}{\sqrt{2\pi}} \int_{-\infty}^{\infty} x e^{-\frac{1}{2}x^2}\, dx = 0,$$

the last equality because the integrand is an odd function. More generally, if $X \sim N(\mu, \sigma^2)$, then

$$\begin{aligned}
\mathbb{E}\, X &= \frac{1}{\sigma\sqrt{2\pi}} \int_{-\infty}^{\infty} x e^{-\frac{1}{2}\left(\frac{x-\mu}{\sigma}\right)^2}\, dx \\
&= \frac{1}{\sqrt{2\pi}} \int_{-\infty}^{\infty} (\sigma y + \mu) e^{-\frac{1}{2}y^2}\, dy \qquad \text{(by the substitution } y = (x-\mu)/\sigma) \\
&= \frac{\sigma}{\sqrt{2\pi}} \int_{-\infty}^{\infty} y e^{-\frac{1}{2}y^2}\, dy + \frac{\mu}{\sqrt{2\pi}} \int_{-\infty}^{\infty} e^{-\frac{1}{2}y^2}\, dy \\
&= \mu.
\end{aligned}$$

◇

The following is a continuous analog of Theorem 6.1.2. For a proof see, for example, [5].

6.2.3 Theorem. *If X is a continuous random variable with density f_X and $h(x)$ is a continuous function, then*

$$\mathbb{E}\, h(X) = \int_{-\infty}^{\infty} h(x) f_X(x)\, dx.$$

6.2.4 Example. Let $X \sim N(0,1)$ and let n be a nonnegative integer. By Theorem 6.1.2,

$$\mathbb{E}\, X^n = \frac{1}{\sqrt{2\pi}} \int_{-\infty}^{\infty} x^n e^{-x^2/2}\, dx.$$

If n is odd, then the integrand is an odd function, hence $\mathbb{E}\, X^n = 0$. If n is even, then

$$\mathbb{E}\, X^n = \frac{2}{\sqrt{2\pi}} \int_{0}^{\infty} x^n e^{-x^2/2}\, dx,$$

and an integration by parts yields

$$\mathbb{E}\, X^n = (n-1)\frac{2}{\sqrt{2\pi}} \int_{0}^{\infty} x^{n-2} e^{-x^2/2}\, dx = (n-1)\mathbb{E}\, X^{n-2}.$$

Iterating, we see that for any even positive integer n,

$$\mathbb{E}\, X^n = (n-1)(n-3)\cdots 3 \cdot 1.$$

In particular, $\mathbb{E}\, X^2 = 1$. The quantity $\mathbb{E}\, X^n$ is called the nth *moment* of X. ◇

Theorem 6.2.3 extends to the case of functions of more than one variable. Here is a version for two variables:

6.2.5 Theorem. *If X and Y are jointly continuous random variables with joint density function $f_{X,Y}$ and $h(x, y)$ is a continuous function, then*

$$\mathbb{E}\, h(X,Y) = \int_{-\infty}^{\infty} \int_{-\infty}^{\infty} h(x,y) f_{X,Y}(x,y)\, dx\, dy.$$

6.2.6 Remark. Theorems 6.1.2, 6.1.4, 6.2.3, and 6.2.5 are sometimes collectively referred to as "the law of the unconscious statistician," reflecting their common use in statistics. ◊

6.3 Basic Properties of Expectation

6.3.1 Theorem. *Let X and Y be either both discrete or both jointly continuous random variables, and let $a, b \in \mathbf{R}$. Then the following hold:*

(a) Unit property: $\mathbb{E}\, 1 = 1$.

(b) Linearity: $\mathbb{E}(aX + bY) = a\mathbb{E}\, X + b\mathbb{E}\, Y$.

(c) Order property: $X \le Y \Rightarrow \mathbb{E}\, X \le \mathbb{E}\, Y$.

(d) Absolute value property: $|\mathbb{E}\, X| \le \mathbb{E}\, |X|$.

Proof. Part (a) is clear. We prove (b) for the discrete case. By Theorem 6.1.4, with $h(x, y) = ax + by$,

$$\begin{aligned}
\mathbb{E}(aX + bY) &= \sum_{x,y} (ax + by) p_{X,Y}(x,y) \\
&= a \sum_x x \sum_y p_{X,Y}(x,y) + b \sum_y y \sum_x p_{X,Y}(x,y) \\
&= a \sum_x x p_X(x) + b \sum_y y p_Y(y) \\
&= a\mathbb{E}\, X + b\mathbb{E}\, Y.
\end{aligned}$$

For the continuous version of (c), set $Z = Y - X$. Then Z is a continuous, nonnegative random variable, hence for $z < 0$

$$\int_{-\infty}^{z} f_Z(t)\, dt = \mathbb{P}(Z \le z) = 0.$$

Differentiating with respect to z, we see that $f_Z(z) = 0$ for $z < 0$. Therefore,

$$\mathbb{E}\, Y - \mathbb{E}\, X = \mathbb{E}\, Z = \int_0^{\infty} z f_Z(z)\, dz \ge 0.$$

Part (d) follows from $\pm\mathbb{E}\, X = \mathbb{E}(\pm X) \le \mathbb{E}\, |X|$, a consequence of part (c). □

6.3.2 Theorem. *Let X and Y be discrete or jointly continuous random variables. If X and Y are independent, then*

$$\mathbb{E}(XY) = (\mathbb{E}\,X)(\mathbb{E}\,Y).$$

Proof. We prove only the jointly continuous case. By Corollary 3.6.2, $f_{X,Y}(x,y) = f_X(x)f_Y(y)$, hence, by Theorem 6.2.5,

$$\begin{aligned}\mathbb{E}(XY) &= \int_{-\infty}^{\infty}\int_{-\infty}^{\infty} xyf_X(x)f_Y(y)\,dx\,dy \\ &= \int_{-\infty}^{\infty} xf_X(x)\,dx \int_{-\infty}^{\infty} yf_Y(y)\,dy \\ &= (\mathbb{E}\,X)(\mathbb{E}\,Y).\end{aligned}$$

□

6.4 Variance of a Random Variable

It is important in many applications to determine the degree to which a random variable deviates from its mean. There are various measures of this, but by far the most important is the variance, defined as follows.

Let X be a discrete or continuous random variable with mean $\mu := \mathbb{E}\,X$. The *variance* and *standard deviation* of X are defined, respectively, by

$$\mathbb{V}\,X = \mathbb{E}(X-\mu)^2 \quad\text{and}\quad \sigma(X) = \sqrt{\mathbb{V}X}.$$

Expanding $(X-\mu)^2$ and using the linearity of expectation we see that

$$\mathbb{V}\,X = \mathbb{E}\,X^2 - 2\mu\mathbb{E}\,X + \mu^2 = \mathbb{E}\,X^2 - \mu^2,$$

that is,

$$\mathbb{V}\,X = \mathbb{E}\,X^2 - \mathbb{E}^2X, \tag{6.3}$$

where we have used the shorthand notation $\mathbb{E}^2X$ for $(\mathbb{E}\,X)^2$.

6.4.1 Theorem. *For real numbers a and b,*

$$\mathbb{V}(aX+b) = a^2\mathbb{V}\,X.$$

Proof. Let $\mu = \mathbb{E}\,X$. By linearity,

$$\mathbb{E}(aX+b)^2 = a^2\mathbb{E}\,X^2 + 2ab\mu + b^2$$

and

$$\mathbb{E}^2(aX+b) = (a\mu+b)^2 = a^2\mu^2 + 2ab\mu + b^2.$$

Subtracting these equations and using (6.3) proves the theorem. □

6.4.2 Theorem. *If X and Y are independent, then* $\mathbb{V}(X+Y) = \mathbb{V}\,X + \mathbb{V}\,Y$.

Proof. By Theorem 6.3.2,

$$\mathbb{E}(X+Y)^2 = \mathbb{E}(X^2 + 2XY + Y^2) = \mathbb{E}\,X^2 + 2(\mathbb{E}\,X)(\mathbb{E}\,Y) + \mathbb{E}\,Y^2.$$

Also,

$$\mathbb{E}^2(X+Y) = (\mathbb{E}\,X + \mathbb{E}\,Y)^2 = \mathbb{E}^2 X + 2(\mathbb{E}\,X)(\mathbb{E}\,Y) + \mathbb{E}^2 Y.$$

Subtracting these equations and using (6.3) proves the theorem. □

6.4.3 Examples.

(a) If X is a Bernoulli random variable with parameter p, then $\mathbb{E}\,X = p = \mathbb{E}\,X^2$, hence

$$\mathbb{V}\,X = p - p^2 = p(1-p).$$

(b) Let $X \sim B(n,p)$. By Example 6.1.1(b), $\mathbb{E}\,X = np$, and by Example 6.1.3, $\mathbb{E}\,X^2 = np(np+q)$. Therefore,

$$\mathbb{V}\,X = np(np+q) - n^2p^2 = npq, \quad q := 1-p.$$

A somewhat simpler argument uses the fact that X has the same pmf as a sum of n independent Bernoulli random variables X_j with parameter p (Example 3.8.1). It then follows from Theorem 6.4.2 and (a) that

$$\mathbb{V}\,X = \mathbb{V}\,X_1 + \mathbb{V}\,X_2 + \cdots + \mathbb{V}\,X_n = np(1-p).$$

(c) Let $Y_1, Y_2, \ldots$ be a sequence of iid random variables such that Y_j takes on the values 1 and -1 with probabilities p and $q := 1-p$, respectively, and set $Z_n = Y_1 + \cdots + Y_n$. Then $\mathbb{E}\,Z_n = n(2p-1)$. Moreover, the random variable $X_j := (Y_j+1)/2$ is Bernoulli and $Z_n = 2\sum_{j=1}^n X_j - n$, so by Theorem 6.4.2, Theorem 6.4.1, and (a)

$$\mathbb{V}\,Z_n = 4npq.$$

The random variable Z_n may be interpreted as the position of a particle starting at the origin of a line and moving one step to the right with probability p, one step to the left with probability q. The stochastic process $(Y_n)_{n\geq 1}$ is called a *random walk.*

(d) If $X \sim N(\mu, \sigma^2)$, then $Y = (X-\mu)/\sigma \sim N(0,1)$, hence, by Example 6.2.4, $\mathbb{E}\,Y = 0$ and $\mathbb{E}\,Y^2 = 1$. Therefore, by Theorem 6.4.1,

$$\mathbb{V}\,X = \mathbb{V}(\sigma Y + \mu) = \sigma^2 \mathbb{V}\,Y = \sigma^2.$$

◊

6.5 Moment Generating Functions

The *moment generating function* (mgf) ϕ_X of a random variable X is defined by

$$\phi_X(\lambda) = \mathbb{E}\, e^{\lambda X}$$

for all real numbers λ for which the expectation is finite. To see how ϕ_X gets its name, expand $e^{\lambda X}$ in a power series and take expectations to obtain

$$\phi_X(\lambda) = \sum_{n=0}^{\infty} \frac{\lambda^n}{n!} \mathbb{E}\, X^n. \tag{6.4}$$

Differentiating we have $\phi_X^{(n)}(0) = \mathbb{E}\, X^n$, which is the *nth moment* of X.

6.5.1 Example. Let $X \sim N(0,1)$. Then

$$\phi_X(\lambda) = \frac{1}{\sqrt{2\pi}} \int_{-\infty}^{\infty} e^{\lambda x - x^2/2}\, dx = \frac{e^{\lambda^2/2}}{\sqrt{2\pi}} \int_{-\infty}^{\infty} e^{-(x-\lambda)^2/2}\, dx = e^{\lambda^2/2}.$$

To find the moments of X write

$$e^{\lambda^2/2} = \sum_{n=0}^{\infty} \frac{\lambda^{2n}}{n! 2^n} \tag{6.5}$$

and compare power series in (6.4) and (6.5) to obtain $\mathbb{E}\, X^{2n+1} = 0$ and

$$\mathbb{E}\, X^{2n} = \frac{(2n)!}{n! 2^n} = (2n-1)(2n-3) \cdots 3 \cdot 1.$$

(See Example 6.2.4, where these moments were found by direct integration.) More generally, if $X \sim N(\mu, \sigma^2)$ then $Y := (X - \mu)/\sigma \sim N(0,1)$ hence

$$\phi_X(\lambda) = \mathbb{E}\, e^{\lambda(\sigma Y + \mu)} = e^{\mu\lambda} \phi_Y(\sigma\lambda) = e^{\mu\lambda + \sigma^2\lambda^2/2}. \qquad \diamond$$

Moment generating functions derive their importance from the following theorem, which asserts that the distribution of a random variable is completely determined by its mgf. A proof may be found in standard texts on probability.

6.5.2 Theorem. *If random variables X and Y have the same mgf, then they have the same cdf.*

6.5.3 Example. Let $X_j \sim N(\mu_j, \sigma_j^2)$, $j = 1, 2$. If X_1 and X_2 are independent, then, by Example 6.5.1,

$$\begin{aligned}\phi_{X_2+X_2}(\lambda) &= \mathbb{E}\left(e^{\lambda X_1} e^{\lambda X_2}\right) = \phi_{X_1}(\lambda)\phi_{X_2}(\lambda) = e^{\mu_1\lambda + (\sigma_1\lambda)^2/2} e^{\mu_2\lambda + (\sigma_2\lambda)^2/2} \\ &= e^{\mu\lambda + (\sigma\lambda)^2/2},\end{aligned}$$

where $\mu = \mu_1 + \mu_2$ and $\sigma^2 = \sigma_1^2 + \sigma_2^2$. By Theorem 6.5.2, $X_1 + X_2 \sim N(\mu, \sigma^2)$, a result obtained in Example 3.8.2 with considerably more effort. $\diamond$

The *moment generating function* $\phi_{\boldsymbol{X}}$ of a random vector $\boldsymbol{X} = (X_1, \ldots, X_n)$ is defined (whenever the expectation exists) by

$$\phi_{\boldsymbol{X}}(\boldsymbol{\lambda}) = \mathbb{E}\left(e^{\boldsymbol{\lambda}\cdot\boldsymbol{X}}\right), \quad \boldsymbol{\lambda} = (\lambda_1, \ldots, \lambda_n),$$

where $\boldsymbol{\lambda} \cdot \boldsymbol{X} := \sum_{j=1}^{n} \lambda_j X_j$. The following result generalizes Theorem 6.5.2 to random vectors.

6.5.4 Theorem. *If* $\boldsymbol{X} = (X_1, \ldots, X_n)$ *and* $\boldsymbol{Y} = (Y_1, \ldots, Y_n)$ *have the same mgf, then they have the same joint cdf.*

6.5.5 Corollary. *Let* $\boldsymbol{X} = (X_1, \ldots, X_n)$*. Then* $X_1, \ldots, X_n$ *are independent iff for all* λ

$$\phi_{\boldsymbol{X}}(\boldsymbol{\lambda}) = \phi_{X_1}(\lambda_1)\phi_{X_2}(\lambda_2)\cdots\phi_{X_n}(\lambda_n). \tag{6.6}$$

Proof. The necessity is clear. To prove the sufficiency, let $Y_1, \ldots, Y_n$ be independent random variables such that $F_{Y_j} = F_{X_j}$ for all j and set $\boldsymbol{Y} = (Y_1, \ldots, Y_n)$.[1] Then $\phi_{Y_j} = \phi_{X_j}$ so, by independence and (6.6),

$$\phi_{\boldsymbol{Y}}(\boldsymbol{\lambda}) = \phi_{Y_1}(\lambda_1)\phi_{Y_2}(\lambda_2)\cdots\phi_{Y_n}(\lambda_n) = \phi_{\boldsymbol{X}}(\boldsymbol{\lambda}).$$

By Theorem 6.5.4, $F_{\boldsymbol{X}} = F_{\boldsymbol{Y}}$. Therefore,

$$\begin{aligned} F_{\boldsymbol{X}}(x_1, x_2, \ldots, x_n) &= F_{\boldsymbol{Y}}(x_1, x_2, \ldots, x_n) \\ &= F_{Y_1}(x_1)F_{Y_2}(x_2)\cdots F_{Y_n}(x_n) \\ &= F_{X_1}(x_1)F_{X_2}(x_2)\cdots F_{X_n}(x_n), \end{aligned}$$

which shows that $X_1, \ldots, X_n$ are independent. □

6.6 The Strong Law of Large Numbers

The strong law of large numbers (SLLN) is the theoretical basis for the observation that the longer one flips a fair coin the closer the relative frequency of heads is to 1/2, the probability of heads on a single toss. Here is the precise statement:

6.6.1 Theorem. *Let* $\{X_n\}$ *be an infinite sequence of iid random variables with* $\mu = \mathbb{E}\, X_n$*. Then with probability one*

$$\lim_{n\to\infty} \frac{X_1 + \cdots + X_n}{n} = \mu.$$

[1] The random variable Y_j is generally taken to be the jth coordinate function on a new probability space $\mathbf{R}^n$, where the probability measure is defined so that the sets $(-\infty, x_1] \times (-\infty, x_2] \times \cdots \times (-\infty, x_n]$ have probability $F_{X_1}(x_1)F_{X_2}(x_2)\cdots F_{X_n}(x_n)$.

In an experiment where the outcome of the nth trial is expressed by a random variable X_n, the expression $(X_1 + \cdots + X_n)/n$ is called a *sample average* of the experiment. Thus the SLLN asserts that for large n the sample average of an experiment approximates the mean of the random variables X_n. The proof may be found in standard texts on advanced probability.

6.6.2 Example. For a generalization of the coin example, consider an experiment consisting of an infinite sequence of independent trials. Let A be any event and define

$$X_n = \begin{cases} 1 & \text{if the event } A \text{ occurred on the } n\text{th trial,} \\ 0 & \text{otherwise.} \end{cases}$$

Then $(X_1 + \cdots + X_n)/n$ is the average number of occurrences of A in n trials and $\mu = \mathbb{P}(A)$. The SLLN asserts that this average approaches $\mathbb{P}(A)$ with probability one. ◊

6.6.3 Example. Consider the experiment of observing a quantity $f(X)$, where X is a number chosen randomly from an interval $[a, b]$ and f is a continuous function. Repeating the experiment produces a sequence of iid variables $f(X_n)$, where $X_n \sim U(a, b)$. By SLLN and Theorem 6.2.3, with probability one we have

$$\lim_{n\to\infty} \frac{f(X_1) + \cdots + f(X_n)}{n} = \frac{1}{b-a} \int_a^b f(x)\,dx.$$

This may be viewed as a technique for approximating $\int_a^b f(x)\,dx$ and illustrates the fundamental idea behind *Monte Carlo simulation*, wherein a computer program calculates sample averages for large n using a random number generator to simulate X_j. The module **LLNUniform** illustrates this for user-defined functions f. See also **LLNBinomial** where analogous calculations are carried out but with binomially distributed X.

◊

6.7 The Central Limit Theorem

The central limit theorem (CLT) is a fundamental result in probability theory that ranks in importance with the law of large numbers and underlies many statistical applications. The theorem conveys the remarkable fact that the distribution of a (suitably normalized) sum of a large number of iid random variables is approximately that of a standard normal random variable. It explains why the data distribution from populations frequently exhibits a bell shape. Proofs of the CLT may be found in standard texts on advanced probability.

6.7.1 Theorem (Central Limit Theorem). *Let $X_1, X_2, \ldots$ be an infinite sequence of iid random variables with mean μ and standard deviation σ, and let $S_n = \sum_{j=1}^n X_j$. Then*

$$\lim_{n\to\infty} \mathbb{P}\left(\frac{S_n - n\mu}{\sigma\sqrt{n}} \le x\right) = \Phi(x). \tag{6.7}$$

It follows that

$$\lim_{n\to\infty} \mathbb{P}\left(a \le \frac{S_n - n\mu}{\sigma\sqrt{n}} \le b\right) = \Phi(b) - \Phi(a). \tag{6.8}$$

6.7.2 Remarks. (a) The CLT asserts that for large n, $(S_n - n\mu)/(\sigma\sqrt{n})$ has a nearly standard normal distribution. Now set

$$\overline{X}_n := \frac{X_1 + \cdots + X_n}{n} = \frac{S_n}{n},$$

Then $\mathbb{E}\,\overline{X}_n = \mu$ and $\mathbb{V}\,\overline{X}_n = \sigma^2/n$, hence

$$\frac{S_n - n\mu}{\sigma\sqrt{n}} = \frac{\overline{X}_n - \mathbb{E}\,\overline{X}_n}{\sigma(\overline{X}_n)}.$$

Thus the CLT may be written

$$\lim_{n\to+\infty} \mathbb{P}\left(\frac{\overline{X}_n - \mathbb{E}\,\overline{X}_n}{\sigma(\overline{X}_n)} \le x\right) = \mathbb{P}\left(\frac{X - \mu}{\sigma} \le x\right),$$

where $X \sim N(\mu, \sigma)$.

(b) In the special case that the random variables X_j are Bernoulli with parameter $p \in (0,1)$, (6.7) becomes

$$\lim_{n\to\infty} \mathbb{P}\left(\frac{S_n - np}{\sqrt{np(1-p)}} \le x\right) = \Phi(x) \tag{6.9}$$

(see Example 6.4.3(a)). This result is known as the *DeMoivre-Laplace Theorem.*

(c) For later reference, we note that the DeMoivre-Laplace theorem has the following extension: Let $p_n \in (0,1)$ and $p_n \to p \in (0,1)$. For each n let $X_{n,1}, X_{n,2}, \ldots X_{n,n}$, be independent Bernoulli random variables with parameter p_n. Set

$$S_n = X_{n,1} + X_{n,2} + \cdots + X_{n,n}.$$

If $a_n \to a$ and $b_n \to b$, then

$$\lim_{n\to\infty} \mathbb{P}\left(a_n \le \frac{S_n - np_n}{\sqrt{np_n(1-p_n)}} \le b_n\right) = \Phi(b) - \Phi(a).$$

(d) One can use (6.9) to obtain the following approximation for the pmf of the binomial random variable S_n: For $k = 0, 1, \ldots, n$,

$$\begin{aligned}\mathbb{P}(S_n = k) &= \mathbb{P}(k - .5 < S_n < k + .5) \\ &= P\left(\frac{k - .5 - np}{\sqrt{npq}} < \frac{S_n - np}{\sqrt{npq}} < \frac{k + .5 - np}{\sqrt{npq}}\right) \\ &\approx \Phi\left(\frac{k + .5 - np}{\sqrt{npq}}\right) - \Phi\left(\frac{k - .5 - np}{\sqrt{npq}}\right).\end{aligned}$$

In the first equality, a correction was made to compensate for using a continuous distribution to approximate a discrete one. In the special case $p = .5$, we have the approximation

$$\mathbb{P}(S_n = k) \approx \Phi\left(\frac{2k + 1 - n}{\sqrt{n}}\right) - \Phi\left(\frac{2k - 1 - n}{\sqrt{n}}\right).$$

For example, suppose we flip a fair coin 50 times. For $k = 25$, the right side (rounded to four decimal places) is

$$\Phi\left(\frac{1}{\sqrt{50}}\right) - \Phi\left(\frac{-1}{\sqrt{50}}\right) = .1125.$$

The actual probability (rounded to four decimal places) is

$$\mathbb{P}(S_{50} = 25) = \binom{50}{25}\left(\frac{1}{2}\right)^{50} = .1123. \qquad \diamond$$

6.8 Exercises

1. A jar contains r red and w white marbles. The marbles are drawn one at a time and replaced. Let Y denote the number of red marbles drawn before the second white one. Find $\mathbb{E}Y$ in terms of r and w. (Use Examples 3.7.1 and 6.1.1(c).)

2. Refer to Exercise 2.18(c) on roulette. Suppose a gambler employs the following betting strategy: She initially bets \$1 on black. If black appears, she takes her profit and quits. If she loses, she bets \$2 on black. If she wins this bet she takes her profit of \$1 and quits. Otherwise, she bets \$4 on the next spin. She continues in this way, quitting if she wins a game, doubling the bet on black otherwise. She decides to quit after the Nth game, win or lose. Show that her expected profit is $1 - (2p)^N < 0$, where $p = 20/38$. What is the expected profit for a roulette wheel with no green pockets? (The general betting strategy of doubling a wager after a loss is called a *martingale*.)

3. Show that the variance of a geometric random variable X with parameter p is q/p^2, where $q = 1 - p$.

4. A *Poisson random variable* X *with parameter* $\lambda > 0$ has distribution

$$p_X(n) = \frac{\lambda^n}{n!} e^{-\lambda}, \quad n = 0, 1, 2, \ldots.$$

Find the expectation and variance of X. (Poisson random variables are used for modeling the random flow of events such as motorists arriving at a toll booth or calls arriving at a service center.)

5. A jar contains r red and w white marbles. Marbles are drawn randomly one at a time until the drawing produces two marbles of the same color. Find the expected number of marbles drawn if, after each draw, the marble is
(a) replaced. (b) discarded.

6. A hat contains a slips of paper labeled with the number 1, and b slips labeled with the number 2. Two slips are drawn at random without replacement. Let X be the number on the first slip drawn and Y the number on the second. Show that

(a) X and Y are identically distributed. (b) $\mathbb{E}(XY) \neq (\mathbb{E}\,X)(\mathbb{E}\,Y)$.

7. Let $A_1, \ldots, A_n$ be independent events and set $X = \sum_{j=1}^{n} a_j I_{A_j}$, $a_j \in \mathbf{R}$. Show that

$$\mathbb{V}\,X = \sum_{j=1}^{n} a_j^2 P(A_j) P(A_j').$$

8. Let Y be a binomial random variable with parameters (n, p). Find $\mathbb{E}\,2^Y$.

9. Let X and Y be independent and uniformly distributed on $[0, 1]$. Calculate

$$\mathbb{E}\left(\frac{XY}{X^2 + Y^2 + 1}\right).$$

10. Find $\mathbb{E}|X|$ if $X \sim N(0, 1)$.

11. Let X and Y be independent random variables with $\mathbb{E}\,X = \mathbb{E}\,Y = 0$. Show that

$$\mathbb{E}\,(X+Y)^2 = \mathbb{E}\,X^2 + \mathbb{E}\,Y^2 \quad \text{and} \quad \mathbb{E}\,(X+Y)^3 = \mathbb{E}\,X^3 + \mathbb{E}\,Y^3.$$

What can you say about higher powers?

12. Let X and Y be independent, continuous random variables, each with an even density function. Show that if n is odd, then $\mathbb{E}\,(X+Y)^n = 0$.

13. Express each of the following integrals in terms of Φ, where $\alpha \neq 0$.

(a) $\displaystyle\int_a^b e^{\alpha x} \varphi(x)\, dx.$ (b) $\displaystyle\int_a^b e^{\alpha x} \Phi(x)\, dx.$

14. A positive random variable X such that $\ln X \sim N(\mu, \sigma^2)$ is said to be *lognormal with parameters* μ *and* σ^2. Show that

$$\mathbb{E}\,X = \exp(\mu + \sigma^2/2) \quad \text{and} \quad \mathbb{E}\,X^2 = \exp(2\mu + 2\sigma^2).$$

15. Find $\mathbb{V} X$ if X is uniformly distributed on the interval (a, b).

16. Let X and Y be independent random variables with X uniformly distributed on $[0,1]$ and $Y \sim N(0,1)$. Find the mean and variance of Ye^{XY}.

17. Show that $\mathbb{E}^2 X \le \mathbb{E} X^2$.

18. Find the nth moment of a random variable X with density $f_X(x) = \frac{1}{2}e^{-|x|}$.

19. Show that, in the notation of (3.4), the expectation of a hypergeometric random variable with parameters (p, z, M) is pz.

20. Find the expected value of the random variables Z and Θ in Exercises 3.13 and 3.14.

21. Use the CLT to estimate the probability that the number Y of heads appearing in 100 tosses of a fair coin

    (a) is exactly 50. (b) lies between 40 and 60.

    (A spreadsheet with a built-in normal cdf function is useful here.) Find the exact probability for part (a).

22. A true-false exam has 54 questions. Use the CLT to approximate the probability of getting a passing score of 35 or more correct answers simply by guessing on each question.

23. Let $X \sim B(n, p)$ and $0 \le m \le n$. Show that

$$\mathbb{E}\left[X(X-1)\cdots(X-m)\right] = n(n-1)\cdots(n-m)p^{m+1}.$$

24. Let $X \sim B(n, p)$. Use Exercise 23 to show that

$$\mathbb{E} X^3 = n(n-1)(n-2)p^3 + 3n(n-1)p^2 - 2np.$$

25. Let X and Y be independent random variables uniformly distributed on $[0,1]$. Find

    (a) $\mathbb{E}(X-Y)^2$. (b) $\mathbb{E}\left(\dfrac{Y}{1+XY}\right)$.

26. Let $X_1, \ldots, X_n$, be independent Bernoulli random variables with parameter p, and let $Y_j = X_1 + \cdots + X_j$. Show that

$$\mathbb{E}\, 2^{Y_1+\cdots+Y_n} = (q+2^n p)(q+2^{n-1}p)\cdots(q+2^2 p)(q+2p)$$

27. Let $X_1, \ldots, X_n$ be independent Bernoulli random variables with parameter p and let let $Y_j = X_1 + \cdots + X_j$. Let p_X denote the pmf of a Bernoulli random variable X with parameter p. Use Exercise 3.29 to show that

$$\mathbb{E} f(Y_1, \ldots, Y_n) = \sum_{\substack{0 \le k_1 \le \cdots \le k_n \le n \\ k_j - k_{j-1} \le 1}} f(k_1, \ldots, k_n) p^{k_n} q^{n-k_n}.$$

28. Let $X_1, \ldots, X_n$ be independent Bernoulli random variables with parameter p and let N be a random variable with range $\{1, \ldots, n\}$. Define the *random sum* Y_N of $X_1, \ldots, X_N$ by

$$Y_N(\omega) = X_1(\omega) + \cdots + X_{N(\omega)}(\omega)$$

and the *random average* by Y_N/N. Suppose that N is independent of $X_1, \ldots, X_N$. Calculate $\mathbb{E}\, Y_N$ and $\mathbb{E}(Y_N/N)$.

29. Find the mgfs of
    (a) a binomial random variable X with parameters (n, p).
    (b) a geometric random variable X with parameter p.

30. Find the mgf of a random variable X uniformly distributed on $[0, 1]$.

Chapter 7

The Binomial Model

In this chapter, we construct the geometric binomial model for stock price movement and use the model to determine the value of a general European claim. An important consequence is the Cox-Ross-Rubinstein formula for the price of a call option. Formulas for path-dependent claims are given in the last section. The valuation techniques in this chapter are based on the notion of self-financing portfolio described in Chapter 5.

7.1 Construction of the Binomial Model

Consider a (non-dividend paying) stock $\mathcal{S}$ with initial price S_0 such that during each discrete time period the price changes either by a factor u with probability p or by a factor d with probability $q := 1 - p$, where $0 < d < u$. The symbols u and d are meant to suggest the words "up" and "down" and we shall use these terms to describe the price movement from one period to the next, even though u may be less than one (prices drift downward) or d may be greater than one (prices drift upward). The model has essentially three components: the stock price formula, the probability law that governs the price, and the flow of information in the form of a filtration.

The Stock Price

The stock's movements are modeled as follows: For a fixed positive integer N, the number of periods over which the stock price is observed, let Ω be the set of all sequences $\omega = (\omega_1, \ldots, \omega_N)$, where $\omega_n = u$ if the stock moved up during the nth time period and $\omega_n = d$ if the stock moved down. We may express Ω as a Cartesian product

$$\Omega = \Omega_1 \times \Omega_2 \times \cdots \times \Omega_N,$$

where $\Omega_n = \{u, d\}$ represents the possible movements of the stock at time n. A member of Ω represents a particular market scenario. For example the sequence $(u, d, *, \ldots, *)$ represents all market scenarios with the property that during the time interval $[0, 1]$ the stock value changed from S_0 to uS_0, and during the time interval $[1, 2]$ the stock value changed from uS_0 to duS_0. In general, for a given scenario $\omega = (\omega_1, \ldots, \omega_N)$, the price $S_n(\omega)$ of the stock at time n satisfies

$$S_n(\omega) = \left\{ \begin{array}{ll} uS_{n-1}(\omega) & \text{if } \omega_n = u \\ dS_{n-1}(\omega) & \text{if } \omega_n = d \end{array} \right\} = \omega_n S_{n-1}(\omega).$$

Iterating, we have

$$S_n(\omega) = \omega_n S_{n-1}(\omega) = \omega_n \omega_{n-1} S_{n-2}(\omega) = \cdots = \omega_n \omega_{n-1} \cdots \omega_1 S_0. \tag{7.1}$$

Now let $X_j = 1$ if the stock goes up in the jth time period and $X_j = 0$ if the stock goes down, that is,

$$X_j(\omega) = \begin{cases} 1 & \text{if } \omega_j = u, \\ 0 & \text{if } \omega_j = d. \end{cases}$$

The function U_n on Ω defined by

$$U_n(\omega) := X_1(\omega) + \cdots + X_n(\omega) \tag{7.2}$$

then counts the number of upticks of the stock in the time period from 0 to n. From (7.1) we obtain the formula

$$S_n = u^{U_n} d^{n-U_n} S_0 = v^{U_n} d^n S_0, \quad v := ud^{-1}. \tag{7.3}$$

For a numerical example, consider the following table, which gives stock values after n days for various values of u and n. The stock is assumed to have an initial value of $S_0 = \$10$. For simplicity, $d = .7$ throughout. The table is based on formula (7.3) by specifying the number of upticks U_n. As the

TABLE 7.1: Stock values.

n	10	10	10	10	100	100	100	100
u	1.3	1.4	1.3	1.4	1.3	1.4	1.3	1.4
U_n	6	6	7	7	57	57	58	58
S_n	11.59	18.08	21.52	36.16	6.82	466.14	12.67	932.27

table suggests, even slight changes in u, d, or the number of upticks can cause significant fluctuations in the price S_n. This is due to the exponential nature of the binomial model. In general, an increase in u or d causes an increase in price, as may be seen from (7.3). In this connection, the reader may wish to run the program **StockPriceGraph–Bin**, which shows the variation of S_N with u, d, and the number of upticks.

The Price Path

The sequence $(S_0(\omega), S_1(\omega), \ldots, S_N(\omega))$ is called the (*price*) *path* of the stock corresponding the scenario ω. Figure 7.5 shows the path for the scenario $(d, u, d, u, d, u, \ldots)$, where $u = 2$ and $d = 1/2$. The path was generated by the module **StockPricePath–Bin**, which also generates *random* paths governed by user-selected values of the uptick probability p.

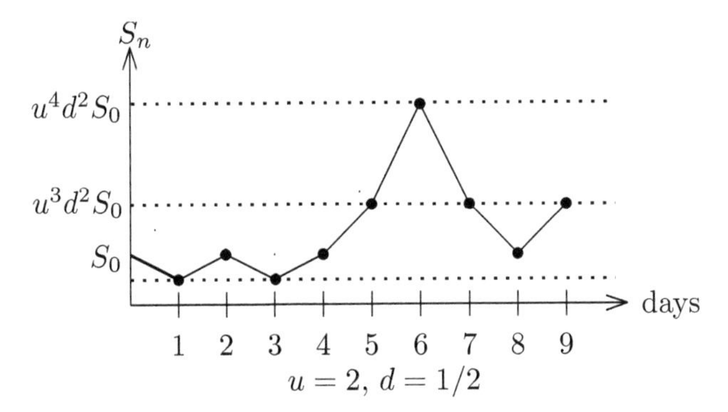

FIGURE 7.1: The stock price path $(d, u, d, u, \cdots)$.

The Probability Law

The random behavior of the stock is modeled as follows. Define a probability measure $\mathbb{P}_n$ on Ω_n by

$$\mathbb{P}_n(u) = p \quad \text{and} \quad \mathbb{P}_n(d) = q.$$

Using the measures $\mathbb{P}_n$, we define a probability measure $\mathbb{P}$ on subsets A of Ω by

$$\mathbb{P}(A) = \sum_{(\omega_1,\ldots,\omega_N)\in A} \mathbb{P}_1(\omega_1)\mathbb{P}_2(\omega_2)\cdots\mathbb{P}_N(\omega_N).$$

For example, the probability that the stock rises during the first period and falls during the next is $\mathbb{P}_1(u)\mathbb{P}_2(d)\mathbb{P}_3(\Omega_3)\cdots\mathbb{P}_N(\Omega_N) = pq$. More generally, if $A_n \subseteq \Omega_n$, $n = 1, 2, \ldots, N$, then

$$\mathbb{P}(A_1 \times A_2 \times \cdots \times A_N) = \mathbb{P}_1(A_1)\mathbb{P}_2(A_2)\cdots\mathbb{P}_N(A_N). \tag{7.4}$$

In particular, setting

$$B_j = \Omega_1 \times \cdots \times \Omega_{j-1} \times A_j \times \Omega_{j+1} \cdots \times \Omega_N$$

we have $\mathbb{P}(B_j) = \mathbb{P}_j(A_j)$, hence

$$\mathbb{P}(B_1B_2\cdots B_N) = \mathbb{P}(B_1)\mathbb{P}(B_2)\cdots\mathbb{P}(B_N).$$

Thus the probability measure $\mathbb{P}$ models the independence of the stock movements. $\mathbb{P}$ is called the *product* of the measures $\mathbb{P}_n$. As usual, we denote the corresponding expectation operator by $\mathbb{E}$.

Under the probability law $\mathbb{P}$, the X_j's are independent Bernoulli random variables on Ω with parameter p, hence the function U_n defined in (7.2) is a binomial random variable with parameters (n, p). The stochastic process $S = (S_n)_{n=0}^N$ is called a *geometric binomial price process.*

Figure 7.2 shows the possible stock movements for three time periods. The

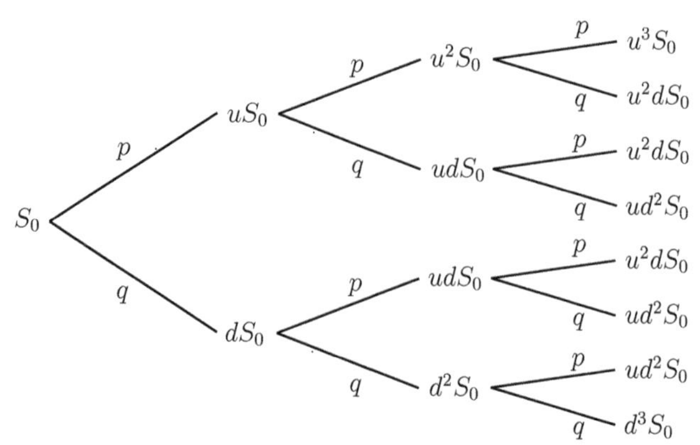

FIGURE 7.2: 3-Step binomial tree.

probabilities along the edges are conditional probabilities; specifically,

$$\mathbb{P}(S_n = ux \mid S_{n-1} = x) = p \text{ and } \mathbb{P}(S_n = dx \mid S_{n-1} = x) = q.$$

Other conditional probabilities may be found by multiplying probabilities along edges and adding. For example,

$$\mathbb{P}(S_n = udx \mid S_{n-2} = x) = 2pq,$$

since there are two paths leading from vertex x to vertex udx.

7.1.1 Example. The probability that the price of the stock at time n is larger than its initial price is found from (7.3) by observing that

$$S_n > S_0 \iff \left(\frac{u}{d}\right)^{U_n} d^n > 1 \iff U_n > b := \frac{n \ln d}{\ln d - \ln u}.$$

If $d > 1$, b is negative and $\mathbb{P}(S_n > S_0) = 1$. If $d < 1$,

$$\mathbb{P}(S_n > S_0) = \sum_{j=m+1}^{n} \mathbb{P}(U_n = j) = \sum_{j=m+1}^{n} \binom{n}{j} p^j q^{n-j},$$

where $m := \lfloor b \rfloor$ is the greatest integer in b. The following table lists these probabilities for $n = 100$ and various values of u, d, and p.

TABLE 7.2: Probabilities that $S_{100} > S_0$.

p	.50	.50	.50	.55	.55	.55	.45	.45	.45
u	1.20	1.21	1.20	1.20	1.21	1.20	1.20	1.21	1.20
d	.80	.80	.79	.80	.80	.79	.80	.80	.79
$\mathbb{P}(S_{100} > S_0)$	.14	.24	.10	.46	.62	.38	.02	.04	.01

As in the case of stock prices, the probabilities are seen to be sensitive to slight variations of the parameters. ◇

The Filtration

We model the flow of stock price information by the filtration $(\mathcal{F}_n)_{n=0}^N$, where $\mathcal{F}_0 = \{\emptyset, \Omega\}$, and for $n \geq 1$ $\mathcal{F}_n$ consists of Ω, $\emptyset$, and all unions of sets of the form

$$A_\eta := \{\eta\} \times \Omega_{n+1} \times \cdots \Omega_N, \quad \eta := (\eta_1, \eta_2, \ldots, \eta_n). \tag{7.5}$$

Here η runs through all market scenarios up to time n. For example, for $N = 4$, $\mathcal{F}_2$ consists of Ω, $\emptyset$, and all unions of the sets

$$\{(u,u)\} \times \Omega_3 \times \Omega_4, \quad \{(u,d)\} \times \Omega_3 \times \Omega_4, \quad \{(d,u)\} \times \Omega_3 \times \Omega_4, \quad \{(d,d)\} \times \Omega_3 \times \Omega_4.$$

Members of $\mathcal{F}_2$ therefore describe exactly what occurred in the first two time periods. One checks that $(\mathcal{F}_n)_{n=0}^N$ is the natural filtration $(\mathcal{F}_n^S)_{n=0}^N$ (Exercise 3).

The following notational convention will be useful: If Z is a $\mathcal{F}_N^S$-random variable on Ω that depends only on the first n coordinates of ω, we will suppress the last $N - n$ coordinates in the notation $Z(\omega_1, \ldots, \omega_N)$ and write instead $Z(\omega_1, \ldots, \omega_n)$. Such a random variable is $\mathcal{F}_n$-measurable, since the event $\{Z = z\}$ is of the form $A \times \Omega_{n+1} \times \cdots \times \Omega_N$ and hence is a union of the sets A_η in (7.5) for $\eta \in A$. Moreover, since $\mathbb{P}_{n+1}(\Omega_{n+1}) = \cdots = \mathbb{P}_N(\Omega_N) = 1$, (7.6) implies the following "truncated" form of the expectation of Z:

$$\mathbb{E}\, Z = \sum_{(\omega_1,\ldots,\omega_n)} Z(\omega_1, \ldots, \omega_n) \mathbb{P}_1(\omega_1) \cdots \mathbb{P}_n(\omega_n). \tag{7.6}$$

7.2 Completeness and Arbitrage in the Binomial Model

In this section, we determine the fair price of a European claim H when the underlying stock $\mathcal{S}$ has a geometric binomial price process $(S_n)_{n=0}^N$, as described in the preceding section. Let $\mathcal{B}$ be a risk-free bond with price process $B_n = (1+i)^n$. According to Theorem 5.5.1, if the binomial model is arbitrage-free, then the proper value of the claim at time n is that of a self-financing, replicating portfolio (ϕ, θ) based on $(\mathcal{B}, \mathcal{S})$ with final value H. For the time being, however, we do *not* make the assumption that the model is arbitrage-free.

We begin with a portfolio (ϕ, θ) based on $(\mathcal{B}, \mathcal{S})$ with value process $(V_n)_{n=0}^N$. Assume that the process is self-financing. By part (c) of Theorem 5.4.1,

$$V_{n+1} = \theta_{n+1} S_{n+1} + a(V_n - \theta_{n+1} S_n), \quad a := 1 + i, \quad n = 0, 1, \ldots, N-1. \tag{7.7}$$

For each scenario $\omega = (\omega_1, \ldots, \omega_n)$, Equation (7.7) may expanded into the system

$$\begin{aligned} \theta_{n+1}(\omega) S_{n+1}(\omega, u) + a[V_n(\omega) - \theta_{n+1}(\omega) S_n(\omega)] &= V_{n+1}(\omega, u) \\ \theta_{n+1}(\omega) S_{n+1}(\omega, d) + a[V_n(\omega) - \theta_{n+1}(\omega) S_n(\omega)] &= V_{n+1}(\omega, d), \end{aligned} \tag{7.8}$$

where we have employed our convention of displaying only the time-relevant coordinates of members of Ω. To solve (7.8) for $\theta_{n+1}(\omega)$, subtract the equations and use

$$S_{n+1}(\omega, \omega_{n+1}) = \omega_{n+1} S_n(\omega)$$

to obtain

$$\theta_{n+1}(\omega) = \frac{V_{n+1}(\omega, u) - V_{n+1}(\omega, d)}{(u-d)S_n(\omega)}, \quad n = 0, 1, \ldots, N-1. \tag{7.9}$$

Equation (7.9) is called the *delta hedging rule*. Solving the first equation in (7.8) for $aV_n(\omega)$ and using the delta hedging rule, we obtain

$$\begin{aligned} aV_n(\omega) &= V_{n+1}(\omega, u) + \theta_{n+1}(\omega)\big(aS_n(\omega) - S_{n+1}(\omega, u)\big) \\ &= V_{n+1}(\omega, u) + \theta_{n+1}(\omega)S_n(\omega)(a-u) \\ &= V_{n+1}(\omega, u) + \big[V_{n+1}(\omega, u) - V_{n+1}(\omega, d)\big]\frac{a-u}{u-d} \\ &= V_{n+1}(\omega, u)\, p^* + V_{n+1}(\omega, d)\, q^*, \end{aligned} \tag{7.10}$$

where

$$p^* := \frac{a-d}{u-d} \quad \text{and} \quad q^* := \frac{u-a}{u-d}, \qquad a := 1+i. \tag{7.11}$$

Dividing (7.10) by a^{n+1} we obtain the following *reverse recursion* formula for the discounted value process $\widetilde{V}$:

$$\widetilde{V}_n(\omega) = \widetilde{V}_{n+1}(\omega, u)p^* + \widetilde{V}_{n+1}(\omega, d)q^*. \tag{7.12}$$

We have proved that (7.7) implies (7.9) and (7.12).

Conversely, assume that θ and V satisfy (7.9) and (7.12). We show that (7.7) holds. It will then follow from Theorem 5.4.1 that (ϕ, θ) is self-financing. Now, dividing by a^{n+1} and rearranging we see that (7.7) may be written as

$$\Delta\widetilde{V}_n(\omega, \omega_{n+1}) = \theta_{n+1}(\omega)\Delta\widetilde{S}_n(\omega, \omega_{n+1}), \quad n = 0, 1, \ldots, N-1. \tag{7.13}$$

By (7.12) and the fact that $p^* + q^* = 1$, the left side of (7.13) is

$$\begin{aligned} \widetilde{V}_{n+1}(\omega, \omega_{n+1}) - \widetilde{V}_n(\omega) &= \widetilde{V}_{n+1}(\omega, \omega_{n+1}) - \widetilde{V}_{n+1}(\omega, u)p^* - \widetilde{V}_{n+1}(\omega, d)q^* \\ &= \big[\widetilde{V}_{n+1}(\omega, \omega_{n+1}) - \widetilde{V}_{n+1}(\omega, u)\big]p^* + \big[\widetilde{V}_{n+1}(\omega, \omega_{n+1}) - \widetilde{V}_{n+1}(\omega, d)\big]q^*. \end{aligned}$$

In particular, taking $\omega_{n+1} = u$ we see from (7.9) and the definition of q^* that

$$\widetilde{V}_{n+1}(\omega, u) - \widetilde{V}_n(\omega) = \big[\widetilde{V}_{n+1}(\omega, u) - \widetilde{V}_{n+1}(\omega, d)\big]q^* = \frac{u-a}{a^{n+1}}\theta_{n+1}(\omega)S_n(\omega).$$

Since $uS_n(\omega) = S_{n+1}(\omega, u)$, we obtain

$$\widetilde{V}_{n+1}(\omega, u) - \widetilde{V}_n(\omega) = \theta_{n+1}(\omega)\frac{S_{n+1}(\omega, u) - aS_n(\omega)}{a^{n+1}} = \widetilde{\theta}_{n+1}(\omega)\Delta\widetilde{S}_n(\omega, u),$$

which is (7.13) for $\omega_{n+1} = u$. A similar argument shows that the equation holds for $\omega_{n+1} = d$ as well. Thus (7.7) holds, as asserted. We have proved

7.2.1 Theorem. *A portfolio (ϕ, θ) in the binomial model with value process V is self-financing iff* (7.9) *and* (7.12) *hold.*

Using the theorem we may prove that the binomial model is complete, that is, every European claim may be hedged:

7.2.2 Corollary. *Given a European claim H in the binomial model there exists a unique self-financing trading strategy (ϕ, θ) with value process V satisfying $V_N = H$.*

Proof. Define a process $(V_n)_{n=0}^N$ by setting $V_N = H$ and using the reverse recursion scheme (7.12) to obtain $V_{N-1}, V_{N-2}, \ldots, V_0$. Next, define θ_{n+1} by (7.9). By the proof of Theorem 7.2.1, (7.7) holds. For $n = 0, 1, \ldots, N-1$ define

$$\phi_{n+1}(\omega) = a^{-n}\big[V_n(\omega) - \theta_{n+1}(\omega)S_n(\omega)\big], \quad a = (1+i). \tag{7.14}$$

Since ϕ_{n+1} and θ_{n+1} depend only on the first n time steps, the process (ϕ, θ) is predictable and hence is a trading strategy. Note that Equations (7.9) and (7.14) define θ_1 and ϕ_1 as constants, in accordance with the portfolio theory of Chapter 5. From (7.7) and (7.14) with n replaced by $n-1$ we have

$$V_n = \theta_n S_n + a[V_{n-1} - \theta_n S_{n-1}] = \theta_n S_n + a^n \phi_n = \theta_n S_n + \phi_n B_n,$$

which shows that V is the value process for the portfolio. Since, by construction, (7.9) and (7.12) hold, (ϕ, θ) is self-financing, We have constructed a unique self-financing, replicating strategy for the claim H, as required. □

Since $p^* + q^* = 1$, (p^*, q^*) is a probability vector iff $p^* > 0$ and $q^* > 0$, that is, iff $0 < d < 1 + i < u$. Thus if the latter inequality holds, then we may construct a probability measure $\mathbb{P}^*$ on Ω in exactly the same manner as $\mathbb{P}$ was constructed, but with p and q replaced, respectively, by p^* and q^*. Denoting the corresponding expectation operator by $\mathbb{E}^*$, we have the following consequence of Corollary 7.2.2, which gives a general formula for the price of a European claim.

7.2.3 Corollary. *The binomial model is arbitrage-free iff $d < 1 + i < u$. In this case, given a European claim H, the unique discounted value process $\widetilde{V}$ with $V_N = H$ satisfies*

$$\mathbb{E}^* \widetilde{V}_n = \mathbb{E}^* \widetilde{V}_{n+1}, \quad n = 0, 1, \ldots, N-1.$$

In particular,

$$V_0 = (1+i)^{-N} \mathbb{E}^* H,$$

which is the arbitrage-free price of the claim.

Proof. If the binomial model is arbitrage-free, then $d < 1 + i < u$ (Example 4.2.1.) Conversely, assume that these inequalities hold. Then, using (7.12) and (7.6) with $\mathbb{P}_j$ replaced by $\mathbb{P}_j^*$, we have

$$\begin{aligned}
\mathbb{E}^*\widetilde{V}_n &= \sum_{\omega=(\omega_1,\ldots,\omega_n)} \widetilde{V}_n(\omega)\mathbb{P}_1^*(\omega_1)\cdots\mathbb{P}_n^*(\omega_n) \\
&= \sum_{\omega=(\omega_1,\ldots,\omega_n)} \left[\widetilde{V}_{n+1}(\omega,u)p^* + \widetilde{V}_{n+1}(\omega,d)q^*\right]\mathbb{P}_1^*(\omega_1)\cdots\mathbb{P}_n^*(\omega_n) \\
&= \sum_{\omega=(\omega_1,\ldots,\omega_{n+1})} \widetilde{V}_{n+1}(\omega)\mathbb{P}_1^*(\omega_1)\cdots\mathbb{P}_{n+1}^*(\omega_{n+1}) \\
&= \mathbb{E}^*\widetilde{V}_{n+1}.
\end{aligned}$$

This verifies the first equality of the corollary. The second equality follows by iteration, noting that $V_N = H$ and $\mathbb{E}^*\widetilde{V}_0 = V_0$.

To show that the binomial model is arbitrage-free, suppose for a contradiction that there exists a self-financing trading strategy with value process $W = (W_n)$ such that $W_0 = 0$, $W_n \geq 0$ for all n, and $\mathbb{P}(W_N > 0) > 0$. Then $\{W_N > 0\} \neq \emptyset$, and since $\mathbb{P}^*(\omega) > 0$ for all ω it follows that $\mathbb{E}^*W_N > 0$. On the other hand, if we take $H = W_N$ in Corollary 7.2.2, then, by uniqueness, W must be the process V constructed there, hence, by what has just been proved, $\mathbb{E}^*W_N = (1+i)^N W_0 = 0$. This contradiction shows that the binomial model must be arbitrage-free. The last assertion of the corollary follows from Theorem 5.5.1. □

7.2.4 Remarks. **(a)** For $d < 1+i < u$, the random variables X_j defined in Section 7.1 are independent Bernoulli random variables with respect to the new probability measure $\mathbb{P}^*$, hence $U_n = X_1 + \cdots + X_n$ is a binomial random variable with parameters (n, p^*). The probabilities p^* and q^* are called the *risk-neutral probabilities* and $\mathbb{P}^*$ is the *risk-neutral probability measure.* Note that, in contrast to the law $\mathbb{P}$ which reflects the perception of the market, $\mathbb{P}^*$ is a purely mathematical construct.

(b) Note that the equation $Z(\omega) := \mathbb{P}^*(\omega)/\mathbb{P}(\omega)$ defines a random variable Z satisfying $\mathbb{P}^* = Z\mathbb{P}$ and $\mathbb{P} = Z^{-1}\mathbb{P}^*$. Because of these relations, the probability laws $\mathbb{P}$ and $\mathbb{P}^*$ are said to be *equivalent.* We shall return to this idea in later chapters.

(c) Using the identity $d(vp^* + q^*) = up^* + dq^* = 1+i$, we see from (7.3) and independence that

$$\mathbb{E}^*S_n = S_0 d^n \mathbb{E}^* v^{U_n} = S_0 d^n \left(\mathbb{E}^* v^{X_1}\right)^n = S_0 d^n \left(vp^* + q^*\right)^n = (1+i)^n S_0.$$

Thus $\mathbb{E}^*S_n$ is the time-n value of a risk-free account earning compound interest at the rate i. A similar calculation shows that

$$\mathbb{E}\, S_n = (up + dq)^n\, S_0 = (1+j)^n S_0,$$

where $j := up + dq - 1$. Since there is risk involved in buying a stock, one would expect a greater average return, that is, $j > i$, which is equivalent to the inequality $p > p^*$. ◊

7.3 Path-Independent Claims

The following theorem is an immediate consequence of Corollary 7.2.3 and the law of the unconscious statistician.

7.3.1 Theorem. *If $d < 1+i < u$ and the claim H is of the form $f(S_N)$ for some function $f(x)$, then the proper price of the claim is*

$$V_0 = (1+i)^{-N}\mathbb{E}^* f(S_N) = (1+i)^{-N}\sum_{j=0}^{N}\binom{N}{j} f\left(S_0 u^j d^{N-j}\right) p^{*j} q^{*N-j}.$$

Application to Forwards

For a forward, take $f(S_N) = S_N - K$ in Theorem 7.3.1 to obtain

$$\begin{aligned}(1+i)^N V_0 &= \sum_{j=0}^{N}\binom{N}{j}(u^j d^{N-j} S_0 - K)p^{*j} q^{*N-j}\\ &= S_0\sum_{j=0}^{N}\binom{N}{j}(up^*)^j (dq^*)^{N-j} - K\sum_{j=0}^{N}\binom{N}{j} p^{*j} q^{*N-j}\\ &= S_0(up^* + dq^*)^N - K(p^* + q^*)^N\\ &= S_0(1+i)^N - K.\end{aligned}$$

Therefore, $V_0 = S_0 - K(1+i)^{-N}$. Recalling that there is no cost to enter a forward contract, we have $K = S_0(1+i)^N$. This is the discrete-time analog of Equation (4.3), which was obtained by a general arbitrage argument.

Application to Calls and Puts

For the price C_0 of a call one takes $f(S_N) = (S_N - K)^+$ in 7.3.1 and for the price P_0 of a put one takes $f(S_N) = (K - S_N)^+$. Table 7.3 below gives C_0 and P_0 for various values of K, u, and d. The initial value of the stock is $S_0 = 20.00$, the nominal rate is $r = .06$, and we consider daily fluctuations of the stock, so $i = .06/365 = .00016$. The options are assumed to expire in $N = 90$ days.

The table suggests that C_0 typically increases as $v = u/d$ (a measure of the volatility of the stock price) increases, a fact that may be directly verified from formula (7.19) below. The table also suggests that C_0 decreases as K increases. This also follows from (7.19) and is to be expected, as a larger K results in a smaller payoff for the holder, making the option less attractive. The put price has the reverse behavior. The prices in Table 7.3 were calculated using the module **EuropeanOptionPrice–Bin**. In this program the user enters an arbitrary payoff function of the form $F(S_N, K)$ and the program

TABLE 7.3: Variation of C_0 and P_0 with K, u, and d.

u	d	K	C_0	P_0	K	C_0	P_0	K	C_0	P_0
1.1	.7	10	14.92	4.77	19	12.59	11.31	30	10.72	20.28
1.4	.7	10	18.61	8.47	19	18.05	16.77	30	17.54	27.10
1.1	.9	10	11.81	1.66	19	7.71	6.43	30	4.97	14.53
1.4	.9	10	14.91	4.76	19	12.82	11.54	30	11.14	20.70

calculates the price. For example, the price of a call option is obtained by taking $F(S_N, K) = \text{MAX}(S_N - K, 0)$, the Excel maximum function.

The following are graphs of the call option price against u and d. The figures were generated by the module **GraphOfEuropeanOptionPrice–Bin**. As in the module **EuropeanOptionPrice–Bin**, the user enters the payoff function.

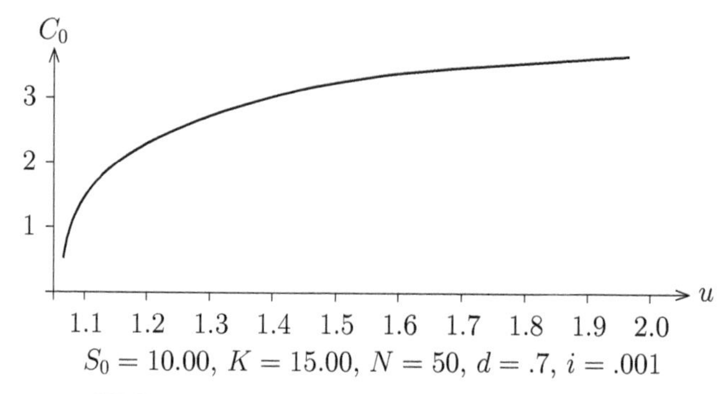

FIGURE 7.3: Call option price against u.

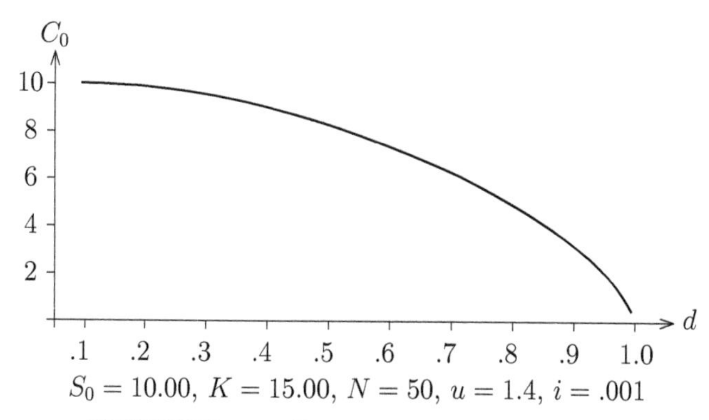

FIGURE 7.4: Call option price against d.

7.3.2 Remark. If $d < 1 + i < u$, then, by using the identity

$$(K - x)^+ = (x - K)^+ + K - x$$

in Theorem 7.3.1, one obtains the relation

$$P_0 = C_0 - S_0 + (1+i)^{-N}K, \tag{7.15}$$

which is the discrete version of the put-call parity formula derived in 4.9.2 by a general arbitrage argument. ◊

The Cox-Ross-Rubinstein Formula

Assume that $d < 1+i < u$. Taking $f(x) = (x-K)^+$ in Theorem 7.3.1 we see that the fair price C_0 of a call option with strike price K is given by

$$C_0 = (1+i)^{-N}\sum_{j=0}^{N}\binom{N}{j}\left(S_0 d^N v^j - K\right)^+ p^{*j} q^{*N-j}.$$

If $S_0 u^N \le K$, then $S_0 d^N v^j = S_0 u^j d^{N-j} \le S_0 u^j u^{N-j} \le K$ for all j and so $C_0 = 0$. Now assume that $S_0 u^N > K$. Then there must be a smallest integer $0 \le m \le N$ for which $S_0 v^m d^N = S_0 u^m d^{N-m} > K$, or, equivalently,

$$m > b := \frac{\ln(K) - \ln(d^N S_0)}{\ln v}. \tag{7.16}$$

It follows that $m - 1 = \lfloor b \rfloor$, the greatest integer in b. Since $v > 1$, $S_0 d^N v^j$ is increasing in j, hence $S_0 v^j d^N > K$ for $j \ge m$. Thus

$$\begin{aligned}
C_0 &= (1+i)^{-N}\sum_{j=m}^{N}\binom{N}{j}\left(S_0 u^j d^{N-j} - K\right)p^{*j}q^{*N-j}\\
&= S_0\sum_{j=m}^{N}\binom{N}{j}\left(\frac{p^*u}{1+i}\right)^j\left(\frac{q^*d}{1+i}\right)^{N-j} - (1+i)^{-N}K\sum_{j=m}^{N}\binom{N}{j}p^{*j}q^{*N-j}\\
&= S_0\sum_{j=m}^{N}\binom{N}{j}\widehat{p}^j\widehat{q}^{N-j} - (1+i)^{-N}K\sum_{j=m}^{N}\binom{N}{j}p^{*j}q^{*N-j}
\end{aligned} \tag{7.17}$$

where

$$\widehat{p} := \frac{p^*u}{1+i}, \quad \text{and} \quad \widehat{q} := 1 - \widehat{p} = \frac{q^*d}{1+i}. \tag{7.18}$$

The inequality $u > 1+i$ implies that $\widehat{p} < 1$. Since $\widehat{p} + \widehat{q} = 1$, $(\widehat{p}, \widehat{q})$ is a probability vector. For arbitrary $0 < \widetilde{p} < 1$, let $\Theta(k, N, \widetilde{p})$ denote the cdf of a binomial random variable with parameters $(\widetilde{p}, N)$ and define

$$\Psi(k, N, \widetilde{p}) = 1 - \Theta(k-1, N, \widetilde{p}) = \sum_{j=k}^{N}\binom{N}{j}\widetilde{p}^j(1-\widetilde{p})^{N-j}, \quad k = 1, 2, \ldots N.$$

The summations in (7.17) may then be written in terms of Ψ as, respectively, $\Psi(m, N, \widehat{p})$ and $\Psi(m, N, p^*)$. We have proved

7.3.3 Theorem (Cox-Ross-Rubinstein (CRR) Formula). *If $d < 1 + i < u$, then the cost C_0 of a call option with strike price K to be exercised after N time steps is*

$$C_0 = S_0\Psi\big(\lfloor b\rfloor + 1, N, \widehat{p}\big) - (1+i)^{-N}K\Psi\big(\lfloor b\rfloor + 1, N, p^*\big) \tag{7.19}$$

where $\widehat{p}$ is given by (7.18) and b by (7.16). If $\lfloor b\rfloor \geq N$ (which occurs iff $K \geq u^N S_0$), the right side of (7.19) is interpreted as zero.

7.3.4 Example. The diagram below was generated by the module **PortfolioProfitDiagram–Bin**, which uses the CRR formula and put-call parity to determine the cost of the portfolio. The input data for the portfolio is given in Table 7.4, which shows that the portfolio is in long in a \$15- and \$30-call, a \$10- and a \$20-put, and 3.2 shares of the stock; and short in cash in the amount of \$50, two \$5-calls, and three \$20-calls.

TABLE 7.4: Initial portfolio data.

S_0	u	d	i	Periods
11.00	1.5	.7	.001	90
Strike prices	Calls	Puts	Shares	Cash
5	−2		3.2	−50.00
10		1		
15	1			
20	−3	1		
30	1			

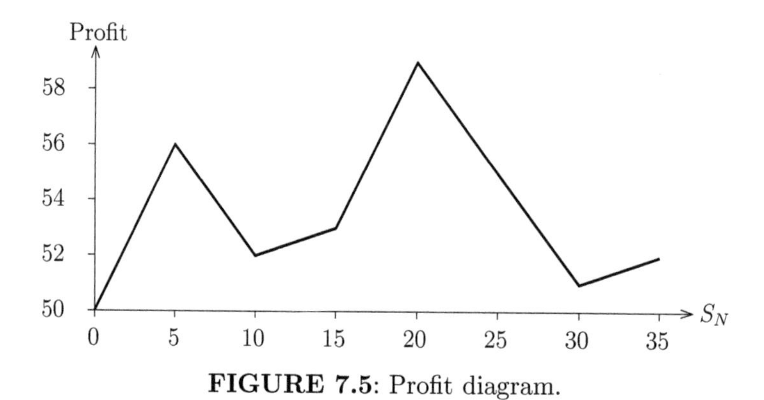

FIGURE 7.5: Profit diagram.

*7.4 Path-Dependent Claims

In this section we consider European derivatives with payoffs of the form $H = f(S_0, S_1, \ldots, S_N)$, where f is an arbitrary function of $N+1$ variables. We assume throughout that $d < 1+i < u$.

Corollary 7.2.3 implies that the arbitrage-free price of the claim is

$$V_0 = (1+i)^{-N}\mathbb{E}^* f(S_0, S_1, \ldots, S_N).$$

To find a concrete formula for V_0, observe first that by (7.3)

$$S_n = S_0 d^n v^{X_1+\cdots+X_n}, \quad v := \frac{u}{d},$$

where X_j is Bernoulli with parameter p^*. Thus the sequence $(S_0, S_1, \ldots, S_N)$ takes on the values

$$(S_0, S_0 d v^{k_1}, S_0 d^2 v^{k_1+k_2}, S_0 d^3 v^{k_1+k_2+k_3}, \ldots, S_0 d^N v^{k_1+k_2+\cdots+k_N})$$

with probability

$$p^{*k_1+\cdots+k_N} q^{*N-(k_1+\cdots+k_N)} = q^{*N} r^{*k_1+\cdots+k_N}, \quad r^* := \frac{p^*}{q^*},$$

where $k_j = 1$ or 0, corresponding as to whether there was an uptick or not on day j. Using (7.6) we then have

7.4.1 Theorem. *Let $d < 1+i < u$. Given a function $f(s_0, s_1, \ldots, s_N)$, the arbitrage-free price of a claim of the form $H = f(S_0, S_1, \ldots, S_N)$ is*

$$V_0 = (1+i)^{-N}\mathbb{E}^* H,$$

where

$$\mathbb{E}^* H = \sum_{(k_1,\ldots,k_N)} f(S_0, S_0 d v^{k_1}, S_0 d^2 v^{k_1+k_2}, \ldots, S_0 d^N v^{k_1+\cdots+k_N}) q^{*N} r^{*k_1+\cdots+k_N},$$

the sum taken over all 2^N sequences $(k_1, \ldots, k_N)$, where $k_j = 0$ or 1.

The above formula for $\mathbb{E}^* H$ does not generally have a simple closed form. However, in many cases of interest it is possible to construct a computer program based on the formula that can be used to calculate V_0. This may be accomplished by either deterministic or probabilistic methods. The deterministic method relies on a somewhat simplified version of the formula in Theorem 7.4.1. The revised formula uses expression

$$\delta(a,b) := \frac{1}{2}\left(1 + (-1)^{\lfloor (a-1)2^{-(b-1)} \rfloor}\right), \quad a = 1, \ldots 2^N, \; b = 1, \ldots N,$$

to calculate zeros and ones as values for the k_j. The values generated by this scheme may be conveniently displayed as a $2^N \times N$ matrix D in "truth table" format with the rows labeled by the a's and the columns by the b's, so that $\delta(a,b)$ appears in row a and column b. For example, for $N = 2$,

$$D = \begin{matrix} \text{scenario 1:} \\ \text{scenario 2:} \\ \text{scenario 3:} \\ \text{scenario 4:} \end{matrix} \begin{pmatrix} 1 & 1 \\ 0 & 1 \\ 1 & 0 \\ 0 & 0 \end{pmatrix} \quad \text{corresponding to} \quad \begin{pmatrix} u & u \\ d & u \\ u & d \\ d & d \end{pmatrix}$$

Now let $s_n(a)$ denote the sum of the first n entries in row a:

$$s_n(a) = \delta(a,1) + \delta(a,2) + \cdots + \delta(a,n).$$

The $s_n(a)$'s may be substituted for the sums $k_1 + \cdots + k_n$ in (7.4.1) resulting in the formula

$$\mathbb{E}^* H = \sum_{a=1}^{2^N} f(S_0, S_0 d v^{s_1(a)}, S_0 d^2 v^{s_2(a)}, \ldots, S_0 d^N v^{s_N(a)}) \times q^{*N} r^{*s_N(a)}.$$

While this may seem to be merely a notational simplification, the revised formula actually has the advantage of being amenable to coding.

The deterministic approach works well for short-term options, but for longer terms, the probabilistic method, known as Monte Carlo simulation, is generally superior. This method uses a random number generator to generate the zeros and ones in the sums $k_1 + \cdots + k_n$. The calculations are repeated many times and V_0 is then taken as the average of the iterations. The validity of this method is based on the law of large numbers. (See §6.6.)

We illustrate both approaches with the options in the list below, where the function $f(S_0, S_1 \ldots, S_N)$ is specified by the notation

$$A := \frac{S_0 + S_1 + \cdots + S_N}{N+1}, \quad G := (S_0 S_1 \cdots S_N)^{N+1}$$
$$M := \max\{S_0, \ldots, S_N\}, \text{ and } m := \min\{S_0, \ldots, S_N\}.$$

- Fixed strike arithmetic average call option: $(A - K)^+$.
- Floating strike arithmetic average call option: $(S_N - A)^+$.
- Fixed strike geometric average call option: $(G - K)^+$.
- Floating strike geometric average call option: $(S_N - G)^+$.
- Fixed strike lookback call option: $(M - K)^+$.
- Down-and-out barrier call option: $(S_N - K)^+ \mathbf{1}_{(c,+\infty)}(m)$.
- Up-and-in barrier call option: $(S_N - K)^+ \mathbf{1}_{(c,+\infty)}(M)$.

The modules **AsianOptionPrice–Bin**, **LookbackOptionPrice–Bin**, and **BarrierOptionPrice–Bin** calculate prices of the above options for a variety of user-input parameters. The reader may wish to compare the prices generated by these modules to those for standard call and put options generated by **EuropeanOptionPrice–Bin**. (See Table 7.3.)

7.5 Exercises

The following exercises refer to the binomial model, where $0 < d < 1 + i < u$.

1. Prove Equation 7.4.

2. Let $\eta \in \Omega_1 \times \ldots \times \Omega_n$. Show that

$$\mathbb{P}(\{\eta\} \times \Omega_{n+1} \times \ldots \times \Omega_N) = p^{U_n(\eta)} q^{n - U_n(\eta)}.$$

3. For $n = 1, \ldots, N$, define Z_n on Ω by $Z_n(\omega) = \omega_n$. Show that

   (a) $Z_n = (u - d)X_n + d$. (b) $S_n = Z_n Z_{n-1} \cdots Z_1 S_0$.

   (c) $X_n = (S_n - dS_{n-1})/(u - d)S_{n-1}$. Conclude from (c) that $\mathcal{F}_n^S = \mathcal{F}_n^X$.

4. Find the probability (in terms of n and p) that the price of the stock in the binomial model goes down at least twice during the first n time periods.

5. In the one-step binomial model, the Cox-Ross-Rubinstein formula reduces to

$$C_0 = (1 + i)^{-1} \left[(S_0 u - K)^+ p^* + (S_0 d - K)^+ q^* \right].$$

   (a) Show that if $S_0 d < K < S_0 u$ then $\frac{\partial C_0}{\partial u} > 0$ and $\frac{\partial C_0}{\partial d} < 0$. Conclude that for $u > K/S_0$ and $d < K/S_0$, C_0 increases as the spread $u - d$ increases.

   (b) Show that C_0, as a function of (u, d), is constant for the range of values $u > 1 + i > d \geq K/S_0$.

6. Suppose in Example 7.1.1 that $d = 1/u$ and $p = 1/2$. Show that

$$\mathbb{P}(S_n > S_0) = \mathbb{P}(S_n < S_0) = \begin{cases} 1/2 & \text{if } n \text{ is odd} \\ (1/2)\left[1 - \binom{n}{n/2} 2^{-n}\right] & \text{if } n \text{ is even.} \end{cases}$$

7. Let $S_0 = \$10$, $p = .5$, and $d = .8$. Fill in the following table.

u	1.1	1.2	1.3	1.5
$\mathbb{P}(S_5 > 20)$				

8. Find the probability that a call option finishes in the money.

9. A *k-run of upticks* is a sequence of k consecutive stock price increases not contained in any larger such sequence. Show that if $N/2 \le k < N$ then the probability of a k-run of upticks in N time periods is

$$p^k[2q + (N-k-1)q^2].$$

10. (*Cash-or-nothing call option*). Let A be a fixed amount of cash. Show that the cost V_0 of a claim with payoff $A\mathbf{1}_{(K,\infty)}(S_N)$ is

$$(1+i)^{-N} A\Psi(m, N, p^*),$$

where m is defined as in Theorem 7.3.3.

11. (*Asset-or-nothing call option*). Show that the cost V_0 of a claim with payoff $S_N\mathbf{1}_{(K,\infty)}(S_N)$ is $S_0\Psi(m, N, \hat{p})$, where m and $\hat{p}$ are defined as in Theorem 7.3.3.

12. Show that a portfolio long in an asset-or-nothing call option (Exercise 11) and short in a cash-or-nothing call option with cash K (Exercise 10) has the same time-n value as a portfolio with a long position in a call option with strike price K.

13. Let $1 \le M < N$. Show that the cost V_0 of a claim with payoff $(S_N - S_M)^+$ is

$$S_0\Psi(k, L, \hat{p}) - (1+i)^{-L}\Psi(k, L, p^*),$$

where

$$L := N - M, \quad k := (\lfloor b \rfloor + 1)^+, \quad b := \frac{L \ln d}{\ln d - \ln u},$$

and $\hat{p}$ is defined as in Theorem 7.3.3.

14. Show that the cost of a claim with payoff $S_N(S_N - K)$ is

$$V_0 = S_0^2 \left(\frac{w}{1+i}\right)^N - KS_0,$$

where $w = (u+d)(1+i) - ud$.

15. Show that the cost of a claim with payoff $S_N(S_N - K)^+$ is

$$V_0 = S_0^2 \left(\frac{w}{1+i}\right)^N \Psi(m, N, \tilde{p}) - KS_0\Psi(m, N, \hat{p}),$$

where m, and $\hat{p}$ are defined as in Theorem 7.3.3, $w = (u+d)(1+i) - ud$, and $\tilde{p} = p^*u^2/v$.

16. Use Exercises 14 and 15 and the law of one price to find the cost V_0 of a claim with payoff $S_N(K - S_N)^+$.

17. Show that the price of a claim with payoff $f(S_m, S_n)$, where $1 \le m < n \le N$, is

$$V_0 = \frac{1}{a^N} \sum_{j=0}^{m} \sum_{k=j}^{n-m+j} \binom{m}{j}\binom{n-m}{k-j} p^{*k} q^{*n-k} f(S_0 u^j d^{m-j}, S_0 u^k d^{n-k}),$$

where $a := (1+i)$.

18. Referring to Example 7.1.1 with $n = 100$, $p = .5$, and $d = .8$, use a spreadsheet to find the smallest value of u that results in $\mathbb{P}(S_n > S_0) \approx .85$.

19. Consider a claim with payoff $\left[\frac{1}{2}(S_1 + S_N) - K\right]^+$. Use Exercise 17 to show that if N is sufficiently large, specifically, if

$$u^{N-1} > \frac{2K - S_0 d}{S_0 d},$$

then there exist nonnegative integers k_1 and k_2 less than N such that the price of the claim is

$$V_0 = \frac{(S_0 d - 2K)q^*}{2(1+i)^N}\Psi(k_1, N-1, p^*) + \frac{S_0 d q^*}{2}\Psi(k_1, N-1, \hat{p}) + \frac{(S_0 u - 2K)p^*}{2(1+i)^N}\Psi(k_2, N-1, p^*) + \frac{S_0 u p^*}{2}\Psi(k_2, N-1, \hat{p}),$$

where $\hat{p}$ is defined as in Theorem 7.3.3. Show that k_1 is the smallest nonnegative integer k such that $S_0 d + S_0 u^k d^{N-k} > 2K$, and k_2 is the smallest nonnegative integer k such that $S_0 u + S_0 u^{k+1} d^{N-k-1} > 2K$.

20. Find the probability that the stock goes up

(a) three times in the first 5 days, given that it went up twice in the first 3 days.

(b) twice in the first 3 days, given that it goes up three times in the first 5 days.

21. Let P denote the probability of a call option on $\mathcal{S}$ of finishing in the money after 100 days, where $S_0 = 20$. Use a spreadsheet to fill in Table 7.5, which shows the variation of P with u, d, p, and K.

TABLE 7.5: Variation of P with u, d, p, and K.

K	u	d	p	P	K	u	d	p	P
10	1.20	.80	.45		30	1.20	.80	.45	
10	1.18	.80	.45		30	1.18	.80	.45	
10	1.20	.82	.45		30	1.20	.82	.45	
10	1.20	.80	.55		30	1.20	.80	.55	
10	1.18	.80	.55		30	1.18	.80	.55	
10	1.20	.82	.55		30	1.20	.82	.55	

22. Let k_j be a nonnegative integer and $x_j = d^j S_0 v^{k_j}$. Use Exercise 3.29 to show that

$$\mathbb{P}(S_1 = x_1,\, S_2 = x_2, \ldots, S_n = x_n) = \prod_{j=1}^{n} p_X(k_j - k_{j-1}),$$

where p_X is the pmf of a Bernoulli random variable X with parameter p.

23. Use Theorem 7.3.1 with $f(s) = (s - K)^+$ and $(K - s)^+$ to obtain the put-call parity relation.

24. (Asian arithmetic forward.) Consider a claim with payoff

$$H = f(S_1, \ldots, S_n) = (S_1 + \cdots + S_N)/N - K.$$

Setting $v := (u/d)$ and $x = d(q^* + vp^*)$ show that

$$E^* H = \frac{1}{N} \mathbb{E}^*(S_1 + \cdots + S_N) - K = \frac{S_0}{N} \sum_{n=1}^{N} x^n - K = \frac{S_0}{N} \frac{x^{N+1} - x}{x - 1} - K.$$

Conclude that the proper forward price is

$$K = \frac{S_0}{N} \frac{x^{N+1} - x}{x - 1}.$$

25. (Asian geometric forward.) Consider a claim with payoff

$$H = f(S_1, \ldots, S_n) = (S_1 S_2 \cdots S_N)^{1/N} - K.$$

Setting $w := (u/d)^{\frac{1}{N}}$ and $a := 1 + i$ show that

$$\mathbb{E}^*(S_1 S_2 \cdots S_N)^{\frac{1}{N}} = S_0 d^{\frac{N+1}{2}} (u - d)^{-N} \prod_{j=1}^{N} \left(u - a + (a - d) w^j\right).$$

Conclude that the proper forward price is

$$K = \frac{S_0 d^{\frac{N+1}{2}}}{(u - d)^N} \prod_{j=1}^{N} \left[u - a + (a - d)(u/d)^{j/N}\right].$$

Chapter 8

Conditional Expectation

Conditional expectation generalizes the notion of expectation by taking into account information provided by a given σ-field. We use conditional expectation in the next chapter to place the binomial model in a broader context. This will ultimately lead to the formulation of the general option valuation models studied in later chapters.

8.1 Definition of Conditional Expectation

Let X be a random variable on a probability space $(\Omega, \mathcal{F}, \mathbb{P})$ and let $\mathcal{G}$ be a σ-field with $\mathcal{G} \subseteq \mathcal{F}$. A $\mathcal{G}$-random variable Y with the property that

$$\mathbb{E}\,(\mathbf{1}_A Y) = \mathbb{E}\,(\mathbf{1}_A X) \quad \text{for all } A \in \mathcal{G} \tag{8.1}$$

is called a *conditional expectation of X given $\mathcal{G}$*. It may be shown that such a random variable Y exists for any random variable X for which $\mathbb{E}\,X$ exists. Further, it may be shown that it suffices to take the sets A in the definition to lie in a collection closed under finite intersections that generates $\mathcal{G}$.[1] Thus to verify that a random variable Y is the conditional expectation of X with respect to $\mathcal{G}$ it is enough to show that

- Y is $\mathcal{G}$-measurable, and
- $\mathbb{E}(\mathbf{1}_A Y) = \mathbb{E}(\mathbf{1}_A X)$ for all A in a collection closed under finite intersections that generates $\mathcal{G}$.

Conditional expectations are not unique. Indeed, if Y satisfies the above conditions, then so does any $\mathcal{G}$-measurable random variable Y' that differs from Y on a set of probability zero. As a consequence, properties regarding conditional expectations generally hold only *almost surely*, that is, only for those ω forming a set of probability one. This is an important technical detail, but for ease of exposition we usually omit the qualifier that a given property holds only almost surely.

[1] This fact, as well as the existence of conditional expectation, is established in Section 8.5 for the special case of a finite sample space.

The standard notation for the conditional expectation of X given $\mathcal{G}$ is $\mathbb{E}(X \mid \mathcal{G})$. In the special case that $\mathcal{G} = \sigma(X_1, \ldots, X_n)$ (see page 81), $\mathbb{E}(X \mid \mathcal{G})$ is called the *conditional expectation of* X *given* $X_1, \ldots, X_n$ and is usually denoted by $\mathbb{E}(X \mid X_1, \ldots, X_n)$. In this case the sets A in the foregoing may be taken to be of the form $\{X_1 \in J_1, \ldots, X_n \in J_n\}$ for intervals J_n.

8.2 Examples of Conditional Expectations

Given the "known" information provided by a σ-field $\mathcal{G}$, the conditional expectation of a random variable X may be seen as the average of the values of X over the remaining "unknown" information. This is demonstrated explicitly in Example 8.2.3 below. The conditional expectation $\mathbb{E}(X \mid \mathcal{G})$ may also be characterized as the best prediction of X given $\mathcal{G}$. (See Exercise 6(e).) The following examples illustrate these ideas.

8.2.1 Example. For the trivial σ-field $\mathcal{G} = \{\emptyset, \Omega\}$, $\mathbb{E}(X \mid \mathcal{G}) = \mathbb{E}(X)$. This follows from the equality $\mathbb{E}(\mathbf{1}_A \mathbb{E}\, X) = \mathbb{E}(\mathbf{1}_A X)$, $A \in \mathcal{G}$, as is easily verified. Thus, the best prediction of X given the information $\mathcal{G}$, which is to say no information at all, is simply the expected value of X. ◊

8.2.2 Example. If $\mathcal{G}$ is the σ-field consisting of all subsets of Ω, then, trivially, $\mathbb{E}(X \mid \mathcal{G}) = X$: the best prediction of X given all possible information is X itself. ◊

8.2.3 Example. Toss a coin N times and observe the outcome heads H or tails T on each toss. Let p be the probability of heads on a single toss and set $q = 1 - p$. The sample space for the experiment is $\Omega = \Omega_1 \times \Omega_2 \times \cdots \times \Omega_N$, where $\Omega_n = \{H, T\}$ is the set of outcomes of the nth toss. The probability law for the experiment may be expressed as

$$\mathbb{P}(\omega) = p^{H(\omega)} q^{T(\omega)}, \quad \omega = (\omega_1, \ldots, \omega_N),$$

where $H(\omega)$ denotes the number of heads in the sequence ω, and $T(\omega)$ the number of tails. Fix $n < N$ and for $\omega \in \Omega$ write

$$\omega = (\omega', \omega''), \quad \omega' \in \Omega_1 \times \cdots \times \Omega_n, \quad \omega'' \in \Omega_{n+1} \times \cdots \times \Omega_N.$$

Let $\mathcal{G}_n$ denote the σ-field generated by the sets

$$A_{\omega'} = \{(\omega', \omega'') \mid \omega'' \in \Omega_{n+1} \times \cdots \times \Omega_N\}. \tag{8.2}$$

Then $\mathcal{G}_n$ represents the information generated by the first n tosses of the coin. We claim that for any random variable X

$$\mathbb{E}(X \mid \mathcal{G}_n)(\omega) = \mathbb{E}(X \mid \mathcal{G}_n)(\omega', \omega'') = \sum_{\eta} p^{H(\eta)} q^{T(\eta)} X(\omega', \eta), \tag{8.3}$$

where the sum on the right is taken over all $\eta \in \Omega_{n+1} \times \cdots \times \Omega_N$. Equation (8.3) asserts that the best prediction of X given the information provided by the first n tosses (the "known") is the average of X over the remaining outcomes η (the "unknown").

To verify (8.3), denote the sum on the right by $Y(\omega) = Y(\omega', \omega'')$ and note that, since Y depends only on the first n tosses, Y is $\mathcal{G}_n$-measurable. It therefore remains to show that $\mathbb{E}(Y\mathbf{1}_A) = \mathbb{E}(X\mathbf{1}_A)$ for the sets $A = A_{\omega'}$ defined in (8.2). Noting that $\sum_{\omega''} p^{H(\omega'')} q^{T(\omega'')} = 1$ (the expectation of the constant function 1 on the probability space $\Omega_{n+1} \times \cdots \times \Omega_N$), we have

$$
\begin{aligned}
\mathbb{E}(\mathbf{1}_{A_{\omega'}} Y) &= \sum_{\omega \in A_{\omega'}} Y(\omega)\mathbb{P}(\omega) \\
&= \sum_{\omega''} Y(\omega', \omega'') p^{H(\omega')+H(\omega'')} q^{T(\omega')+T(\omega'')} \\
&= p^{H(\omega')} q^{T(\omega')} \sum_{\omega''} p^{H(\omega'')} q^{T(\omega'')} \sum_{\eta} p^{H(\eta)} q^{T(\eta)} X(\omega', \eta) \\
&= p^{H(\omega')} q^{T(\omega')} \sum_{\eta} p^{H(\eta)} q^{T(\eta)} X(\omega', \eta) \\
&= \sum_{\eta} p^{H(\omega')+H(\eta)} q^{T(\omega')+T(\eta)} X(\omega', \eta) \\
&= \sum_{\omega \in A_{\omega'}} p^{H(\omega)} q^{T(\omega)} X(\omega) \\
&= \mathbb{E}(\mathbf{1}_{A_{\omega'}} X),
\end{aligned}
$$

as required. ◊

8.2.4 Example. Consider the geometric binomial price process S with its natural filtration $(\mathcal{F}_n^S)$. We show that for $0 \le n < m \le N$ and any real-valued function $f(x)$,

$$
\mathbb{E}\left(f(S_m) \mid \mathcal{F}_n^S\right) = \sum_{j=0}^{k} \binom{k}{j} p^j q^{k-j} f\left(u^j d^{k-j} S_n\right), \quad k := m - n.
$$

Let Y denote the sum on the right. Since Y is obviously $\mathcal{F}_n^S$-measurable it remains to show that $\mathbb{E}\left(f(S_m)\mathbf{1}_A\right) = \mathbb{E}(Y\mathbf{1}_A)$ for all $A \in \mathcal{F}_n^S$. For this we may assume that A is of the form

$$
A = \{\eta\} \times \Omega_{n+1} \times \cdots \times \Omega_N,
$$

where $\eta \in \Omega_1 \times \cdots \times \Omega_n$, since these sets generate $\mathcal{F}_n^S$. Noting that

$$
\mathbb{P}(A) = \mathbb{P}_1(\eta_1)\mathbb{P}_2(\eta_2) \cdots \mathbb{P}_n(\eta_n)
$$

and

$$
\sum_{(\omega_{m+1}, \ldots, \omega_N)} \mathbb{P}_{m+1}(\omega_{m+1}) \cdots \mathbb{P}_N(\omega_N) = 1,
$$

we have

$$\begin{aligned}
\mathbb{E}\left[f(S_m)\mathbf{1}_A\right] &= \sum_{\omega\in A} f\left(S_m(\omega)\right)\mathbb{P}(\omega), \qquad \omega = (\eta_1,\dots,\eta_n,\omega_{n+1},\dots,\omega_N) \\
&= \sum_{\omega\in A} f\left(\omega_{n+1}\cdots\omega_m S_m(\eta)\right)\mathbb{P}_1(\eta_1)\cdots\mathbb{P}_n(\eta_n)\mathbb{P}_{n+1}(\omega_{n+1})\cdots\mathbb{P}_N(\omega_N) \\
&= \sum_{\omega\in A} f\left(\omega_{n+1}\cdots\omega_m S_m(\eta)\right)\mathbb{P}_1(\eta_1)\cdots\mathbb{P}_n(\eta_n)\mathbb{P}_{n+1}(\omega_{n+1})\cdots\mathbb{P}_m(\omega_m) \\
&= \mathbb{P}(A)\sum_{(\omega_{n+1},\dots,\omega_m)} f\left(\omega_{n+1}\cdots\omega_m S_n(\eta)\right)\mathbb{P}_{n+1}(\omega_{n+1})\cdots\mathbb{P}_m(\omega_m).
\end{aligned}$$

Collecting together all terms in the last sum for which the sequence $(\omega_{n+1},\dots,\omega_m)$ has exactly j u's ($j = 0, 1, \dots, k := m - n$), we see that

$$\mathbb{E}\left(f(S_m)\mathbf{1}_A\right) = \mathbb{P}(A)\sum_{j=0}^{k}\binom{k}{j}p^j q^{k-j} f\left(u^j d^{k-j} S_n(\eta)\right). \tag{$\dagger$}$$

Similarly, for each $j \le k$ we have

$$\mathbb{E}\left(\mathbf{1}_A f(u^j d^{k-j} S_n)\right) = \sum_{\omega\in A} f(u^j d^{k-j} S_n(\omega))\mathbb{P}(\omega) \quad = \mathbb{P}(A) f(u^j d^{k-j} S_n(\eta)).$$

Recalling the definition of Y we see from these equations that

$$\begin{aligned}
\mathbb{E}(\mathbf{1}_A Y) &= \sum_{j=0}^{k}\binom{k}{j}p^j q^{k-j}\mathbb{E}(\mathbf{1}_A f(u^j d^{k-j} S_n)) \\
&= \mathbb{P}(A)\sum_{j=0}^{k}\binom{k}{j}p^j q^{k-j} f\left(u^j d^{k-j} S_n(\eta)\right).
\end{aligned}$$

By (†) the last expression is $\mathbb{E}\left(f(S_m)\mathbf{1}_A\right)$, as required. ◊

8.3 Properties of Conditional Expectation

In this section we consider those properties of conditional expectation that will be needed for the remainder of the book. The reader will note that some of these properties are identical or analogous to those of ordinary expectation, while others have no such counterpart. Throughout this section, X and Y are random variables with finite expectation on a probability space $(\Omega, \mathcal{F}, \mathbb{P})$ and $\mathcal{G}$ is a σ-field with $\mathcal{G} \subseteq \mathcal{F}$.

The first property follows directly from the definition of conditional expectation; the proof is left to the reader.

8.3.1 Theorem (Preservation of constants). *For any constant c, $\mathbb{E}(c \mid \mathcal{G}) = c$.*

8.3.2 Theorem (Linearity). *For a, $b \in \mathbf{R}$,*

$$\mathbb{E}(aX + bY \mid \mathcal{G}) = a\,\mathbb{E}(X \mid \mathcal{G}) + b\,\mathbb{E}(Y \mid \mathcal{G}).$$

Proof. Let Z denote the $\mathcal{G}$-random variable $a\,\mathbb{E}(X \mid \mathcal{G}) + b\,\mathbb{E}(Y \mid \mathcal{G})$. By linearity of expectation, for all $A \in \mathcal{G}$ we have

$$\begin{aligned}\mathbb{E}(\mathbf{1}_A Z) &= a\,\mathbb{E}\left[\mathbf{1}_A \mathbb{E}(X \mid \mathcal{G})\right] + b\,\mathbb{E}\left[\mathbf{1}_A \mathbb{E}(Y \mid \mathcal{G})\right] \\ &= a\,\mathbb{E}(\mathbf{1}_A X) + b\,\mathbb{E}(\mathbf{1}_A Y) \\ &= \mathbb{E}\left[\mathbf{1}_A (aX + bY)\right].\end{aligned}$$

Therefore Z is the conditional expectation $\mathbb{E}(aX + bY \mid \mathcal{G})$. □

8.3.3 Theorem (Order Property). *If $X \le Y$, then $\mathbb{E}(X \mid \mathcal{G}) \le \mathbb{E}(Y \mid \mathcal{G})$.*

Proof. The assertion of the theorem may be shown to be a consequence of the inequality.

$$\mathbb{E}\left[\mathbf{1}_A \mathbb{E}(X \mid \mathcal{G})\right] = \mathbb{E}(\mathbf{1}_A X) \le \mathbb{E}(\mathbf{1}_A Y) = \mathbb{E}\left[\mathbf{1}_A \mathbb{E}(Y \mid \mathcal{G})\right] \quad (A \in \mathcal{G}).$$

A verification of this inequality for the special case of a finite sample space is given in §8.5. □

8.3.4 Corollary. $|\,\mathbb{E}(X \mid \mathcal{G})\,| \le \mathbb{E}(|X| \mid \mathcal{G})$.

Proof. This follows from $\pm X \le |X|$, hence, by the theorem, $\pm\mathbb{E}(X \mid \mathcal{G}) = \mathbb{E}(\pm X \mid \mathcal{G}) \le \mathbb{E}(|X| \mid \mathcal{G})$. □

The next theorem asserts that known factors may be moved outside the conditional expectation operator.

8.3.5 Theorem (Factor Property). *If X is a $\mathcal{G}$-random variable, then the equation $\mathbb{E}(XY \mid \mathcal{G}) = X\mathbb{E}(Y \mid \mathcal{G})$ holds. In particular, $\mathbb{E}(X \mid \mathcal{G}) = X$.*

Proof. Since $X\mathbb{E}(Y \mid \mathcal{G})$ is $\mathcal{G}$-measurable, for the first assertion it suffices to show that it has the conditional expectation property $\mathbb{E}\left[\mathbf{1}_A X\mathbb{E}(Y \mid \mathcal{G})\right] = \mathbb{E}(\mathbf{1}_A XY)$ $(A \in \mathcal{G})$. We prove this for the special case that the range of X is a finite set $\{x_1, x_2, \ldots, x_n\}$. Set $A_j = \{X = x_j\}$. Then $A_j \in \mathcal{G}$ and

$$\mathbf{1}_A X = \sum_{j=1}^{n} x_j \mathbf{1}_{AA_j}.$$

Multiplying this equation by $\mathbb{E}(Y \mid \mathcal{G})$ and using the linearity of conditional expectation we have

$$\begin{aligned}\mathbb{E}\left[\mathbf{1}_A X\mathbb{E}(Y \mid \mathcal{G})\right] &= \sum_{j=1}^{n} x_j \mathbb{E}\left[\mathbf{1}_{AA_j}\mathbb{E}(Y \mid \mathcal{G})\right] = \sum_{j=1}^{n} x_j \mathbb{E}(\mathbf{1}_{AA_j} Y) \\ &= \mathbb{E}(\mathbf{1}_A XY),\end{aligned}$$

as required. The second assertion follows by taking $Y = 1$. □

The next theorem asserts that if X is independent of $\mathcal{G}$, then, in calculating $\mathbb{E}(X \mid \mathcal{G})$, nothing is gained by incorporating the information provided by $\mathcal{G}$.

8.3.6 Theorem (Independence Property). *If X is independent of $\mathcal{G}$, that is, if X and $\mathbf{1}_A$ are independent for all $A \in \mathcal{G}$, then $\mathbb{E}(X \mid \mathcal{G}) = \mathbb{E}(X)$.*

Proof. Obviously the constant random variable $Z = \mathbb{E}\,X$ is $\mathcal{G}$-measurable. Moreover, if $A \in \mathcal{G}$, then, by independence of X and $\mathbf{1}_A$,

$$\mathbb{E}(\mathbf{1}_A X) = (\mathbb{E}\,\mathbf{1}_A)(\mathbb{E}\,X) = \mathbb{E}\big(\mathbf{1}_A \mathbb{E}(X)\big) = \mathbb{E}(\mathbf{1}_A Z).$$

Thus Z has the required properties of conditional expectation. □

The following theorem asserts that successive predictions of X based on nested levels of information produce the same result as a single prediction using the least information.

8.3.7 Theorem (Iterated Conditioning Property). *Let $\mathcal{H}$ be a σ-field with $\mathcal{H} \subseteq \mathcal{G}$. Then*

$$\mathbb{E}\left[\mathbb{E}(X \mid \mathcal{G}) \mid \mathcal{H}\right] = \mathbb{E}(X \mid \mathcal{H}).$$

Proof. Set $Z := \mathbb{E}(X \mid \mathcal{G})$. To prove that $\mathbb{E}(Z \mid \mathcal{H}) = \mathbb{E}(X \mid \mathcal{H})$, it must be shown that

$$\mathbb{E}\left[\mathbf{1}_A \mathbb{E}(Z \mid \mathcal{H})\right] = \mathbb{E}(\mathbf{1}_A X) \quad \text{for all } A \in \mathcal{H}.$$

But, by the defining property of conditional expectation with respect to $\mathcal{H}$, the last equation reduces to $\mathbb{E}(\mathbf{1}_A Z) = \mathbb{E}(\mathbf{1}_A X)$, which holds by the defining property of conditional expectation of X with respect to $\mathcal{G}$. □

8.4 Special Cases

There are two special constructions that lead to concrete representations of conditional expectation.

The Conditional Probability Mass Function

Let X and Y be discrete random variables on $(\Omega, \mathcal{F}, \mathbb{P})$ with $p_Y(y) > 0$ for all y in the range of Y. The function $p_{X|Y}$ defined by

$$p_{X|Y}(x \mid y) := \mathbb{P}(X = x \mid Y = y) = \frac{p_{X,Y}(x, y)}{p_Y(y)}$$

is called the *conditional probability mass function (pmf) of X given Y*. The following proposition gives a concrete representation of the conditional expectation $\mathbb{E}(X \mid Y)$ in terms of this function.

8.4.1 Proposition. *Let $p_{X|Y}(x \mid Y)$ denote the random variable defined by*

$$p_{X|Y}(x \mid Y)(\omega) := p_{X|Y}(x \mid Y(\omega)), \quad \omega \in \Omega.$$

Then

$$\mathbb{E}(X \mid Y) = \sum_x x\, p_{X|Y}(x \mid Y),$$

where the sum is taken over all x in the range of X.

Proof. Set $g(y) := p_{X|Y}(x \mid y)$. Then $p_{X|Y}(x \mid Y) = g \circ Y$, which is measurable with respect to the σ-field generated by Y. Since the sets $\{Y = y_0\}$ generate $\sigma(Y)$, it remains to show that

$$\mathbb{E}\left(\mathbf{1}_{\{Y=y_0\}} g(Y)\right) = \mathbb{E}\left(\mathbf{1}_{\{Y=y_0\}} X\right)$$

for all y_0 in the range of Y. But by 6.1.2, the left side of the equation is

$$\sum_y \mathbf{1}_{\{y_0\}}(y) g(y) p_Y(y) = g(y_0) p_Y(y_0) = \sum_x x\, p_{X,Y}(x, y_0),$$

and the right side is

$$\sum_{x,\,y} x \mathbf{1}_{\{y_0\}}(y) p_{X,Y}(x, y) = \sum_x x p_{X,Y}(x, y_0). \qquad \square$$

The Conditional Density Function

Let X and Y be jointly distributed continuous random variables on a probability space $(\Omega, \mathcal{F}, \mathbb{P})$ with $f_Y(y) > 0$ for all y. Assume that $f_{X,Y}$ is continuous. The function $f_{X|Y}$ defined by

$$f_{X|Y}(x \mid y) = \frac{f_{X,Y}(x, y)}{f_Y(y)}.$$

is called the *conditional density of X given Y*. Note that $f_{X|Y}(x \mid y)$ is continuous in (x, y). We claim that

$$\mathbb{E}(X \mid Y) = \int_{-\infty}^{\infty} x f_{X|Y}(x \mid Y)\, dx \tag{8.4}$$

where

$$f_{X|Y}(x \mid Y)(\omega) := f_{X|Y}(x \mid Y(\omega)), \quad \omega \in \Omega.$$

To see this, let

$$g(y) = \int_{-\infty}^{\infty} x f_{X|Y}(x \mid y)\, dx.$$

Then g is continuous (by the continuity of $f_{X,Y}$) and it follows that $g(Y)$, which is the integral in (8.4), is measurable with respect to the σ-field generated by Y. It remains to show that for any interval J

$$\mathbb{E}\left(\mathbf{1}_{\{Y \in J\}} g(Y)\right) = \mathbb{E}\left(\mathbf{1}_{\{Y \in J\}} X\right).$$

But by 6.2.3, the left side of the equation is

$$\mathbb{E}\left(\mathbf{1}_J(Y)g(Y)\right) = \int_{-\infty}^{\infty} \mathbf{1}_J(y)g(y)f_Y(y)\,dy = \int_{-\infty}^{\infty}\int_{-\infty}^{\infty} \mathbf{1}_J(y)xf_{X,Y}(x,y)\,dx\,dy,$$

which, by 6.2.5, is the right side.

The reader should calculate $f_{X|Y}(x \mid y)$ if X and Y are independent.

---

*8.5 Existence of Conditional Expectation

In this section, $(\Omega, \mathcal{F}, \mathbb{P})$ is a probability space with finite sample space Ω such that $\mathbb{P}(\omega) > 0$ for all ω, and $\mathcal{G}$ is a σ-field contained in $\mathcal{F}$. Recall that, by Proposition 5.1.2, $\mathcal{G}$ is necessarily generated by a partition.

8.5.1 Lemma. *Let $\mathcal{G}$ be a σ-field of subsets of Ω and let X and Y be $\mathcal{G}$-random variables on Ω. If $\{A_1, \ldots, A_n\}$ is a generating partition for $\mathcal{G}$, then*

(a) *$X \geq Y$ iff $\mathbb{E}(X\mathbf{1}_{A_j}) \geq \mathbb{E}(Y\mathbf{1}_{A_j})$ for all j.*

(b) *$X = Y$ iff $\mathbb{E}(X\mathbf{1}_{A_j}) = \mathbb{E}(Y\mathbf{1}_{A_j})$ for all j.*

Proof. The necessity of (a) is clear and (b) follows from (a). To prove the sufficiency of (a), set $Z = X - Y$ so that, by hypothesis, $\mathbb{E}(Z\mathbf{1}_{A_j}) \geq 0$ for all j. We show that the set $A = \{Z < 0\}$ is empty.

Suppose to the contrary that $A \neq \emptyset$. Since Z is a $\mathcal{G}$-random variable, $A \in \mathcal{G}$, hence there exists a subset $J \subseteq \{1, 2, \ldots, n\}$ such that $A = \bigcup_{j \in J} A_j$. Since the sets A_j are pairwise disjoint,

$$\mathbb{E}(Z\mathbf{1}_A) = \sum_{j \in J} \mathbb{E}(Z\mathbf{1}_{A_j}) \geq 0.$$

On the other hand, by definition of A,

$$\mathbb{E}(Z\mathbf{1}_A) = \sum_{\omega \in \Omega} Z(\omega)\mathbf{1}_A(\omega)\mathbb{P}(\omega) = \sum_{\omega \in A} Z(\omega)\mathbb{P}(\omega) < 0.$$

This contradiction shows that A must be empty, hence $X \geq Y$. □

We may now prove the following special case of the existence and uniqueness of conditional expectation.

8.5.2 Theorem. *If X is a random variable on $(\Omega, \mathcal{F}, \mathbb{P})$, then there exists a $\mathcal{G}$-random variable Y such that*

$$\mathbb{E}\left(\mathbf{1}_A Y\right) = \mathbb{E}\left(\mathbf{1}_A X\right) \tag{8.5}$$

for all $A \in \mathcal{G}$. Moreover, if Y' is a $\mathcal{G}$-random variable such that (8.5) holds for all A in a generating set for $\mathcal{G}$, then $Y' = Y$. In particular, Y is unique.

Proof. Let $\{A_1, \ldots, A_m\}$ be a partition of Ω generating $\mathcal{G}$. Define a $\mathcal{G}$-random variable Y on Ω by

$$Y = \sum_{j=1}^{m} a_j \mathbf{1}_{A_j}, \quad a_j := \frac{\mathbb{E}(\mathbf{1}_{A_j} X)}{\mathbb{P}(A_j)}. \tag{8.6}$$

Since $A_j A_k = \emptyset$ for $j \neq k$,

$$\mathbf{1}_{A_k} Y = \sum_{j=1}^{m} a_j \mathbf{1}_{A_j} \mathbf{1}_{A_k} = a_k \mathbf{1}_{A_k},,$$

hence

$$\mathbb{E}\left(\mathbf{1}_{A_k} Y\right) = a_k \mathbb{P}(A_k) = \mathbb{E}(\mathbf{1}_{A_k} X).$$

Since any member A of $\mathcal{G}$ is a disjoint union of sets A_k, (8.5) holds.

For the last part we have $\mathbb{E}\left(\mathbf{1}_A Y'\right) = \mathbb{E}\left(\mathbf{1}_A Y\right)$ for all A in a generating set, hence $Y' = Y$ by Lemma 8.5.1. □

8.5.3 Corollary. *Under the conditions of the theorem, if X, X_1, …, X_n are random variables on $(\Omega, \mathcal{F}, \mathbb{P})$, then there exists a function $g_X(x_1, x_2, \ldots, x_n)$ such that*

$$\mathbb{E}(X \mid X_1, \ldots, X_n) = g_X(X_1, \ldots, X_n).$$

Proof. The σ-field $\mathcal{G} := \sigma(X_1, \ldots, X_n)$ is generated by the partition consisting of sets of the form

$$A(\boldsymbol{x}) = \{X_1 = x_1, \ldots, X_n = x_n\}, \quad \boldsymbol{x} := (x_1, \ldots, x_n). \tag{8.7}$$

Define

$$g_X(\boldsymbol{x}) = \begin{cases} \dfrac{\mathbb{E}(\mathbf{1}_{A(\boldsymbol{x})} X)}{\mathbb{P}(A(\boldsymbol{x}))} & \text{if } A(\boldsymbol{x}) \neq \emptyset, \\ 0 & \text{otherwise.} \end{cases}$$

By (8.6),

$$\mathbb{E}(X \mid X_1, X_2, \ldots, X_n) = \sum_{\boldsymbol{x}} g_X(\boldsymbol{x}) \mathbf{1}_{A(\boldsymbol{x})}.$$

If $\omega \in A(\boldsymbol{x})$, then $\boldsymbol{x} = (X_1(\omega), X_2(\omega), \ldots, X_n(\omega))$, hence

$$\begin{aligned} \mathbb{E}(X \mid X_1, X_2, \ldots, X_n)(\omega) &= g_X(x) \mathbf{1}_{A(x)}(\omega) \\ &= g_X(X_1(\omega), X_2(\omega), \ldots X_n(\omega)). \end{aligned}$$

Since every $\omega \in \Omega$ is in some $A(\boldsymbol{x})$, the last equation holds for all $\omega \in \Omega$. □

8.6 Exercises

1. Let $\mathcal{G}$ be a σ-field with $\mathcal{G} \subseteq \mathcal{F}$, let X be a $\mathcal{G}$-random variable, and let Y be an $\mathcal{F}$-random variable independent of $\mathcal{G}$. Show that $\mathbb{E}(XY) = \mathbb{E}(X)\mathbb{E}(Y)$. (*Hint.* Condition on $\mathcal{G}$.)

2. Let $\mathcal{G}$ be a σ-field with $\mathcal{G} \subseteq \mathcal{F}$, X a $\mathcal{G}$-random variable, and Y an $\mathcal{F}$-random variable with $X - Y$ independent of $\mathcal{G}$. Show that if either $\mathbb{E}\,X = 0$ or $\mathbb{E}\,X = \mathbb{E}\,Y$, then
$$\mathbb{E}(XY) = \mathbb{E}\,X^2 \quad \text{and} \quad \mathbb{E}(Y - X)^2 = \mathbb{E}\,Y^2 - \mathbb{E}\,X^2.$$

3. Let X and Y be independent Bernoulli random variables with parameter p. Show that
$$\mathbb{E}(X \mid X + Y) = \mathbf{1}_{\{2\}}(X + Y) + \tfrac{1}{2}\mathbf{1}_{\{1\}}(X + Y).$$

4. Let (X_n) be a sequence of independent Bernoulli random variables with parameter p and set $Y_k = X_1 + \cdots + X_k$. For all cases find

   (a) $\mathbb{E}(Y_m \mid Y_n)$. (b) $\mathbb{E}(X_j \mid Y_n)$. (c) $\mathbb{E}(X_k \mid X_j)$. (d) $\mathbb{E}(Y_n \mid X_j)$.

5. Let X and Y be discrete independent random variables. Show that
$$\mathbb{E}(X \mid X + Y) = \frac{\sum_x x p_X(x) p_Y(X + Y - x)}{\sum_{x'} p_X(x') p_Y(X + Y - x')}$$

6. Let Y be a $\mathcal{G}$-random variable. Verify the following:

   (a) $\mathbb{E}\big(\mathbb{E}(X \mid \mathcal{G})Y\big) = \mathbb{E}(XY)$.

   (b) $\mathbb{E}\big(\mathbb{E}(X \mid \mathcal{G})^2\big) = \mathbb{E}\big(X\mathbb{E}(X \mid \mathcal{G})\big)$.

   (c) $\mathbb{E}\big[(X - \mathbb{E}(X \mid \mathcal{G}))Y\big] = 0$.

   (d) $\mathbb{E}\big[X - \mathbb{E}(X \mid \mathcal{G})\big]^2 + \mathbb{E}\big[Y - \mathbb{E}(X \mid \mathcal{G})\big]^2 = \mathbb{E}(X - Y)^2$.

   (e) $\mathbb{E}\big[X - \mathbb{E}(X \mid \mathcal{G})\big]^2 \le \mathbb{E}(X - Y)^2$.

   Part (e) asserts that $\mathbb{E}(X \mid \mathcal{G})$ is the $\mathcal{G}$-random variable that best approximates X in the *mean square* sense.

7. Let $h(x)$ be a continuous function. Verify the law of the unconscious statistician for conditional expectation:
$$\mathbb{E}(h(X) \mid Y) = \int_{-\infty}^{\infty} h(x) f_{X|Y}(x \mid Y)\, dx.$$

Chapter 9

Martingales in Discrete Time Markets

Martingale theory unifies many concepts in finance and provides techniques for analyzing more general markets. In this chapter we consider those aspects of (discrete-time) martingale theory that bear directly on these issues and show how martingales arise naturally in the binomial model in both the dividend and non-dividend-paying cases. We also show how martingales may be used to price derivatives in general markets. The material in this chapter foreshadows results on continuous-time martingales discussed in later chapters.

9.1 Discrete Time Martingales

Definition and Examples

Let $(\Omega, \mathcal{F}, \mathbb{P})$ be a probability space and let $(M_n)_{n\geq 0}$ be a (finite or infinite) stochastic process adapted to a filtration $(\mathcal{F}_n)_{n\geq 0}$. The sequence (M_n) is said to be a *martingale* (with respect to $\mathbb{P}$ and $(\mathcal{F}_n)$) if

$$\mathbb{E}(M_{n+1} \mid \mathcal{F}_n) = M_n, \quad n = 0, 1, \ldots .^{1} \tag{9.1}$$

A d-dimensional stochastic process is a *martingale* if every component is a martingale (with respect to the same probability law and filtration).

If (M_n) is a martingale, then, because M_n is $\mathcal{F}_n$-measurable, $\mathbb{E}(M_n \mid \mathcal{F}_n) = M_n$, hence the martingale property may be written as

$$\mathbb{E}(M_{n+1} - M_n \mid \mathcal{F}_n) = 0, \qquad n = 0, 1, \ldots .$$

This condition has the following gambling interpretation: Let M_n represent the accumulated winnings of a gambler up through the nth play. The martingale property is the hallmark of a fair game, that is, one that requires that the best prediction of the gain $M_{n+1} - M_n$ on the next play, based on the information obtained during the first n plays, be zero.

The following are examples of martingales on an arbitrary probability space $(\Omega, \mathcal{F}, \mathbb{P})$.

[1]We shall also consider martingales that start at the index one.

9.1.1 Examples. (a) Let $X_1, X_2, \ldots$ be a sequence of independent random variables with mean 1, and set $M_n = X_1 X_2 \cdots X_n$. By the factor and independence properties,

$$\begin{aligned}\mathbb{E}(M_{n+1} - M_n \mid M_1, \ldots, M_n) &= M_n \mathbb{E}(X_{n+1} - 1 \mid M_1, \ldots, M_n) \\ &= M_n \mathbb{E}(X_{n+1} - 1) \\ &= 0.\end{aligned}$$

Therefore, (M_n) is a martingale with respect to the natural filtration.

(b) Let $X_1, X_2, \ldots$ be a sequence of independent random variables with mean p and set $M_n = X_1 + \cdots + X_n - np$. By the independence property,

$$\begin{aligned}\mathbb{E}(M_{n+1} - M_n \mid M_1, \ldots, M_n) &= \mathbb{E}(X_{n+1} - p \mid M_1, \ldots, M_n) \\ &= \mathbb{E}(X_{n+1} - p) \\ &= 0,\end{aligned}$$

hence (M_n) is a martingale with respect to the natural filtration.

(c) Let X be a random variable and let $(\mathcal{F}_n)$ be a filtration on Ω. Define $M_n = \mathbb{E}(X \mid \mathcal{F}_n)$, $n = 0, 1, \ldots$. By the iterated conditioning property,

$$\mathbb{E}(M_{n+1} \mid \mathcal{F}_n) = \mathbb{E}\left[\mathbb{E}(X \mid \mathcal{F}_{n+1}) \mid \mathcal{F}_n\right] = \mathbb{E}(X \mid \mathcal{F}_n) = M_n,$$

hence (M_n) is a martingale with respect to the filtration $(\mathcal{F}_n)$. ◊

Properties of Martingales

9.1.2 Theorem (Multistep Property). *For any $m > n$*

$$\mathbb{E}(M_m \mid \mathcal{F}_n) = M_n.$$

Proof. By the iterated conditioning property,

$$\mathbb{E}(M_m \mid \mathcal{F}_n) = \mathbb{E}[\mathbb{E}(M_m \mid \mathcal{F}_{m-1}) \mid \mathcal{F}_n] = \mathbb{E}(M_{m-1} \mid \mathcal{F}_n).$$

Repeating this process we arrive at $\mathbb{E}(M_m \mid \mathcal{F}_n) = \mathbb{E}(M_n \mid \mathcal{F}_n) = M_n$. □

9.1.3 Theorem (Linearity). *If (M_n) and (M'_n) are $(\mathcal{F}_n)$-martingales, then so is $(aM_n + a'M'_n)$ $(a, a' \in \mathbf{R})$.*

Proof. This follows from the linearity of conditional expectation:

$$\begin{aligned}\mathbb{E}\big(aM_{n+1} + a'M'_{n+1} \mid \mathcal{F}_n\big) &= a\mathbb{E}\big(M_{n+1} \mid \mathcal{F}_n\big) + a'\mathbb{E}\big(M'_{n+1} \mid \mathcal{F}_n\big) \\ &= aM_n + a'M'_n.\end{aligned}$$ □

The next theorem asserts that reducing the amount of information provided by a filtration preserves the martingale property. (The same conclusion does not necessarily hold if information is *increased*.)

9.1.4 Theorem (Change of Filtration). *Let $\mathcal{G}_n$ and $\mathcal{F}_n$ be filtrations with $\mathcal{G}_n \subseteq \mathcal{F}_n \subseteq \mathcal{F}$, $n = 0, 1, \ldots$. If (M_n) is adapted to $(\mathcal{G}_n)$ and is a martingale with respect to $(\mathcal{F}_n)$, then it is also a martingale with respect to $(\mathcal{G}_n)$. In particular, an $(\mathcal{F}_n)$-martingale $M = (M_n)$ is an $(\mathcal{F}_n^M)$-martingale.*

Proof. This follows from the iterated conditioning property:

$$\mathbb{E}(M_{n+1} \mid \mathcal{G}_n) = \mathbb{E}\big(\mathbb{E}(M_{n+1} \mid \mathcal{F}_n) \mid \mathcal{G}_n\big) = \mathbb{E}(M_n \mid \mathcal{G}_n) = M_n. \qquad \square$$

9.2 The Value Process as a Martingale

Consider a discrete-time market consisting of a risky asset $\mathcal{S}$ and a risk-free bond $\mathcal{B}$ earning compound interest at a rate of i per period. Let $S = (S_n)_{n=0}^N$ denote the price process of $\mathcal{S}$ on some probability space $(\Omega, \mathcal{F}, \mathbb{P}^*)$ and let (ϕ, θ) be a trading strategy with value process V.

9.2.1 Theorem. *Suppose that the trading strategy is self-financing. If the discounted price process $\widetilde{S}$ is a martingale with respect to $\mathbb{P}^*$ and the natural filtration $(\mathcal{F}_n^S)_{n=0}^N$ for $\mathcal{S}$, then so is the discounted value process $\widetilde{V}$. Thus*

$$\mathbb{E}^*(V_m \mid \mathcal{F}_n^S) = (1+i)^{m-n} V_n, \quad 0 \le n \le m \le N, \tag{9.2}$$

hence the time-n value of a contingent claim H is

$$V_n = (1+i)^{n-N} \mathbb{E}^*\left(H \mid \mathcal{F}_n\right), \quad 0 \le n \le N. \tag{9.3}$$

In particular

$$V_0 = (1+i)^{-N} \mathbb{E}^*(H). \tag{9.4}$$

Proof. It is clear from the definition of value process that $(\widetilde{V}_n)$ is adapted to the filtration $(\mathcal{F}_n^S)$. Recalling that a self-financing portfolio has the property

$$V_{n+1} - aV_n = \theta_{n+1}(S_{n+1} - aS_n) \quad n = 0, 1, \ldots, N-1, \quad a := 1+i,$$

(Theorem (5.4.1)), we have (using the predictability of the trading strategy)

$$\mathbb{E}^*(V_{n+1} - aV_n \mid \mathcal{F}_n^S) = \theta_{n+1} \mathbb{E}^*(S_{n+1} - aS_n \mid \mathcal{F}_n^S).$$

Dividing by a^{n+1} yields

$$\mathbb{E}^*(\widetilde{V}_{n+1} - \widetilde{V}_n \mid \mathcal{F}_n^S) = \theta_{n+1} \mathbb{E}^*(\widetilde{S}_{n+1} - \widetilde{S}_n \mid \mathcal{F}_n^S).$$

Since $\tilde{S}$ is a martingale, the right side is zero, proving that $\widetilde{V}$ is a martingale. Equation (9.2) is a consequence of the multi-step property for martingales. $\square$

9.3 A Martingale View of the Binomial Model

Recall that in the binomial model the price S of the security each period goes up by a factor of u with probability p or down by a factor of d with probability $q = 1 - p$, where $0 < d < u$. The price of $\mathcal{S}$ at time n satisfies

$$S_n = S_{n-1}u^{X_n}d^{1-X_n} = S_0 u^{U_n} d^{n-U_n} = S_0 d^n \left(\frac{u}{d}\right)^{U_n}. \tag{9.5}$$

where $X_1, \ldots, X_N$ are iid Bernoulli random variables and $U_n = X_1 + \cdots + X_n$ is the number upticks of the price in time n. The flow of information is given by the filtration $(\mathcal{F}_n)_{n=0}^N$, where

$$\mathcal{F}_n = \sigma(S_0, S_1, \ldots, S_n) = \sigma(X_1, \ldots, X_n) = \mathcal{F}_n^X.$$

All martingales considered in this section are relative to this filtration. The following theorems show how the inequalities $d < 1 + i < u$ emerge naturally from the martingale property.

9.3.1 Theorem. *If $d < 1 + i < u$, then $(\widetilde{S}_n)$ is a martingale with respect to the risk-neutral probability measure $\mathbb{P}^*$ in the binomial model.*

Proof. From (9.5)

$$\begin{aligned} \mathbb{E}^*(S_n \mid \mathcal{F}_{n-1}) &= S_{n-1}\mathbb{E}^*(u^{X_n}d^{1-X_n} \mid \mathcal{F}_{n-1}) \\ &= \big(u\,\mathbb{P}^*(X_n = 1) + d\,\mathbb{P}^*(X_n = 0)\big)S_{n-1} \\ &= (up^* + dq^*)S_{n-1}. \end{aligned}$$

Dividing by $(1 + i)^n$ and recalling that $up^* + dq^* = 1 + i$, we see that $(\widetilde{S}_n)$ is a martingale with respect to $\mathbb{P}^*$. □

From Theorem 9.2.1, we have

9.3.2 Corollary. *Let (ϕ, θ) be a trading strategy for $(\mathcal{B}, \mathcal{S})$. Then the value process is a martingale with respect to $\mathbb{P}^*$.*

The next theorem asserts that $\mathbb{P}^*$ is the only probability law with respect to which $(\widetilde{S}_n)$ is a martingale.

9.3.3 Theorem. *Let $\widehat{\mathbb{P}}$ be a probability measure on $(\Omega, \mathcal{F})$ with respect to which $X_1, \ldots, X_N$ are iid Bernoulli random variables. Define*

$$\widehat{p} := \widehat{\mathbb{P}}(X_n = 1) \quad \text{and} \quad \widehat{q} := \widehat{\mathbb{P}}(X_n = 0) = 1 - \widehat{p}.$$

If $(\widetilde{S}_n)$ is a martingale with respect to $\widehat{\mathbb{P}}$, then $d < 1 + i < u$ and $\widehat{\mathbb{P}}$ is the risk-neutral measure $\mathbb{P}^*$.

Proof. Let $\widehat{\mathbb{E}}$ denote the expectation operator in the probability space $(\Omega, \mathcal{F}, \widehat{\mathbb{P}})$. From (9.5) we have

$$\begin{aligned}\widehat{\mathbb{E}}(S_n \mid \mathcal{F}_{n-1}) &= S_{n-1}\widehat{\mathbb{E}}(u^{X_n}d^{1-X_n} \mid \mathcal{F}_{n-1}) \\ &= S_{n-1}\big[u\,\widehat{\mathbb{P}}(X_n = 1) + d\,\widehat{\mathbb{P}}(X_n = 0)\big] \\ &= S_{n-1}\big[u\widehat{p} + d\widehat{q}\big].\end{aligned}$$

Dividing by $(1+i)^n$, we see that $(\widetilde{S}_n)$ is a martingale with respect to $\widehat{\mathbb{P}}$ if and only if $u\widehat{p} + d\widehat{q} = 1 + i$. Combining this with the equation $\widehat{p} + \widehat{q} = 1$ we obtain, by Cramer's rule, the unique solution

$$\widehat{p} = \frac{1+i-d}{u-d} \quad \text{and} \quad \widehat{q} = \frac{u-(1+i)}{u-d}.$$

These equations imply that $d < 1 + i < u$ and that $(\widehat{p}, \widehat{q})$ is the risk-neutral probability vector (p^*, q^*). Thus $\widehat{\mathbb{P}} = \mathbb{P}^*$. □

Dividends in the Binomial Model

In the non-dividend case, the binomial price process (S_n) satisfies the recursion equation

$$S_{n+1} = Z_{n+1}S_n, \quad Z_{n+1} := d\left(\frac{u}{d}\right)^{X_{n+1}}.$$

By 5.4.1(c), the value process (V_n) of a self-financing portfolio based on the stock may then be expressed as

$$V_{n+1} = \theta_{n+1}Z_{n+1}S_n + (1+i)(V_n - \theta_{n+1}S_n), \quad n = 0, 1, \ldots, N-1.$$

Now suppose that at each of the times $n = 1, 2, \ldots, N$ the stock pays a dividend which is a fraction $\delta_n \in (0, 1)$ of the value of $\mathcal{S}$ at that time. We assume that δ_n is a random variable and that the process $(\delta_n)_{n=1}^N$ is adapted to the price process filtration. An arbitrage argument shows that after the dividend is paid the value of the stock is reduced by exactly the amount of the dividend (see Section 4.10). Thus at time $n+1$, just after payment of the dividend $\delta_{n+1}Z_{n+1}S_n$, the value of the stock becomes

$$S_{n+1} = (1 - \delta_{n+1})Z_{n+1}S_n, \quad n = 0, 1, \ldots, N-1. \tag{9.6}$$

Since dividends contribute to the portfolio these must be added to the value process, hence

$$\begin{aligned}V_{n+1} &= \theta_{n+1}S_{n+1} + (1+i)(V_n - \theta_{n+1}S_n) + \theta_{n+1}\delta_{n+1}Z_{n+1}S_n \\ &= \theta_{n+1}(1-\delta_{n+1})Z_{n+1}S_n + (1+i)(V_n - \theta_{n+1}S_n) + \theta_{n+1}\delta_{n+1}Z_{n+1}S_n \\ &= \theta_{n+1}Z_{n+1}S_n + (1+i)(V_n - \theta_{n+1}S_n). \end{aligned} \tag{9.7}$$

Thus the value process in the dividend-paying case satisfies the same recursion equation as in the non-dividend-paying case. The methods of §7.2 may then be applied to show that, given a European claim H, there exists a unique self-financing trading strategy (ϕ, θ) with value process V such that $V_N = H$.

One easily checks that in the dividend-paying case the discounted price process is no longer a martingale. (In this connection, see Exercise 9.) Nevertheless, as in the non-dividend-paying case, we have

9.3.4 Theorem. *The discounted value process* $(\widetilde{V}_n := (1+i)^{-n} V_n)_{n=0}^N$ *of a self-financing portfolio* (ϕ, θ) *based on a risk-free bond and a dividend-paying stock is a martingale with respect to* $\mathbb{P}^*$. *In particular, Equations* (9.3) *and* (9.4) *hold.*

Proof. Using (9.7), noting that θ_{n+1}, S_n, and V_n are $\mathcal{F}_n$-measurable, we have

$$\begin{aligned}\mathbb{E}^*(V_{n+1} \mid \mathcal{F}_n) &= \theta_{n+1} S_n \mathbb{E}^*(Z_{n+1} \mid \mathcal{F}_n) + (1+i)(V_n - \theta_{n+1} S_n) \\ &= (1+i)V_n + \theta_{n+1} S_n \left[\mathbb{E}^*(Z_{n+1} \mid \mathcal{F}_n) - (1+i)\right]\end{aligned}$$

Since Z_{n+1} is independent of $\mathcal{F}_n$,

$$\mathbb{E}^*(Z_{n+1} \mid \mathcal{F}_n) = \mathbb{E}^*(Z_{n+1}) = up^* + dq^* = 1 + i. \tag{9.8}$$

Therefore, $\mathbb{E}^*(V_{n+1} \mid \mathcal{F}_n) = (1+i)V_n$. Dividing by $(1+i)^{n+1}$ gives the martingale property. □

9.4 The Fundamental Theorems of Asset Pricing

Consider again a discrete-time market consisting of a risky asset $\mathcal{S}$ and a risk-free bond $\mathcal{B}$ earning compound interest at a rate of i per period. In this section we show how martingales may be used to characterize arbitrage and completeness in the market and to price a derivative, thus generalizing the binomial model. Throughout, $S = (S_n)_{n=0}^N$ denotes the price process of $\mathcal{S}$ on some probability space $(\Omega, \mathcal{F}, \mathbb{P})$ and (ϕ, θ) a trading strategy with value process V.

The First Fundamental Theorem of Asset Pricing

For the theorems in this subsection and the next we need the following definition: We say that probability measures $\mathbb{P}_1$ and $\mathbb{P}_2$ on Ω are *equivalent* if they satisfy the condition

$$\mathbb{P}_1(\omega) > 0 \text{ if and only if } \mathbb{P}_2(\omega) > 0.$$

Thus equivalent measures have precisely the same sets of probability zero. The following result uses this notion to give a martingale characterization of arbitrage.

9.4.1 Theorem. *Assume the sample space Ω of market scenarios is finite. Then the market is arbitrage-free iff the discounted price process $\widetilde{S}$ of the stock is a martingale with respect to some probability measure $\mathbb{P}^*$ equivalent to $\mathbb{P}$.*

Proof. We prove the theorem only in one direction. Assume that $\widetilde{S}$ is a martingale with respect to some probability measure $\mathbb{P}^*$ equivalent to $\mathbb{P}$. Consider any self-financing trading strategy with value process V satisfying $V_0 = 0$ and $\mathbb{P}(V_N \geq 0) = 1$. We show that $\mathbb{P}(V_N > 0) = 0$ (see §5.5). Now, if $V_N(\omega) < 0$, then, because $\mathbb{P}(V_N \geq 0) = 1$, it must be the case that $\mathbb{P}(\omega) = 0$ hence also $\mathbb{P}^*(\omega) = 0$. It follows that

$$\mathbb{E}^*(V_N) = \sum_{V_N(\omega)>0} V_N(\omega)\mathbb{P}^*(\omega).$$

By Theorem 9.2.1, $(\widetilde{V}_n)$ is a martingale, hence $\mathbb{E}^*(V_N) = (1+i)^N \mathbb{E}^*(V_0) = 0$. Therefore

$$\sum_{V_N(\omega)>0} V_N(\omega)\mathbb{P}^*(\omega) = 0.$$

Since the terms in the sum are nonnegative, $V_N(\omega) > 0$ implies that $\mathbb{P}^*(\omega) = 0$, hence also $\mathbb{P}(\omega) = 0$. Thus

$$\mathbb{P}(V_N > 0) = \sum_{V_N(\omega)>0} \mathbb{P}(\omega) = 0,$$

proving that the market is arbitrage-free. The proof of the converse may be found in [2, 4]. □

The Second Fundamental Theorem of Asset Pricing

The next theorem brings in the notion of completeness.

9.4.2 Theorem. *Assume the sample space Ω of market scenarios is finite. Then the market is arbitrage-free and complete iff the discounted price process of the stock is a martingale with respect to a* unique *probability measure $\mathbb{P}^*$ equivalent to $\mathbb{P}$.*

Proof. We prove the theorem only in one direction. Assume the market is arbitrage-free and complete. By the first fundamental theorem of asset pricing, there exists a probability measure $\mathbb{P}^*$ equivalent to $\mathbb{P}$ relative to which the discounted price process of the stock is a martingale. Assume, for a contradiction, that there is another such measure, say $\mathbb{P}^{**}$. Let H be a European claim and let (ϕ, θ) be a self-financing trading strategy with value process V that replicates the claim. By Theorem 5.4.1, $\Delta\widetilde{V}_n = \theta_{n+1}\Delta\widetilde{S}_n$ $(n = 0, 1, \ldots, N-1)$. Summing over n, we have

$$\widetilde{H} = \widetilde{V}_N = V_0 + \sum_{n=0}^{N-1} \theta_{n+1}\Delta\widetilde{S}_n.$$

Since $\widetilde{S}$ is a martingale with respect to $\mathbb{P}^*$ and θ is predictable,

$$\mathbb{E}^*\left(\theta_{n+1}\Delta\widetilde{S}_n \mid \mathcal{F}_n\right) = \theta_{n+1}\mathbb{E}^*\left(\Delta\widetilde{S}_n \mid \mathcal{F}_n\right) = 0.$$

Taking expectations, we have $\mathbb{E}^*\left(\theta_{n+1}\Delta\widetilde{S}_n\right) = 0$ for all n. Therefore, $\mathbb{E}^*\widetilde{H} = V_0$. Similarly $\mathbb{E}^{**}\widetilde{H} = V_0$. We now have

$$\mathbb{E}^* H = \mathbb{E}^{**} H \quad \text{for all claims } H.$$

In particular, this holds for claims of the form $H = \mathbf{1}_A$, $A \in \mathcal{F}$. Therefore, $\mathbb{P}^*(A) = \mathbb{P}^{**}(A)$ for all $A \in \mathcal{F}$, that is, $\mathbb{P}^* = \mathbb{P}^{**}$, proving the necessity of the theorem. For the sufficiency, see [2, 4]. □

Pricing a Derivative

The probability measure $\mathbb{P}^*$ in Theorem 9.4.2 is called the *risk-neutral measure* for the market. Using $\mathbb{P}^*$ we can find a fair price for a claim H in a finite market model that is arbitrage-free and complete. Indeed, for such a market there exists a self-financing portfolio (ϕ, θ) with value process V that replicates the claim and a unique risk-neutral measure $\mathbb{P}^*$. By Theorem 9.4.2, $\widetilde{S}$ is a martingale with respect to $\mathbb{P}^*$. Therefore the fair price V_0 is given by equation (9.4).

---

*9.5 Change of Probability

Assume that the market is complete and arbitrage-free. Theorem 9.2.1 expresses V_n as a conditional expectation relative to the risk-neutral probability measure $\mathbb{P}^*$. It is also possible to express V_n as a conditional expectation relative to the *original* probability measure $\mathbb{P}$. This is the content of the following theorem.

9.5.1 Theorem. *There exists a positive random variable Z on $(\Omega, \mathcal{F}, \mathbb{P})$ with $\mathbb{E}(Z) = 1$ such that*

$$V_n = (1+i)^{n-m}\frac{\mathbb{E}(Z_m V_m \mid \mathcal{F}_n^S)}{Z_n}, \quad 0 \le n \le m \le N,$$

where $Z_n := \mathbb{E}(Z \mid \mathcal{F}_n^S)$.

The core of the proof consists of the following three lemmas.

9.5.2 Lemma. *There exists a unique positive random variable Z on $(\Omega, \mathcal{F}, \mathbb{P})$ with $\mathbb{E}\, Z = 1$ such that, for any random variable X,*

$$\mathbb{E}^* X = \mathbb{E}(XZ). \tag{9.9}$$

In particular,

$$\mathbb{P}^*(A) = \mathbb{E}(\mathbf{1}_A Z). \tag{9.10}$$

Proof. Define Z by

$$Z(\omega) = \frac{\mathbb{P}^*(\omega)}{\mathbb{P}(\omega)}.$$

For any random variable X on $(\Omega, \mathcal{F}, \mathbb{P})$ we have

$$\mathbb{E}(XZ) = \sum_{\omega} X(\omega)Z(\omega)\mathbb{P}(\omega) = \sum_{\omega} X(\omega)\mathbb{P}^*(\omega) = \mathbb{E}^*(X).$$

To see that Z is unique, take $A = \{\omega\}$ in (9.10) to obtain $\mathbb{P}^*(\omega) = Z(\omega)\mathbb{P}(\omega)$. □

The random variable Z in Lemma 9.5.2 is called the *Radon-Nikodym derivative of* $\mathbb{P}^*$ *with respect to* $\mathbb{P}$. It provides a connection between the expectations $\mathbb{E}^*$ and $\mathbb{E}$. The next lemma shows that Z provides an analogous connection between the corresponding conditional expectations.

9.5.3 Lemma. *For any σ-fields* $\mathcal{H} \subseteq \mathcal{G} \subseteq \mathcal{F}$ *and any* $\mathcal{G}$*-random variable* X,

$$\mathbb{E}^*(X \mid \mathcal{H}) = \frac{\mathbb{E}(XZ \mid \mathcal{H})}{\mathbb{E}(Z \mid \mathcal{H})}. \tag{9.11}$$

Proof. Let $Y = \mathbb{E}(Z \mid \mathcal{H})$. We show first that $Y > 0$. Let $A = \{Y \le 0\}$. Then $A \in \mathcal{H}$ and

$$\mathbb{E}(\mathbf{1}_A Y) = \mathbb{E}(\mathbf{1}_A Z) = \mathbb{E}^*(\mathbf{1}_A) = \mathbb{P}^*(A),$$

where we have used the defining property of conditional expectations in the first equality and the defining property of the Radon-Nikodym derivative in the second. Since $\mathbf{1}_A Y \le 0$, $\mathbb{P}^*(A) = 0$. Therefore, $A = \emptyset$ and $Y > 0$.

To prove (9.11) we show that the $\mathcal{H}$-random variable $U := Y^{-1}\mathbb{E}(XZ|\mathcal{H})$ on the right has the defining property of conditional expectation with respect to $\mathbb{P}^*$, namely,

$$\mathbb{E}^*(\mathbf{1}_A U) = \mathbb{E}^*(\mathbf{1}_A X) \quad \text{for all} \quad A \in \mathcal{H}.$$

Indeed, by the factor property,

$$\begin{aligned}\mathbb{E}^*(\mathbf{1}_A U) &= \mathbb{E}(\mathbf{1}_A UZ) = \mathbb{E}\left[\mathbb{E}(\mathbf{1}_A UZ \mid \mathcal{H})\right] = \mathbb{E}\left(\mathbf{1}_A UY\right) \\ &= \mathbb{E}\left[\mathbf{1}_A \mathbb{E}(XZ \mid \mathcal{H})\right] = \mathbb{E}(\mathbf{1}_A XZ) \\ &= \mathbb{E}^*(\mathbf{1}_A X),\end{aligned}$$

as required. □

The following result is a martingale version of the preceding lemma.

9.5.4 Lemma. *Given a filtration* $(\mathcal{G}_n)_{n=0}^N$ *contained in* $\mathcal{F}$ *set*

$$Z_n = \mathbb{E}(Z \mid \mathcal{G}_n), \quad n = 0, 1, \ldots, N.$$

Then (Z_n) *is a martingale with respect to* $(\mathcal{G}_n)$, *and for any* $\mathcal{G}_m$*-random variable* X,

$$\mathbb{E}^*(X \mid \mathcal{G}_n) = \frac{\mathbb{E}(XZ_m \mid \mathcal{G}_n)}{Z_n}, \quad 0 \le n \le m \le N.$$

Proof. That (Z_n) is a martingale is the content of Example 9.1.1(c). By the iterated conditioning and factor properties, for $m \ge n$ we have

$$\begin{aligned}\mathbb{E}(XZ \mid \mathcal{G}_n) &= \mathbb{E}\left[\mathbb{E}(XZ \mid \mathcal{G}_m) \mid \mathcal{G}_n\right] = \mathbb{E}\left[X\mathbb{E}(Z \mid \mathcal{G}_m) \mid \mathcal{G}_n\right] \\ &= \mathbb{E}\left[XZ_m \mid \mathcal{G}_n\right].\end{aligned}$$

Applying Lemma 9.5.3 with $\mathcal{G} = \mathcal{G}_m$ and $\mathcal{H} = \mathcal{G}_n$, we see that

$$\mathbb{E}^*(X \mid \mathcal{G}_n) = \frac{\mathbb{E}(XZ \mid \mathcal{G}_n)}{\mathbb{E}(Z \mid \mathcal{G}_n)} = \frac{\mathbb{E}(XZ_m \mid \mathcal{G}_n)}{Z_n}. \qquad \square$$

For the proof Theorem 9.5.1, take $X = V_m$ and $(\mathcal{G}_n) = (\mathcal{F}_n^S)$ in Lemma 9.5.4 so that

$$\mathbb{E}^*(V_m \mid \mathcal{F}_n^S) = Z_n^{-1}\mathbb{E}(V_m Z_m \mid \mathcal{F}_n^S), \quad 0 \le n \le m \le N.$$

Recalling that

$$V_n = (1+i)^{n-m}\mathbb{E}^*\left(V_m \mid \mathcal{F}_n^S\right), \quad 0 \le n \le m \le N,$$

(Theorem 9.2.1), we obtain the desired equation.

9.5.5 Remark. Lemma 9.5.2 has the following converse: Given a positive random variable Z with $\mathbb{E}\,Z = 1$, the equation $\mathbb{P}^*(A) = \mathbb{E}(\mathbf{1}_A Z)$ defines a probability measure such that (9.9) holds for any random variable X. This may be proved from basic properties of expectation. ◊

9.6 Exercises

1. Show that if (M_n) is a martingale, then $\mathbb{E}(M_n) = \mathbb{E}(M_0)$, $n = 1, 2 \ldots$.

2. Let $X = (X_n)$ be a sequence of independent random variables on $(\Omega, \mathcal{F}, \mathbb{P})$ with mean 0 and variance σ^2, and let $Y_n := X_1 + \cdots + X_n$. Show that

$$M_n := Y_n^2 - n\sigma^2, \quad n = 1, 2, \ldots,$$

defines a martingale with respect to the filtration $(\mathcal{F}_n^X)$.

3. Let $X = (X_n)$ be a sequence of independent random variables on $(\Omega, \mathcal{F}, \mathbb{P})$ with

$$\mathbb{P}(X_n = 1) = p \quad \text{and} \quad \mathbb{P}(X_n = -1) = q := 1 - p,$$

and set $Y_n = \sum_{j=1}^n X_j$. For $a > 0$ define

$$M_n := e^{aY_n}\left(pe^a + qe^{-a}\right)^{-n}, \quad n = 1, 2, \ldots N.$$

Show that (M_n) is a martingale with respect to the filtration $(\mathcal{F}_n^X)$.

4. Let (X_n), (Y_n), and $(\mathcal{F}_n)$ be as in Exercise 3, $0 < p < 1$, and $r := (1-p)/p$. Show that $\left(r^{Y_n}\right)$ is an $(\mathcal{F}_n^X)$-martingale.

5. Let (X_n) and (Y_n) be sequences of independent random variables on $(\Omega, \mathcal{F}, \mathbb{P})$, each adapted to a filtration $(\mathcal{F}_n)$ such that, for each $n \geq 1$, X_n and Y_n are independent of each other and also of $\mathcal{F}_{n-1}$, where $\mathcal{F}_0 = \{\emptyset, \Omega\}$. Suppose also that $\mathbb{E}(X_n) = \mathbb{E}(Y_n) = 0$ for all n. Set
$$A_n = X_1 + \cdots + X_n \quad \text{and} \quad B_n = Y_1 + \cdots + Y_n.$$
Show that $(A_n B_n)$ is an $(\mathcal{F}_n)$-martingale.

6. Let (A_n) and (B_n) be $(\mathcal{F}_n)$-martingales on $(\Omega, \mathcal{F}, \mathbb{P})$ and let $C_n = A_n^2 - B_n$. Show that
$$\mathbb{E}\left[(A_m - A_n)^2 | \mathcal{F}_n\right] = \mathbb{E}\left[C_m - C_n | \mathcal{F}_n\right], \quad 0 \leq n \leq m.$$

7. Let (M_n) be an $(\mathcal{F}_n)$-martingale on $(\Omega, \mathcal{F}, \mathbb{P})$. Show that
$$\mathbb{E}[(M_n - M_m)M_k] = 0, \quad 0 \leq k \leq m \leq n.$$

8. (Doob decomposition). Let $(\mathcal{F}_n)$ be a filtration on $(\Omega, \mathcal{F}, \mathbb{P})$ and (X_n) an adapted process. Define
$$A_0 = 0, \quad \text{and} \quad A_n = A_{n-1} + \mathbb{E}(X_n - X_{n-1} | \mathcal{F}_{n-1}), \quad n = 1, 2, \ldots.$$
Show that, with respect to $(\mathcal{F}_n)$, (A_n) is predictable and $(X_n - A_n)$ is a martingale. Conclude that every adapted process is a sum of a martingale and a predictable process.

9. Referring to Section 9.3, assume that the process (δ_n) is predictable with respect to the price process filtration. Define
$$\xi_n := \prod_{j=1}^{n} (1 - \delta_j), \quad n = 1, 2, \ldots, N.$$
Show that the process
$$\widehat{S}_n := (1+i)^{-n} \xi_n^{-1} S_n, \quad 0 \leq n \leq N,$$
is a martingale with respect to $\mathbb{P}^*$.

10. Referring to Section 9.3, assume that $\delta_n = \delta$ for each n, where $0 < \delta < 1$ is a constant.

(a) Prove the dividend-paying analog of Theorem 7.3.1, namely,
$$V_0 = a^{-N} \sum_{j=0}^{N} \binom{N}{j} p^{*j} q^{*N-j} f\left(b^N u^j d^{N-j} S_0\right),$$
where $a := (1+i)$, $b := 1 - \delta$, and $0 \leq n \leq m \leq N$.

(b) Use (a) to show that the cost C_0 of a call option in the dividend-paying case is
$$C_0 = S_0(1-\delta)^N \Psi(m, N, \hat{p}) - (1+i)^{-N} K \Psi(m, N, p^*),$$
where $\hat{p} = (1+i)^{-1} p^* u$ and m is the smallest nonnegative integer for which $S_0(1-\delta)^N u^m d^{N-m} > K$.

11. Referring to Section 9.5, define

$$Z_n = \left(\frac{p^*}{p}\right)^{Y_n} \left(\frac{q^*}{q}\right)^{n-Y_n}$$

so that, in the notation of Lemma 9.5.2, $Z_N = Z$. Use the fact that $(\mathcal{F}_n^S) = (\mathcal{F}_n^X)$ to show directly that

$$\mathbb{E}(Z_m|\mathcal{F}_n^S) = Z_n, \quad 1 \le n \le m.$$

Chapter 10

American Claims in Discrete-Time Markets

American claims are generally more difficult to analyze than their European counterparts. Indeed, while the writer of a European claim needs to construct a hedge to cover her obligation only at maturity N, the writer of an American claim must devise a hedge that covers the obligation at any exercise time $n \le N$. Moreover, the holder of an American claim has the problem, absent in the European case, of deciding when to exercise the claim so as to obtain the largest possible payoff. In this chapter, we construct a suitable hedge for the writer of an American claim and develop the tools needed to find the holder's optimal exercise time. We assume throughout that the market $(\mathcal{B}, \mathcal{S})$ is complete and arbitrage-free, so that there exists a unique risk-neutral probability measure $\mathbb{P}^*$ on Ω. (Theorem 9.4.2.)

10.1 Hedging an American Claim

Consider an American claim with underlying risky asset $\mathcal{S}$. Let $(\mathcal{F}_n)_{n=0}^N$ denote the natural filtration of the price process $(S_n)_{n=0}^N$ on a probability space $(\Omega, \mathcal{F}, \mathbb{P})$. We assume that the payoff at time n is of the form $f(S_n)$, where $f(x)$ is a nonnegative function. For example in the case of an American put, $f(x) = (K - x)^+$. At time N, a hedge needs to cover the amount $f(S_N)$, hence the value process (V_n) of the portfolio must satisfy

$$V_N = f(S_N).$$

At time $N - 1$, there are two possibilities: If the claim is exercised, then the amount $f(S_{N-1})$ must be covered, so in this case we need

$$V_{N-1} \ge f(S_{N-1}).$$

If the claim is not exercised, then the portfolio must have a value sufficient to cover the claim V_N at time N. By risk-neutral pricing, that value is

$$a^{-1}\mathbb{E}^*(V_N \mid \mathcal{F}_{N-1}), \quad a := 1 + i.$$

Therefore, in this case we should have

$$V_{N-1} \geq a^{-1}\mathbb{E}^*(V_N \mid \mathcal{F}_{N-1}).$$

Both cases may be optimally satisfied by requiring that

$$V_{N-1} = \max\big(f(S_{N-1}), a^{-1}\mathbb{E}^*(V_N \mid \mathcal{F}_{N-1})\big).$$

The same argument may be made at each stage of the process. This leads to the backward recursion scheme

$$\begin{aligned} V_N &= f(S_N), \\ V_n &= \max\{f(S_n), a^{-1}\mathbb{E}^*(V_{n+1} \mid \mathcal{F}_n)\}, \;\; n = N-1, N-2, \ldots, 0. \end{aligned} \tag{10.1}$$

The process $V = (V_n)$ is clearly adapted to $(\mathcal{F}_n)$. By Theorem 5.4.1, for V to be the value process of a self-financing trading strategy (ϕ, θ), it must be the case that

$$\Delta\widetilde{V}_n = \theta_{n+1}\Delta\widetilde{S}_n \;\text{ and }\; \phi_{n+1} = V_0 - \sum_{k=0}^{n} \widetilde{S}_k \Delta\theta_k, \;\; 0 \leq n \leq N-1. \tag{10.2}$$

If $\Delta\widetilde{S}_n \neq 0$, that is, if $S_{n+1} \neq (1+i)S_n$, which is the case, for example, in the binomial model, then the preceding equations may be used to define a self-financing trading strategy (ϕ, θ) with value process V. Thus we have

10.1.1 Theorem. *If $\Delta\widetilde{S}_n \neq 0$ for all n, then the process $V = (V_n)$ defined by (10.1) is the value process for a self-financing portfolio with trading strategy (ϕ, θ). The portfolio covers the claim at any time n, that is,*

$$V_N = f(S_N) \;\; \text{and} \;\; V_n \geq f(S_n), \;\; n = 1, 2, \ldots, N-1.$$

Hence, the proper price of the claim is the initial cost V_0 of setting up the portfolio.

10.1.2 Remark. In contrast to the case of a European claim, the discounted value process $\widetilde{V}$ is *not* a martingale. However, it follows from (10.1) that

$$V_n \geq (1+i)^{-1}\mathbb{E}^*(V_{n+1} \mid \mathcal{F}_n^S),$$

and multiplying this inequality by $(1+i)^{-n}$ yields

$$\widetilde{V}_n \geq \mathbb{E}^*(\widetilde{V}_{n+1} \mid \mathcal{F}_n^S), \;\; 0 \leq n < N.$$

A process with this property is called a *supermartingale*. We return to this notion in Section 10.3. ◊

10.2 Stopping Times

We have shown how the writer of an American claim may construct a hedge to cover her obligation at any exercise time $n \leq N$. The *holder* of the claim has a different concern, namely, when to exercise the claim to obtain the largest possible payoff. In this section, we develop the tools needed to determine the holder's optimal exercise time.

Definition and Examples

It is clear that the optimal exercise time of a claim depends on the price of the underlying asset and is therefore a random variable whose value must be determined using only present or past information. This leads to the formal notion of a stopping time, defined as follows: Let $(\mathcal{F}_n) = (\mathcal{F}_n)$ be a filtration on Ω and let $\mathcal{F}_\infty$ be the σ-field generated by $\bigcup_n \mathcal{F}_n$. An $(\mathcal{F}_n)$-*stopping time* is a random variable τ with values in the set $\{0, 1, \ldots, \infty\}$ such that

$$\{\tau = n\} \in \mathcal{F}_n, \quad n = 0, 1, \ldots$$

If there is no possibility of ambiguity, we omit reference to the filtration. Note that if τ is a stopping time, then the set $\{\tau \leq n\}$, as a union of the sets $\{\tau = j\} \in \mathcal{F}_j$, $j \leq n$, is a member of $\mathcal{F}_n$. It follows that $\{\tau \geq n+1\} = \{\tau \leq n\}'$ also lies in $\mathcal{F}_n$.

The constant function $\tau = m$, where m is a fixed integer in $\{0, 1, \ldots, N\}$, is easily seen to be a stopping time. Also, if τ and σ are stopping times, then so is $\tau \wedge \sigma$, where

$$(\tau \wedge \sigma)(\omega) := \min(\tau(\omega), \sigma(\omega)).$$

This follows immediately from the calculation

$$\{\tau \wedge \sigma = n\} = \{\tau = n, \sigma \geq n\} \cup \{\sigma = n, \tau \geq n\}.$$

Combining these observations we see that $\tau \wedge m$ is a stopping time for any positive integer m. Here are some additional examples.

10.2.1 Examples. (a) The first time a stock falls below a value x or exceeds a value y is described mathematically by the formula

$$\tau(\omega) = \begin{cases} \min\{n \mid S_n(\omega) \in A\} & \text{if } \{n \mid S_n(\omega) \in A\} \neq \emptyset \\ \infty & \text{otherwise,} \end{cases}$$

where $A = (-\infty, x) \cup (y, \infty)$. That τ is a stopping time relative to the natural filtration of (S_n) may be seen from the calculations

$$\{\tau = n\} = \{S_n \in A\} \cap \bigcap_{j=0}^{n-1} \{S_j \notin A\} \, (n < N) \text{ and } \{\tau = \infty\} = \bigcap_{j=0}^{N-1} \{S_j \notin A\}.$$

(b) For a related example, consider a process (I_n), say a stock market index like the S&P 500 or the Nikkei 225, and a filtration $(\mathcal{F}_n)$ to which both (S_n) and (I_n) are adapted. Define

$$\tau(\omega) = \begin{cases} \min\{n \mid S_n(\omega) > I_n(\omega)\} & \text{if } \{n \mid S_n(\omega) > I_n(\omega)\} \neq \emptyset \\ N & \text{otherwise.} \end{cases}$$

Then τ is a stopping time, as may be seen from calculations similar to those in (a). Such a stopping time could result from an investment decision to sell the stock the first time it exceeds the index value.

(c) It is easy to see that the function

$$\tau(\omega) = \begin{cases} \max\{n \mid S_n(\omega) \in A\} & \text{if } \{n \mid S_n(\omega) \in A\} \neq \emptyset \\ \infty & \text{otherwise} \end{cases}$$

is *not* a stopping time. This is a mathematical formulation of the obvious fact that without foresight (or insider information) an investor cannot determine when the stock's value will lie in a set A for the last time. ◊

Optional Stopping

Let (X_n) be a stochastic process adapted to a filtration $(\mathcal{F}_n)$ and let τ be a stopping time with $\tau(\omega) \leq N$ for all ω. The random variable X_τ defined by

$$X_\tau = \sum_{j=0}^{N} \mathbf{1}_{\{\tau=j\}} X_j$$

is called an *optional stopping* of the process X. From the definition, we see that $X_\tau(\omega) = X_j(\omega)$ for any ω for which $\tau(\omega) = j$. It follows immediately that if $X_j \leq Y_j$ for all j then $X_\tau \leq Y_\tau$ and $\widetilde{X}_\tau \leq \widetilde{Y}_\tau$, where

$$\widetilde{X}_\tau := (1+i)^{-\tau} X_\tau.$$

Stopped Process

For a given stopping time τ, the stochastic process $(X_{\tau \wedge n})_{n=0}^{N}$ is called the *stopped process for* τ. As we shall see, the notion of stopped process is an essential tool in determining the optimal time to exercise a claim.

For a concrete example, let τ be the first time a stock's value reaches a specified value a. Then $S_\tau = a$, and the path of the stopped price process for a given scenario ω is

$$S_0(\omega), S_1(\omega), \ldots, S_{\tau-1}(\omega), a.$$

10.3 Submartingales and Supermartingales

To find the optimal exercise time for an American put, we need the following generalization of a martingale: A stochastic process (M_n) adapted to a filtration $(\mathcal{F}_n)$ is said to be a *submartingale*, respectively, *supermartingale* (with respect to $\mathbb{P}$ and $(\mathcal{F}_n)$) if

$$\mathbb{E}(M_{n+1} \mid \mathcal{F}_n) \geq M_n, \text{ respectively, } \mathbb{E}(M_{n+1} \mid \mathcal{F}_n) \leq M_n, \quad n = 0, 1, \ldots.$$

If $\mathcal{F}_n = \sigma(M_0, \ldots M_n)$, we omit reference to the filtration. Note that (M_n) is a supermartingale iff $(-M_n)$ is a submartingale, and (M_n) is a martingale iff it is both a supermartingale and a submartingale. Moreover, a supermartingale has the multistep property

$$\mathbb{E}(M_m \mid \mathcal{F}_n) \leq M_n, \quad 0 \leq n < m,$$

as may be seen by iterated conditioning. Submartingales have the analogous property with the inequality reversed.

For a gambling interpretation, suppose that M_n denotes a gambler's accumulated winnings at the completion of the nth game. The submartingale property then asserts that the game favors the player, since the best prediction of his gain $M_{n+1} - M_n$ in the next game, relative to the information accrued in the first n games, is nonnegative. Similarly, the supermartingale property describes a game that favors the house.

For Proposition 10.3.2 below we need the following lemma.

10.3.1 Lemma. *Let (Y_n) be a process adapted to a filtration $(\mathcal{F}_n)$ and let τ be a stopping time. Then the stopped process $(Y_{\tau\wedge n})$ is adapted to $(\mathcal{F}_n)$ and*

$$Y_{\tau\wedge(n+1)} - Y_{\tau\wedge n} = (Y_{n+1} - Y_n)\mathbf{1}_{\{n+1\leq\tau\}}. \tag{10.3}$$

Proof. Note first that

$$\sum_{j=1}^{\tau\wedge(n+1)} (Y_j - Y_{j-1}) = Y_{\tau\wedge(n+1)} - Y_0,$$

hence

$$\begin{aligned} Y_{\tau\wedge(n+1)} &= Y_0 + \sum_{j=1}^{\tau\wedge(n+1)} (Y_j - Y_{j-1}) \\ &= Y_0 + \sum_{j=1}^{n+1} (Y_j - Y_{j-1})\mathbf{1}_{\{j\leq\tau\}}, \quad (0 \leq n < N). \end{aligned} \tag{10.4}$$

Since the terms on the right in (10.4) are $\mathcal{F}_{n+1}$-measurable, $(Y_{\tau\wedge n})$ is an adapted process. Subtracting from (10.4) the analogous equation with n replaced by $n-1$ yields (10.3). □

The following proposition implies that a game favorable to the player (house) is still favorable to the player (house) when the game is stopped according to a rule that does not require prescience.

10.3.2 Proposition. *If* (M_n) *is a martingale (submartingale, supermartingale) with respect to* $(\mathcal{F}_n)$ *and if* τ *is an* $(\mathcal{F}_n)$*-stopping time, then the stopped process* $(M_{\tau\wedge n})$ *is a martingale (submartingale, supermartingale) with respect to* $(\mathcal{F}_n)$.

Proof. By Lemma 10.3.1, $(M_{\tau\wedge n})$ is adapted to $(\mathcal{F}_n)$. Also, from (10.3),

$$M_{\tau\wedge(n+1)} - M_{\tau\wedge n} = (M_{n+1} - M_n)\mathbf{1}_{\{n+1\le\tau\}},$$

hence, by the $\mathcal{F}_n$-measurability of $\mathbf{1}_{\{n+1\le\tau\}}$,

$$\mathbb{E}\left(M_{\tau\wedge(n+1)} - M_{\tau\wedge n} \mid \mathcal{F}_n\right) = \mathbf{1}_{\{n+1\le\tau\}}\mathbb{E}(M_{n+1} - M_n \mid \mathcal{F}_n). \tag{10.5}$$

The conclusion of the proposition is immediate from (10.5). For example, if (M_n) is a submartingale, then the right side of (10.5) is nonnegative, hence so is the left side. □

10.4 Optimal Exercise of an American Claim

In Section 10.1 we found that the writer of an American claim can cover her obligations with a hedge whose value process $V = (V_n)$ is given by the backward recursive scheme

$$V_N = H_N, \quad V_n = \max\{H_n, (1+i)^{-1}\mathbb{E}^*(V_{n+1} \mid \mathcal{F}_n)\}, \quad 0 \le n < N,$$

where $H_n := f(S_n)$ denotes the payoff of the claim at time n, $(\mathcal{F}_n)$ is the natural filtration. In terms of the corresponding discounted processes,

$$\widetilde{V}_N = \widetilde{H}_N, \quad \widetilde{V}_n = \max\{\widetilde{H}_n, \mathbb{E}^*(\widetilde{V}_{n+1} \mid \mathcal{F}_n)\}, \quad 0 \le n < N, \tag{10.6}$$

which expresses $(\widetilde{V}_n)$ as the so-called *Snell envelope* of the process $(\widetilde{H}_n)$. In this section, we use the value process V to find the optimal time for the holder to exercise the claim. Specifically, we show that the holder should exercise the claim the first time $H_n = V_n$.

For $m = 0, 1, \ldots, N$, let $\mathcal{T}_m$ denote the set of all $(\mathcal{F}^S_n)$-stopping times with values in the set $\{m, m+1, \ldots, N\}$. Define $\tau_m \in \mathcal{T}_m$ by

$$\tau_m = \min\{j \ge m \mid V_j = H_j\} = \min\{j \ge m \mid \widetilde{V}_j = \widetilde{H}_j\}.$$

Thus τ_m is the first time after m that the value process equals the claim. Note that the set in the definition is not empty (since $V_N = H_N$), hence τ_m is well defined.

Now, for an arbitrary stopping time τ, because $\widetilde{V}$ is a $(\mathcal{F}_n^S)$-supermartingale (Remark 10.1.2), so is the stopped process $(\widetilde{V}_{\tau\wedge n})$, where

$$\widetilde{V}_{\tau\wedge n} = (1+i)^{-\tau\wedge n} V_{\tau\wedge n}$$

(Proposition 10.3.2). The following lemma asserts that, for the special stopping time $\tau = \tau_m$, much more can be said.

10.4.1 Lemma. *For each* $m = 0, 1, \ldots, N-1$, *the stopped process* $(\widetilde{V}_{\tau_m\wedge n})_{n=m}^N$ *is a martingale with respect to* $(\mathcal{F}_n^S)_{n=m}^N$.

Proof. By Lemma 10.3.1, $(\widetilde{V}_{\tau_m\wedge n})$ is adapted to the natural filtration and

$$\widetilde{V}_{\tau_m\wedge(n+1)} - \widetilde{V}_{\tau_m\wedge n} = (\widetilde{V}_{n+1} - \widetilde{V}_n)\mathbf{1}_{\{n+1\le\tau_m\}}.$$

Therefore, it suffices to show that, for $m \le n < N$,

$$\mathbb{E}^*\left[(\widetilde{V}_{n+1} - \widetilde{V}_n)\mathbf{1}_{\{n+1\le\tau_m\}} \mid \mathcal{F}_n\right] = 0.$$

Since $\mathbf{1}_{\{n+1\le\tau_m\}}$ and $\widetilde{V}_n$ are $\mathcal{F}_n^S$-measurable, the last equation is equivalent to

$$\mathbf{1}_{\{n+1\le\tau_m\}}\mathbb{E}^*\left(\widetilde{V}_{n+1} \mid \mathcal{F}_n^S\right) = \mathbf{1}_{\{n+1\le\tau_m\}}\widetilde{V}_n. \tag{10.7}$$

To verify (10.7), fix $\omega \in \Omega$ and consider two cases: If $\tau_m(\omega) < n+1$, then the indicator functions are zero, hence (10.7) is trivially satisfied. If on the other hand $\tau_m(\omega) \ge n+1$, then $\widetilde{V}_n(\omega) \ne \widetilde{H}_n(\omega)$, hence, from (10.6), $\widetilde{V}_n(\omega) = \mathbb{E}^*(\widetilde{V}_{n+1} \mid \mathcal{F}_n)(\omega)$. Therefore, (10.7) holds at each ω. □

The following theorem is the main result of the section. It asserts that, after time m, the optimal time to exercise an American claim is τ_m, in the sense that the largest expected discounted payoff, given the available information, occurs at that time.

10.4.2 Theorem. *For any* $m \in \{0, 1, \ldots, N\}$,

$$\mathbb{E}^*(\widetilde{H}_{\tau_m} \mid \mathcal{F}_m) = \max_{\tau\in\mathcal{T}_m} \mathbb{E}^*(\widetilde{H}_\tau \mid \mathcal{F}_m) = \widetilde{V}_m.$$

In particular,

$$\mathbb{E}^*\widetilde{H}_{\tau_0} = \max_{\tau\in\mathcal{T}_0} \mathbb{E}^*\widetilde{H}_\tau = V_0.$$

Proof. Since $(\widetilde{V}_{\tau_m\wedge n})_{n=m}^N$ is a martingale (Lemma 10.4.1) and $\widetilde{V}_{\tau_m} = \widetilde{H}_{\tau_m}$,

$$\widetilde{V}_m = \widetilde{V}_{\tau_m\wedge m} = \mathbb{E}^*(\widetilde{V}_{\tau_m\wedge N} \mid \mathcal{F}_m) = \mathbb{E}^*(\widetilde{V}_{\tau_m} \mid \mathcal{F}_m) = \mathbb{E}^*(\widetilde{H}_{\tau_m} \mid \mathcal{F}_m).$$

Now let $\tau \in \mathcal{T}_m$. Since $(\widetilde{V}_{\tau\wedge n})$ is a supermartingale (Proposition 10.3.2) and $\widetilde{V}_\tau \ge \widetilde{H}_\tau$,

$$\widetilde{V}_m = \widetilde{V}_{\tau\wedge m} \ge \mathbb{E}^*(\widetilde{V}_{\tau\wedge N} \mid \mathcal{F}_m) = \mathbb{E}^*(\widetilde{V}_\tau \mid \mathcal{F}_m^S) \ge \mathbb{E}^*(\widetilde{H}_\tau \mid \mathcal{F}_m).$$

Therefore, $\widetilde{V}_m = \max_{\tau\in\mathcal{T}_m} \mathbb{E}^*(\widetilde{H}_\tau \mid \mathcal{F}_m)$, completing the proof. □

10.5 Hedging in the Binomial Model

For the binomial model, the scheme (10.1) may be formulated as an algorithm for calculating V_0. First note that, by Example 8.2.4, for any function g we have

$$\mathbb{E}^*\left(g(S_n) \mid \mathcal{F}_{n-1}\right) = p^* g(uS_{n-1}) + q^* g(dS_{n-1}). \tag{10.8}$$

Now let (v_n) be the sequence of functions defined by

$$\begin{aligned} v_N(s) &= f(s) \text{ and} \\ v_n(s) &= \max\left\{f(s), \alpha v_{n+1}(us) + \beta v_{n+1}(ds)\right\}, \quad n = N-1, \ldots, 0, \end{aligned} \tag{10.9}$$

where

$$\alpha := a^{-1}p^* \text{ and } \beta := a^{-1}q^*, \quad a := 1 + i.$$

Taking $g = v_N, v_{N-1}, \ldots$ in (10.8), we have

$$\begin{aligned} V_N &= f(S_N) = v(S_N) \\ V_{N-1} &= \max\left\{f(S_{N-1}), a^{-1}\mathbb{E}^*(V_N \mid \mathcal{F}_{N-1})\right\} \\ &= \max\left\{f(S_{N-1}), a^{-1}\mathbb{E}^*(v_N(S_N) \mid \mathcal{F}_{N-1})\right\} \\ &= \max\left\{f(S_{N-1}), a^{-1}\left(p^* v_N(uS_{N-1}) + q^* v_N(dS_{N-1})\right)\right\} \\ &= v_{N-1}(S_{N-1}), \\ V_{N-2} &= \max\left\{f(S_{N-2}), a^{-1}\mathbb{E}^*(V_{N-1} \mid \mathcal{F}_{N-2})\right\} \\ &= \max\left\{f(S_{N-1}), a^{-1}\mathbb{E}^*(v_{N-1}(S_{N-1}) \mid \mathcal{F}_{N-2})\right\} \\ &= \max\left\{f(S_{N-1}), a^{-1}\left(p^* v_{N-1}(uS_{N-2}) + q^* v_{N-1}(dS_{N-2})\right)\right\} \\ &= v_{N-2}(S_{N-2}), \end{aligned}$$

etc. It follows by backward induction that

$$V_n = v_n(S_n), \quad n = 0, 1, \ldots, N.$$

In particular, for any ω we have

$$V_n(\omega) = \max\left(f\left(u^k d^{n-k} S_0\right), \beta v_{n+1}\left(u^{k+1} d^{n-k} S_0\right) + \beta v_{n+1}\left(u^k d^{n+1-k} S_0\right)\right),$$

where $k := U_n(\omega)$ is the number of upticks during the first n time periods.

For small values of N, the scheme in (10.9) may be readily implemented on a spreadsheet. For example, for $N = 3$ the algorithm may be explicitly rendered as

$$\begin{array}{lclclcl} v_3(s) & = & f(s) & & & & \\ v_2(u^2 S_0) & = & \max\{ f(u^2 S_0), & \alpha v_3(u^3 S_0) & + & \beta v_3(u^2 d S_0) \} \\ v_2(udS_0) & = & \max\{ f(udS_0), & \alpha v_3(u^2 d S_0) & + & \beta v_3(ud^2 S_0) \} \\ v_2(d^2 S_0) & = & \max\{ f(d^2 S_0), & \alpha v_3(ud^2 S_0) & + & \beta v_3(d^3 S_0) \} \\ v_1(uS_0) & = & \max\{ f(uS_0), & \alpha v_2(u^2 S_0) & + & \beta v_2(udS_0) \} \\ v_1(dS_0) & = & \max\{ f(dS_0), & \alpha v_2(udS_0) & + & \beta v_2(d^2 S_0) \} \\ v_0(S_0) & = & \max\{ f(S_0), & \alpha v_1(uS_0) & + & \beta v_1(dS_0) \}. \end{array}$$

For larger values of N, a spreadsheet is impractical and a computer program is needed. The data in Table 10.1 was generated by the VBA modules **AmericanOptionPrice–Bin** and **EuropeanOptionPrice–Bin**. (We included the European option prices for comparison.) Here $S_0 = \$10.00$, the annual rate is 3% and the term is 90 days.

TABLE 10.1: American and European put prices.

K	u	d	P_0^a	P_0^e	K	u	d	P_0^a	P_0^e
8.00	1.1	.9	2.34	2.27	12.00	1.1	.9	5.02	4.89
8.00	1.1	.7	4.58	4.48	12.00	1.1	.7	7.76	7.60
8.00	1.3	.9	4.11	4.03	12.00	1.3	.9	7.24	7.10
8.00	1.3	.7	6.64	6.53	12.00	1.3	.7	10.31	10.13

As with the module **European Option Price–Bin**, **American Option Price–Bin** allows the user to enter a payoff function $F(S_n, K)$, e.g. $(K - S_n)^+$ for puts or $(S_n - K)^+$ for calls. The reader may wish to use the program to confirm that the price of an American call is the same as that of a European call, a fact established in a general context in Section 4.9. In this connection, the reader may also wish to use the VBA module **GraphOfAmericanOptionPrice–Bin**, which is the analog of the module **GraphOfEuropeanOptionPrice–Bin**.

Additional material on American claims in the binomial model, including a version of the hedge that allows consumption at each time n, may be found in [18].

10.6 Optimal Exercise in the Binomial Model

Theorem 10.4.2 asserts that it is optimal to exercise an American claim the first time $f(S_n) = v_n(S_n)$. This leads to the following simple algorithm which may be used to find $\tau_0(\omega)$ for any scenario $\omega = (\omega_1, \omega_2, \ldots, \omega_N)$:

$$\begin{array}{lll}
\text{if} & f(S_0) = V_0, & \tau_0(\omega) = 0, \\
\text{else if} & f(S_0\omega_1) = v_1(S_0\omega_1), & \tau_0(\omega) = 1, \\
\text{else if} & f(S_0\omega_1\omega_2) = v_2(S_0\omega_1\omega_2), & \tau_0(\omega) = 2, \\
\vdots & & \vdots \\
\text{else if} & f(S_0\omega_1\cdots\omega_{N-1}) = v_{N-1}(S_0\omega_1\cdots\omega_{N-1}), & \tau_0(\omega) = N-1, \\
\text{else} & & \tau_0(\omega) = N.
\end{array}$$

The algorithm also gives the stopped scenarios for which $\tau_0 = j$. For small values of N, the algorithm is readily implemented on a spreadsheet by comparing the values of f and v along paths. For large values of N a computer program is needed.

10.6.1 Example. Consider an American put that matures in four periods, where $S_0 = \$10.00$ and $i = .1$. Table 10.2 gives optimal exercise scenarios and payoffs (displayed parenthetically) for various values of K, u, and d. The fourth column gives the prices P_0^a of the put. The table was generated by the module **OptimalStoppingPut–Bin** (See also **OptimalStoppingGeneral–Bin** for a version that allows a user to enter arbitrary payoff functions.)

TABLE 10.2: Put prices and stopping scenarios.

K	u	d	P_0^a	Optimal Stopping Scenarios and Payoffs
20	3	.3	\$13.49	d (17.00); udd (17.30); $uudd$, or $udud$ (11.90)
20	2	.3	\$11.77	d (17.00); ud (14.00); uud (8.00)
12	3	.3	\$7.01	d (9.00); udd (9.30); $uudd$, or $udud$ (3.90)
12	3	.6	\$5.13	dd (8.40); $dudd$, or $uddd$ (5.52)
12	2	.3	\$6.05	d (9.00); udd (10.20); $uudd$, or $udud$ (8.40)
12	2	.6	\$4.04	d (6.00); udd (4.80)
8	3	.3	\$4.07	dd (7.10); dud, or udd (5.30)
8	3	.6	\$2.50	ddd (5.84); $uddd$, $dudd$, or $ddud$ (1.52)

For example, in row 2 we see that the claim should be exercised after one time unit if the stock first goes down, after two time units if the stock goes up then down, and after three time units if the stock goes up twice in succession then down. Scenarios missing from the table are either not optimal or result in zero payoffs. For example, in row 1 it is never optimal to exercise at time 2, and the missing optimal scenarios uuu, $uudu$, $uduu$ all have zero payoffs. In row 4, it is never optimal to exercise at time 1, and the missing optimal scenarios uu, udu, duu, $uddu$, $dudu$ all have zero payoffs. Note that if an optimal scenario $\omega_1, \omega_2, \ldots, \omega_n$ has a zero payoff at time n, then there is no hope of ever obtaining a nonzero payoff; all later scenarios $\omega_1, \ldots, \omega_n, \omega_{n+1}, \ldots$ will also have zero payoffs (see Exercise 3). ◊

For additional material regarding stopping times and American claims in the binomial model, including the case of path-dependent payoffs, see, for example, [18].

10.7 Exercises

1. Show that the following are stopping times:

   (a) $\tau_a :=$ the first time the stock price exceeds the average of all of its previous values.

(b) $\tau_b :=$ the first time the stock price exceeds all of its previous values.

(c) $\tau_c :=$ the first time the stock price exceeds at least one of its previous values.

2. Show that the first time the stock price increases twice in succession is a stopping time.

3. Let $\omega \in \Omega$ and $n = \tau_0(\omega)$. Show that if $f(S_n(\omega)) = 0$ then $f(S_k(\omega)) = 0$ for all $k \geq n$.

Chapter 11

Stochastic Calculus

Stochastic calculus extends classical differential and integral calculus to functions with random components. The fundamental construct in stochastic calculus is the Ito integral, a random variable that is a probabilistic extension of the Riemann integral. In this chapter we construct the integral, examine its properties, and describe its role in solving stochastic differential equations (SDEs).

11.1 Continuous-Time Stochastic Processes

Recall that a discrete-time stochastic process is a sequence of random variables on some probability space. As we have seen, such a construct is useful in modeling experiments consisting of a sequence of trials. However, while effective in many contexts, discrete-time models are not always sufficiently rich to capture all the important dynamical features of an experiment. Furthermore, discrete-time processes have inherent mathematical limitations, notably the unavailability of tools from the calculus. Continuous-time processes frequently offer a more realistic way to model experiments unfolding in time and allow the introduction of powerful techniques from stochastic calculus.

Definition of a Process

A *continuous-time stochastic process* on a probability space $(\Omega, \mathcal{F}, \mathbb{P})$ is a real-valued function X on $D \times \Omega$, where D is an interval of real numbers, such that the function $X(t, \cdot)$ is an $\mathcal{F}$-random variable for each $t \in D$. We shall also refer to X as a *random process* or simply a *process*. The set D is called the *index set* for the process, and the functions $X(\cdot, \omega)$, $\omega \in \Omega$, are the *paths* of the process. If X does not depend on ω, then the process is said to be *deterministic*. If X^1, X^2, ..., X^d are stochastic processes with the same index set, then the d-tuple $\boldsymbol{X} = (X^1, \dots, X^d)$ is called a *d-dimensional* stochastic process.

Depending on context, we will use the notations X_t or $X(t)$ for the random variable $X(t, \cdot)$ and denote the process X by $(X_t)_{t \in D}$ or just (X_t). The latter notation reflects the interpretation of a stochastic process as a collection of random variables indexed by D and hence as a mathematical description of a random system changing in time. The interval D is usually of the form $[0,T]$ or $[0,\infty)$.

An example of a continuous-time process is the price of a stock, a path being the price history for a particular market scenario. For another example, consider the three-dimensional stochastic process which describes the position at time t of a particle randomly bombarded by the molecules of a liquid in which it is suspended. Surprisingly, there is a connection between these seemingly unrelated examples, one that will emerge later.

Continuous-Time Filtrations

As noted above, a continuous-time stochastic process may be viewed as a mathematical description of an evolving experiment. Related to this notion is the flow of information revealed by the experiment. As in the discrete-time case, the evolution of such information may be modeled by a filtration, defined as follows:

A *continuous-time filtration*, or simply *filtration*, on $(\Omega, \mathcal{F}, \mathbb{P})$ is a collection $(\mathcal{F}_t)_{t\in D}$ of σ-fields $\mathcal{F}_t$ indexed by members t of an interval $D \subseteq \mathbf{R}$ such that

$$\mathcal{F}_s \subseteq \mathcal{F}_t \subseteq \mathcal{F} \quad \text{for all} \quad s, t \in D \quad \text{with} \quad s \le t.$$

A stochastic process $(X_t)_{t\in D}$ is said to be *adapted to a filtration* $(\mathcal{F}_t)_{t\in D}$ if X_t is a $\mathcal{F}_t$-random variable for each $t \in D$. A d-dimensional process is *adapted to a filtration* if each coordinate process is adapted. As with stochastic processes, we frequently omit reference to the symbol D and denote the filtration by $(\mathcal{F}_t)$.

A particularly important filtration is the *natural filtration* $\mathcal{F}^X$ for a process $X = (X_t)$. Here, $\mathcal{F}^X_t$ is the intersection of all σ-fields containing events of the form $\{X_s \in J\}$, where J is an arbitrary interval and $s \le t$. $\mathcal{F}^X$ is also called the *filtration generated by the process* (X_t). The natural filtration is the smallest filtration to which X is adapted. It's importance derives from the fact that it contains all time-dependent information related solely to X. An important example of a natural filtration is the filtration generated by a Brownian motion, the fundamental process used in the construction of the stochastic integral. Brownian motion, described in the next section, forms the basis of the pricing models discussed in later chapters.

11.2 Brownian Motion

In 1827, Scottish botanist Robert Brown observed that pollen particles suspended in a liquid exhibited highly irregular motion. Later, it was determined that this motion resulted from random collisions of the particles with molecules in the ambient liquid. In 1900, L. Bachelier noted the same irregular variation in stock prices and attempted to describe this behavior mathematically. In 1905, Albert Einstein used Brownian motion, as the phenomenon came to be called, in his work on measuring Avogadro's number. A rigorous mathematical model of Brownian motion was developed in the 1920s by Norbert Wiener. Since then, the mathematics of Brownian motion and its generalizations have become one of the most active and important areas of probability theory.

Definition of Brownian Motion

Let $(\Omega, \mathcal{F}, \mathbb{P})$ be a probability space. A *Brownian motion* or *Wiener process* is a stochastic process W on $(\Omega, \mathcal{F})$ that has the following properties:

(a) $W(0) = 0$.

(b) $W(t) - W(s) \sim N(0, t-s), \ 0 \le s < t$.

(c) The paths of W are continuous.

(d) For any indices $0 < t_1 < t_2 < \cdots < t_n$, the random variables $W(t_1), W(t_2) - W(t_1), \ldots, W(t_n) - W(t_{n-1})$ are independent.

The *Brownian filtration* on $(\Omega, \mathcal{F}, \mathbb{P})$ is the natural filtration $\mathcal{F}^W = (\mathcal{F}_t^W)_{t \ge 0}$.

Part (a) of the definition says that the process W starts out at zero, the hallmark of so-called *standard Brownian motion.* Part (b) asserts that the increments $W(t) - W(s)$ $(t > s)$ of the process are normally distributed with mean zero and variance $t - s$. In particular, $W(t) = W(t) - W(0)$ is normally distributed with mean zero and variance t. Part (d) asserts that $W(t)$ has *independent increments.*

Brownian motion has the unusual property that while the paths are continuous they are nowhere differentiable. This corresponds to Brown's observation that the paths of the pollen particles seemed to have no tangents. Also, Brownian motion looks the same at any scale; that is, it is a *fractal.* These properties partially account for the usefulness of Brownian motion in modeling systems with noise.

Brownian Motion as a Limit of Random Walks

Rigorous proofs of the existence of Brownian motion may be found in advanced texts on probability theory. The most common of these proofs uses Kolmogorov's Extension Theorem. Another constructs Brownian motion from wavelets. For these the interested reader is referred to [10], [20], or [21]. In this subsection we give a heuristic construction that results in a prototype of (one-dimensional) Brownian motion.

Suppose that every Δt seconds a particle starting at the origin moves right or left a distance Δx, each move with probability 1/2. Let $Z_t^{(n)}$ denote the position of the particle at time $t = n\Delta t$, that is, after n moves. The process $(Z_t^{(n)})$ is called a *random walk.* We assume that Δt and Δx are related by the equation $(\Delta x)^2 = \Delta t$.[1] Let $X_j = 1$ if the jth move is to the right and $X_j = 0$ otherwise. The X_j's are independent Bernoulli variables with parameter $p = 1/2$ and $Y_n := \sum_{j=1}^{n} X_j \sim B(p, n)$ is the number of moves to the right during the time interval $[0, t]$. Thus

$$Z_t^{(n)} = Y_n \Delta x + (n - Y_n)(-\Delta x) = (2Y_n - n)\Delta x = \left(\frac{Y_n - n/2}{\sqrt{n/4}}\right)\sqrt{t},$$

the last equality because $t = n(\Delta x)^2$. By the Central Limit Theorem,

$$\lim_{n\to\infty} \mathbb{P}(Z_t^{(n)} \le z) = \lim_{n\to\infty} \mathbb{P}\left(\frac{Y_n - n/2}{\sqrt{n/4}} \le \frac{z}{\sqrt{t}}\right) = \Phi\left(\frac{z}{\sqrt{t}}\right).$$

Thus, as the step length and step time tend to zero via the relation $\Delta x = \sqrt{\Delta t}$, the sequence $Z_t^{(n)}$ *tends in distribution* to a random variable $Z_t \sim N(0, t)$. A similar argument shows that $Z_t - Z_s$ is the limit in distribution of random walks over the time interval $[s, t]$ and that $Z_t - Z_s \sim N(0, t - s)$.

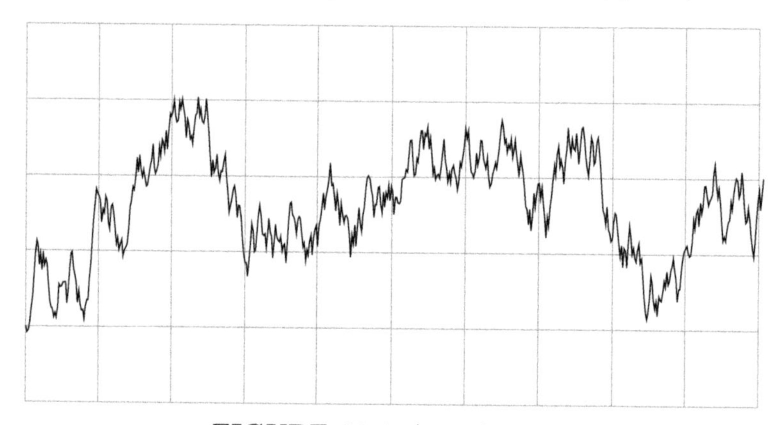

FIGURE 11.1: A random walk.

Variation of Brownian Paths

A useful way to measure the seemingly erratic behavior of Brownian motion is by the *variation* of its paths. For this we need the notion of *mth variation of a function*, defined as follows:

Let $\mathcal{P} = \{a = t_0 < t_1 < \cdots < t_n = b\}$ be a partition of the interval $[a, b]$ and let m be a positive integer. For a real-valued function w on $[a, b]$ define

$$V_{\mathcal{P}}^{(m)}(w) = \sum_{j=0}^{n-1} |\Delta w_j|^m, \quad \Delta w_j := w(t_{j+1}) - w(t_j).$$

[1] This is needed to produce the desired result that the limit Z_t of the random variables $Z_t^{(n)}$ is $N(0, t)$.

The function w is said to have *bounded (unbounded) mth variation on $[a,b]$* if the quantities $V_{\mathcal{P}}^{(m)}(w)$, taken over all partitions $\mathcal{P}$, form a bounded (unbounded) set of real numbers.

11.2.1 Example. Let $\alpha > 0$. It may be shown that the continuous function w on $[0,a]$ defined by

$$w(t) := \begin{cases} t^{\alpha} \sin(1/t) & \text{if } 0 < t \le a, \\ 0 & \text{if } t = 0 \end{cases}$$

has bounded first variation iff $\alpha > 1$. (See [10]). ◊

If f is a stochastic process, then the notion of mth variation may be applied to the paths $f(\cdot,\omega)$, in which case $V_{\mathcal{P}}^{(m)}(f)$ is a random variable. In particular, for Brownian motion we have the definition

$$V_{\mathcal{P}}^{(m)}(W)(\omega) := \sum_{j=0}^{n-1} |\Delta W_j(\omega)|^m, \quad \Delta W_j = W(t_{j+1}) - W(t_j).$$

It may be shown that the paths of Brownian motion have unbounded first variation in every interval. (See, for example, [6].)

The following theorem describes one of several properties of Brownian motion that underlie the differences between stochastic calculus and classical calculus.

11.2.2 Theorem. *Let $\mathcal{P} = \{a = t_0 < t_1 < \cdots < t_n = b\}$ be a partition of the interval $[a,b]$ and set $\|\mathcal{P}\| = \max_j \Delta t_j$, where $\Delta t_j := t_{j+1} - t_j$. Then*

$$\lim_{\|\mathcal{P}\| \to 0} \mathbb{E}\left(V_{\mathcal{P}}^{(2)}(W) - (b-a)\right)^2 = 0.$$

Proof. Define $A_{\mathcal{P}} = V_{\mathcal{P}}^{(2)}(W) - (b-a) = \sum_{j=0}^{n-1} \left[(\Delta W_j)^2 - \Delta t_j\right]$, so that

$$\mathbb{E}(A_{\mathcal{P}}^2) = \sum_{j=0}^{n-1}\sum_{k=0}^{n-1} \mathbb{E}\left\{\left[(\Delta W_j)^2 - \Delta t_j\right]\left[(\Delta W_k)^2 - \Delta t_k\right]\right\}. \tag{11.1}$$

If $j \ne k$, then, by independent increments,

$$\mathbb{E}\left[(\Delta W_j)^2 - \Delta t_j\right]\left[(\Delta W_k)^2 - \Delta t_k\right] = \mathbb{E}\left[(\Delta W_j)^2 - \Delta t_j\right]\mathbb{E}\left[(\Delta W_k)^2 - \Delta t_k\right].$$

Since $W(t) - W(s) \sim N(0, t-s)$, the term on the right is zero, hence

$$\mathbb{E}(A_{\mathcal{P}}^2) = \sum_{j=0}^{n-1} \mathbb{E}\left[(\Delta W_j)^2 - \Delta t_j\right]^2 = \sum_{j=0}^{n-1} \mathbb{E}(Z_j^2 - 1)^2 (\Delta t_j)^2, \tag{11.2}$$

where

$$Z_j := \frac{\Delta W_j}{\sqrt{\Delta t_j}} \sim N(0,1).$$

The quantity $c := \mathbb{E}(Z_j^2 - 1)^2$ is finite and does not depend on j (see Example 6.2.4). Therefore,

$$0 \le \mathbb{E}(A_{\mathcal{P}}^2) \le c\|\mathcal{P}\| \sum_{j=0}^{n-1} \Delta t_j = c\|\mathcal{P}\|(b-a).$$

Letting $\|\mathcal{P}\| \to 0$ forces $\mathbb{E}(A_{\mathcal{P}}^2) \to 0$, which is the assertion of the theorem. □

11.2.3 Remarks. The theorem asserts that the *mean square limit* of $V_{\mathcal{P}}^{(2)}(W)$ as $\|\mathcal{P}\| \to 0$ is $b-a$. The limit is called the *quadratic variation* of Brownian motion on $[a,b]$. That Brownian motion has nonzero quadratic variation on any interval is in stark contrast to the situations one normally encounters in Newtonian calculus. (See Exercise 2 in this regard.)

For $m \ge 3$, the mean square limit of $V_{\mathcal{P}}^{(m)}(W)$ is zero. This follows from the continuity of $W(t)$ and the inequality

$$V_{\mathcal{P}}^{(m)}(W) = \sum_{j=0}^{n-1} |\Delta W_j|^{m-2}|\Delta W_j|^2 \le \max_j |\Delta W_j|^{m-2} V_{\mathcal{P}}^{(2)}(W). \qquad \diamond$$

11.3 Stochastic Integrals

To motivate the construction of the Ito integral, we first give a brief overview of the Riemann-Stieltjes integral.

Riemann-Stieltjes Integrals

Let f and w be bounded functions on an interval $[a,b]$. A *Riemann-Stieltjes sum of f with respect to w* is a sum of the form

$$R_{\mathcal{P}} = \sum_{j=0}^{n-1} f(t_j^*)\Delta w_j, \quad \Delta w_j := w(t_{j+1}) - w(t_j),$$

where $\mathcal{P} = \{a = t_0 < t_1 < \cdots < t_n = b\}$ is a partition of $[a,b]$ and t_j^* is an arbitrary member of $[t_j, t_{j+1}]$. The *Riemann-Stieltjes integral of f with respect to w* is defined as the limit

$$\int_a^b f(t)\,dw(t) = \lim_{\|\mathcal{P}\| \to 0} R_{\mathcal{P}},$$

where $\|\mathcal{P}\| = \max_j \Delta t_j$. The limit is required to be independent of the choice of the t_j^*'s. The integral may be shown to exist if f is continuous and w has bounded first variation on $[a, b]$. The Riemann integral is obtained as a special case by taking $w(t) = t$. More generally, if w is continuously differentiable, then

$$\int_a^b f\,dw = \int_a^b f(t)w'(t)\,dt.$$

The Riemann-Stieltjes integral has many of the familiar properties of the Riemann integral, notably

$$\int_a^b (\alpha f + \beta g)\,dw = \alpha \int_a^b f\,dw + \beta \int_a^b g\,dw, \quad \alpha, \beta \in \mathbf{R}, \quad \text{and}$$

$$\int_a^b f\,dw = \int_a^c f\,dw + \int_c^b f\,dw, \quad a < c < b.$$

For details, the reader is referred to [9] or [17].

Construction of the Ito Integral

For the remainder of the chapter, W denotes a Brownian motion on a probability space $(\Omega, \mathcal{F}, \mathbb{P})$. In this section, we construct the Ito integral

$$I(F) = \int_a^b F(t)\,dW(t), \tag{11.3}$$

where $F(t)$ is a stochastic process on $(\Omega, \mathcal{F}, \mathbb{P})$ with continuous paths. Given such a process F, for each $\omega \in \Omega$ we may form the ordinary Riemann integral

$$\int_a^b F(t, \omega)^2\,dt.$$

For technical reasons, we shall assume that the resulting random variable $\int_a^b F(t)^2\,dt$ has finite expectation:

$$\mathbb{E}\left(\int_a^b F(t)^2\,dt\right) < \infty.$$

To construct the Ito integral, we begin with sums

$$\sum_{j=0}^{n-1} F(t_j^*, \omega)\Delta W_j(\omega), \tag{11.4}$$

over a partition $\mathcal{P} := \{a = t_0 < t_1 < \cdots t_n = b\}$ of $[a, b]$, where $t_j^* \in [t_j, t_{j+1}]$. In light of the discussion on the Riemann-Stieltjes integral, it might seem reasonable to define $I(F)(\omega)$ as the limit of such sums as $\|\mathcal{P}\| \to 0$. However, this fails for several reasons. First, the paths of W do not have bounded first variation, so we can't expect the sums in (11.4) to converge in the usual sense. What is needed instead is mean square convergence. Second, even with the appropriate mode of convergence, the limit of the sums in (11.4) generally depends on the choice of the intermediate points t_j^*. To eliminate this problem, we shall always take the point t_j^* to be the *left* endpoint t_j of the interval $[t_j, t_{j+1}]$. These restrictions, however, are not sufficient to ensure a useful theory. We shall also require that the random variable $F(s)$ be independent of the increment $W(t) - W(s)$ for $0 \le s < t$ and depend only on the information provided by $W(r)$ for $r \le s$. Both conditions are realized by requiring that the process F be adapted to the Brownian filtration. Under these conditions we define the *Ito integral* of F in (11.3) to be the limit of the *Ito sums*

$$I_{\mathcal{P}}(F) := \sum_{j=0}^{n-1} F(t_j)\Delta W_j,$$

where the convergence is in the mean square sense:

$$\lim_{\|\mathcal{P}\|\to 0} \mathbb{E}\big(I_{\mathcal{P}}(F) - I(F)\big)^2 = 0.$$

It may be shown that this limit exists for all continuous processes F satisfying the conditions described above, hereafter referred to as the "usual conditions." In the discussions that follow, we shall assume, usually without explicit mention, that these conditions hold.

11.3.1 Example. If $F(t)$ is *deterministic*, that is, does not depend on ω, and if F has bounded variation, then the following integration by parts formula is valid:

$$\int_a^b F(t)\,dW(t) = F(b)W(b) - F(a)W(a) - \int_a^b W(t)\,dF(t). \tag{11.5}$$

Here the integral on the right, evaluated at any ω, is interpreted as a Riemann-Stieltjes integral.

To verify (11.5), let $\mathcal{P}$ be a partition of $[a, b]$ and write

$$\begin{aligned}\sum_{j=0}^{n-1} F(t_j)\Delta W_j(t) &= \sum_{j=1}^{n} F(t_{j-1})W(t_j) - \sum_{j=0}^{n-1} F(t_j)W(t_j)\\ &= F(t_{n-1})W(b) - F(a)W(a) - \sum_{j=1}^{n-1} W(t_j)\Delta F_{j-1}. \end{aligned}\tag{11.6}$$

As $\|\mathcal{P}\| \to 0$, the sum in (11.6) converges to the Riemann-Stieltjes integral (both pointwise in ω and in the mean square sense) and $F(t_{n-1})$ converges to $F(b)$, verifying the formula.

If F' is continuous, then (11.5) takes the form

$$\int_a^b F(t)\,dW(t) = F(b)W(b) - F(a)W(a) - \int_a^b W(t)F'(t)\,dt, \tag{11.7}$$

as may be seen by applying the Mean Value Theorem to the increments ΔF_j. Because F is deterministic, the random variable $\int_a^b F(t)\,dW(t)$ is normal with mean zero and variance $\int_a^b F^2(t)\,dt$ (Corollary 11.3.4, below). It follows from (11.7) that the random variable

$$\int_a^b W(t)F'(t)\,dt + F(a)W(a) - F(b)W(b)$$

is also normal with mean zero and variance $\int_a^b F^2(t)\,dt$. In particular, taking $F(t) = t - b$, we see that

$$\int_a^b [W(t) - W(a)]\,dt$$

is normal with mean 0 and variance $\int_a^b (t-b)^2\,dt = (b-a)^3/3$. ◊

11.3.2 Example. Theorem 11.2.2 may be used to derive the formula

$$\int_a^b W(t)\,dW(t) = \frac{W^2(b) - W^2(a)}{2} - \frac{b-a}{2}. \tag{11.8}$$

Note first that for any sequence of real numbers x_j,

$$2\sum_{j=0}^{n-1} x_j(x_{j+1} - x_j) = x_n^2 - x_0^2 - \sum_{j=0}^{n-1}(x_{j+1} - x_j)^2,$$

as may be seen by direct expansion of both sides. Now let

$$\mathcal{P} = \{a = t_0 < t_1 < \cdots < t_n = b\}$$

be an arbitrary partition of $[a, b]$. Setting $x_j = W(t_j, \omega)$, we have

$$2\sum_{j=0}^{n-1} W(t_j)\Delta W_j = W^2(b) - W^2(a) - V_{\mathcal{P}}^{(2)}(W).$$

By Theorem 11.2.2, $\lim_{\|\mathcal{P}\|\to 0} V_{\mathcal{P}}^{(2)}(W) = b - a$ in the mean square sense, verifying Equation (11.8). ◊

Remarks. (a) The term $\frac{1}{2}(b-a)$ in (11.8) arises because of the particular choice of t_j^* in the definition of (11.3) as the left endpoint of the interval $[t_j, t_{j+1}]$. If we had instead used *midpoints* (thus producing what is called the *Stratonovich integral*), then the term $\frac{1}{2}(b-a)$ would not appear and the result would conform to the familiar one of classical calculus. The choice of left endpoints is dictated by technical considerations, including the fact that this choice makes the Ito integral a martingale, a result of fundamental importance in both the theory and applications of stochastic calculus. One can also explain the choice of the left endpoint heuristically: Consider the parameter t in $F(t)$ and $W(t)$ to represent time. If t_j represents the present, then we should use the known value $F(t_j, \omega)$ in the jth term of the approximation (11.4) of the integral rather than a value $F(t_j^*, \omega)$, $t_j^* > t_j$, which may be viewed as anticipating the future.

(b) The crucial step in Example 11.3.2 is the result proved in Theorem 11.2.2 that the sums $\sum_{j=0}^{n-1} (\Delta W_j)^2$ converge in the mean square sense to $b - a$. This fact, which is sometimes written symbolically as

$$(dW)^2 = dt,$$

is largely responsible for the difference between the Ito calculus and Newtonian calculus. Similar formulas arise later and will form the basis of a "symbolic calculus." ◊

Properties of the Ito integral

The following theorem summarizes the main properties of the Ito integral. Some of these may be seen as analogous to properties of the classical Riemann integral, while others have no classical counterpart. The processes F and G in the statement of the theorem are assumed to satisfy the usual conditions described in the preceding subsection.

11.3.3 Theorem. *Let $\alpha, \beta \in \mathbf{R}$ and $0 \leq a < c < b$. Then*

(a) $\int_a^b [\alpha F(t) + \beta G(t)]\, dW(t) = \alpha \int_a^b F(t)\, dW(t) + \beta \int_a^b G(t)\, dW(t)$.

(b) $\int_a^b F(t)\, dW(t) = \int_a^c F(t)\, dW(t) + \int_c^b F(t)\, dW(t)$.

(c) $\mathbb{E}\left(\int_a^b F(t)\, dW(t)\right) = 0$.

(d) $\mathbb{E}\left(\int_a^b F(t)\, dW(t)\right)^2 = \int_a^b \mathbb{E}\,(F(t))^2\, dt$.

(e) $\mathbb{E}\left(\int_a^b F(t)\, dW(t) \int_a^b G(t)\, dW(t)\right) = \int_a^b \mathbb{E}\,(F(t)G(t))\, dt$.

(f) $\mathbb{E}\left(\int_a^b F(t)\, dt\right) = \int_a^b \mathbb{E}\,(F(t))\, dt$.

Proof. For part (a), set $H = \alpha F + \beta G$ and $X = \alpha I(F) + \beta I(G)$. Using the inequality

$$(x + y)^2 = 2(x^2 + y^2) - (x - y)^2 \leq 2(x^2 + y^2)$$

and the fact that

$$I_{\mathcal{P}}(H) = \alpha I_{\mathcal{P}}(F) + \beta I_{\mathcal{P}}(G),$$

we have

$$\begin{aligned}
&[I(H) - X]^2 \\
&\quad = |I(H) - I_{\mathcal{P}}(H) + I_{\mathcal{P}}(H) - X|^2 \\
&\quad \le 2\,|I(H) - I_{\mathcal{P}}(H)|^2 + 2\,|I_{\mathcal{P}}(H) - X|^2 \\
&\quad \le 2\,|I(H) - I_{\mathcal{P}}(H)|^2 + 4\alpha^2\,|I_{\mathcal{P}}(F) - I(F)|^2 + 4\beta^2\,|I_{\mathcal{P}}(G) - I(G)|^2 .
\end{aligned}$$

Letting $\|\mathcal{P}\| \to 0$ verifies (a).

For (b), note that a partition $\mathcal{P}$ of $[a, b]$ containing the intermediate point c is the union of partitions $\mathcal{P}_1$ of $[a, c]$ and $\mathcal{P}_2$ of $[c, b]$, hence

$$I_{\mathcal{P}}(F) = I_{\mathcal{P}_1}(F) + I_{\mathcal{P}_2}(F).$$

For partitions that do not contain c, a relation of this sort holds approximately, the approximation improving as $\|\mathcal{P}\| \to 0$. In the limit one obtains (b).

Part (c) follows from

$$\mathbb{E}\, I_{\mathcal{P}}(F) = \sum_{j=0}^{n-1} \mathbb{E}(F(t_j)\Delta W_j) = \sum_{j=0}^{n-1} \big(\mathbb{E}\, F(t_j)\big)\big(\mathbb{E}\, \Delta W_j\big) = 0,$$

where in the last sum we have used the independence of $F(t_j)$ and ΔW_j.

To prove part (d), note that the terms in the double sum

$$\mathbb{E}\,[I_{\mathcal{P}}(F)]^2 = \sum_{j=0}^{n-1}\sum_{k=0}^{n-1} \mathbb{E}\,[F(t_j)\Delta W_j F(t_k)\Delta W_k]$$

for which $j \ne k$ evaluate to zero. Indeed, if $j < k$, then $F(t_j)\Delta W_j F(t_k)$ is $\mathcal{F}^W_{t_k}$-measurable, and since ΔW_k is independent of $\mathcal{F}^W_{t_k}$ it follows that [2]

$$\mathbb{E}\,[F(t_j)\Delta W_j F(t_k)\Delta W_k] = \mathbb{E}\,[F(t_j)\Delta W_j F(t_k)]\,\mathbb{E}\,(\Delta W_k) = 0.$$

Also, because $F(t_j)$ and ΔW_j are independent,

$$\mathbb{E}\,(F(t_j)\Delta W_j)^2 = \mathbb{E}\left(F^2(t_j)\right)\mathbb{E}\,(\Delta W_j)^2 = \mathbb{E}\left(F^2(t_j)\right)(t_{j+1} - t_j).$$

Therefore, the above double sum reduces to

$$\mathbb{E}\,[I_{\mathcal{P}}(F)]^2 = \sum_{j=0}^{n-1} \mathbb{E}\left(F^2(t_j)\right)(\Delta t_j),$$

which is a Riemann sum for the integral $\int_a^b \mathbb{E}\left(F^2(t)\right)\,dt$. Letting $\|\mathcal{P}\| \to 0$ yields (d).

[2] The assertion requires a conditioning argument which may be based on results from Section 8.3. For a discrete version of the calculation, see Exercise 9.1.1.

For (e), define

$$[F,G] = \mathbb{E}\left(\int_a^b F(t)\,dW(t)\int_a^b G(t)\,dW(t)\right)$$

and

$$\langle F,G\rangle = \int_a^b \mathbb{E}\left(F(t)G(t)\right)\,dt.$$

The bracket functions are linear in each argument separately and by (d) yield the same value when $F = G$. Since

$$4\big[F,G\big] = \big[F+G,F+G\big] - \big[F-G,F-G\big],$$

and

$$4\big\langle F,G\big\rangle = \big\langle F+G,F+G\big\rangle - \big\langle F-G,F-G\big\rangle,$$

it follows that $\big[F,G\big] = \big\langle F,G\big\rangle$.

Part (f) is a consequence of *Fubini's Theorem*, which gives general conditions under which integral and expectation may be interchanged. A proof may be found in standard texts on real analysis. □

11.3.4 Corollary. *If $F(t)$ is a deterministic process, then the Ito integral $\int_a^b F(t)\,dW(t)$ is normal with mean zero and variance $\int_a^b F^2(t)\,dt$.*

Proof. That $I(F)$ has mean 0 and variance $\int_a^b F^2(t)\,dt$ follows from (c) and (d) of the theorem. To see that $I(F)$ is normal, note that $I_{\mathcal{P}}(F)$, as sum of independent normal random variables $F(t_j)\Delta W_j$, is itself normal (Example 3.8.2). A standard result in probability theory implies that $I(F)$, as a mean square limit of normal random variables, is also normal. □

11.4 The Ito-Doeblin Formula

The Ito-Doeblin formula, described in various forms in this section, is useful for generating stochastic differential equations, the subject of the next section. First we need the notion of Ito process.

Ito Process

An *Ito process on* $[a,b]$ is a stochastic process X of the form

$$X_t = X_a + \int_a^t F(s)\,dW(s) + \int_a^t G(s)\,ds, \quad a \le t \le b, \tag{11.9}$$

where F and G are continuous processes adapted to $(\mathcal{F}_t^W)$ and

$$\mathbb{E}\left(\int_a^b F^2(t)\,dt\right) + \mathbb{E}\left(\int_a^b |G(t)|\,dt\right) < +\infty.$$

Such processes arise as solutions to stochastic differential equations, special cases of which occur in the Black-Scholes option pricing model discussed later.

Equation (11.9) is usually written in differential notation as

$$dX = F\,dW + G\,dt.$$

For example, if we take $b = t$ in Equation (11.7) and rewrite the resulting equation as

$$F(t)W(t) = F(a)W(a) + \int_a^t F(s)\,dW(s) + \int_a^t W(s)F'(s)\,ds,$$

then FW is seen to be an Ito process with differential

$$d(FW) = F\,dW + WF'\,dt.$$

Similarly, we can rewrite Equation (11.8) as

$$W_t^2 = W_a^2 + 2\int_a^t W(s)\,dW(s) + \int_a^t 1\,ds,$$

which shows that W^2 is an Ito process with differential

$$dW^2 = 2W dW + dt.$$

Note that if X is a deterministic function with continuous derivative then

$$X_t = X_a + \int_a^t X'(s)\,ds.$$

Thus, by the above convention, $dX = X'(t)dt$, in agreement with the classical definition of differential.

Ito-Doeblin Formula, Version 1

Here is the simplest version of the formula.

11.4.1 Theorem. *Let $f(x)$ have continuous first and second derivatives. Then the process $f(W)$ has differential*

$$d\,f(W) = f'(W)\,dW + \tfrac{1}{2}f''(W)\,dt.$$

In integral form,

$$f(W(t)) = f(W(a)) + \int_a^t f'(W(s))\,dW(s) + \frac{1}{2}\int_a^t f''(W(s))\,ds.$$

Proof. We sketch the proof under the assumption that f has a Taylor series expansion. A detailed proof may be found in [11].

Let $\mathcal{P}_n$ be a partition of $[a,t]$ by points where $t_j = a + jt/n$, $j = 0, 1, \ldots, n$. Setting $W_j = W(t_j)$ we have

$$f(W(t)) - f(W(a)) = \sum_{j=0}^{n-1} f(W_{j+1}) - f(W_j)$$

and

$$\begin{aligned} f(W_{j+1}) - f(W_j) &= \sum_{n=1}^{\infty} \frac{1}{n!} f^{(n)}(W_j)(\Delta W_j)^n \\ &= f'(W_j)\Delta W_j + \tfrac{1}{2} f''(W_j)(\Delta W_j)^2 + (\Delta W_j)^3 R_j \\ &= f'(W_j)\Delta t_j + \tfrac{1}{2} f''(W_j)\Delta t_j + \tfrac{1}{2} f''(W_j)\big[(\Delta W_j)^2 - \Delta t_j\big] + (\Delta W_j)^3 R_j \end{aligned}$$

for a suitable remainder term R_j. Thus,

$$f(W(t)) - f(W(a)) = A_n + \tfrac{1}{2} B_n + \tfrac{1}{2} C_n + D_n,$$

where

$$A_n = \sum_{j=0}^{n-1} f'(W(t_j))\Delta W_j \qquad B_n = \sum_{j=0}^{n-1} f''(W_j)\Delta t_j$$

$$C_n = \sum_{j=0}^{n-1} f''(W_j)\big[(\Delta W_j)^2 - \Delta t_j\big] \qquad D_n = \sum_{j=0}^{n-1} (\Delta W_j)^3 R_j.$$

We now consider what happens to A_n, B_n, C_n, and D_n as $n \to \infty$.

By definition of the Ito integral,

$$A_n \to \int_a^t f'(W(s))\, dW(s)$$

in the mean square sense, hence a subsequence converges to the integral on a set of probability one. Furthermore, since $f''(W_t(\omega))$ is continuous in t, B_n converges pointwise to the Riemann integral $\int_a^t f''(W_s(\omega))\, ds$. It thus remains to show that subsequences of (C_n) and (D_n) converge to zero on a set of probability one.

For each n set $C_{nj} = f''(W_j)[(\Delta W_j)^2 - \Delta t_j]$, so that

$$\mathbb{E}(C_n^2) = \sum_{j \neq k} \mathbb{E}(C_{nj} C_{nk}) + \sum_{j=0}^{n-1} \mathbb{E}(C_{nj}^2).$$

By independence, $\mathbb{E}(C_{nj} C_{nk}) = \mathbb{E}(C_{nj})\mathbb{E}(C_{nk})$ and

$$\mathbb{E}(C_{nk}) = \mathbb{E}(f''(W_j)\mathbb{E}[(\Delta W_j)^2 - \Delta t_j])\mathbb{E}(f''(W_k)\mathbb{E}[(\Delta W_k)^2 - \Delta t_k]) = 0,$$

hence

$$\mathbb{E}(C_n^2) = \sum_{j=0}^{n-1} \mathbb{E}(C_{nj}^2) = \sum_{j=0}^{n-1} \mathbb{E}\{f''(W_j)[(\Delta W_j)^2 - \Delta t_j]\}^2.$$

Now assume that f'' is bounded, say $|f''| \leq M$. Then

$$\mathbb{E}(C_n^2) \leq M \sum_{j=0}^{n-1} \mathbb{E}[(\Delta W_j)^2 - \Delta t_j]^2.$$

Moreover, because $\Delta W_j \sim N(0, t/n)$ and $\Delta t_j = t/n$,

$$\mathbb{E}[(\Delta W_j)^2 - \Delta t_j]^2 = \mathbb{E}\left[\left(\sqrt{\frac{t}{n}}X\right)^2 - \frac{t}{n}\right]^2 = \left(\frac{t}{n}\right)^2 \mathbb{E}(X^2 - 1)^2,$$

where $X \sim N(0,1)$. By Example 6.2.4, $\mathbb{E}(X^2 - 1)^2 = 2$, hence the last expression is $2t/n^2$. Therefore, $\mathbb{E}(C_n^2) \leq 2Mt^2/n$ and consequently $\mathbb{E}(C_n^2) \to 0$. It then follows that a subsequence of (C_n) tends to zero on a set of probability one. If f'' is unbounded, the proof is more delicate, requiring one to invoke a certain growth property of Brownian motion.

Finally, using the fact that the order m variation of Brownian motion is zero for $m \geq 3$ (Remarks 11.2.3), one shows that $D_n \to 0$ in the mean square sense, hence a subsequence tends to zero on a set of probability one. Putting all these results together (but using nested subsequences throughout), we obtain the desired conclusion. □

11.4.2 Remark. The equation in Theorem 11.4.1 may be expressed in integral form as

$$\int_a^t f'(W(s))\,dW(s) = f(W(t)) - f(W(a)) - \frac{1}{2}\int_a^t f''(W(s))\,ds,$$

which is Ito's version of the fundamental theorem of calculus. The presence of the integral on the right is a consequence of the nonzero quadratic variation of W. ◊

11.4.3 Example. Applying Theorem 11.4.1 to the function $f(x) = x^k$, $k \geq 2$, we have

$$dW^k = kW^{k-1}\,dW + \tfrac{1}{2}k(k-1)W^{k-2}\,dt,$$

which has integral form

$$W^k(t) = W^k(a) + k\int_a^t W^{k-1}(s)\,dW(s) + \tfrac{1}{2}k(k-1)\int_a^t W^{k-2}(s)\,ds.$$

Rearranging we have

$$\int_a^t W^{k-1}(s)\,dW(s) = \frac{W^k(t) - W^k(a)}{k} - \frac{(k-1)}{2}\int_a^t W^{k-2}(s)\,ds,$$

which may be seen as an evaluation of the Ito integral on the left. The special case $k = 2$ is the content of Example 11.3.2. ◊

We state without proof three additional versions of the Ito-Doeblin Formula, each of which considers differentials of functions of several variables.

Ito-Doeblin Formula, Version 2

The following is a time-dependent version of formula one.

11.4.4 Theorem. *Suppose $f(t,x)$ is continuous with continuous partial derivatives f_t, f_x, and f_{xx}. Then*

$$df(t,W) = f_x(t,W)\,dW + f_t(t,W)\,dt + \tfrac{1}{2}f_{xx}(t,W)\,dt,$$

where we have suppressed the variable t in the notation $W(t)$. In integral form,

$$f(t,W(t)) = f(a,W(a)) + \int_a^t f_x(s,W(s))\,dW(s) + \int_a^t \left[f_t(s,W(s)) + \tfrac{1}{2}f_{xx}(s,W(s))\right] ds.$$

Ito-Doeblin Formula, Version 3

Versions 1 and 2 of the Ito-Doeblin Formula deal only with functions of the process W. The following version treats functions of a general Ito process X.

11.4.5 Theorem. *Suppose $f(t,x)$ is continuous with continuous partial derivatives f_t, f_x, and f_{xx}. Let X be an Ito process with differential $dX = F\,dW + G\,dt$. Then*

$$\begin{aligned} df(t,X) &= f_t(t,X)\,dt + f_x(t,X)\,dX + \tfrac{1}{2}f_{xx}(t,X)\,(dX)^2 \\ &= f_x(t,X)\,F\,dW + \left[f_t(t,X) + f_x(t,X)\,G + \tfrac{1}{2}f_{xx}(t,X)\,F^2\right] dt. \end{aligned}$$

11.4.6 Remark. The second equality in the formula may be obtained from the first by substituting $F\,dW + G\,dt$ for dX and using the formal multiplication rules summarized in Table 11.1. The rules reflect the mean square limit

TABLE 11.1: Symbol Table One.

$\cdot$	dt	dW
dt	0	0
dW	0	dt

properties

$$\sum_{j=0}^{n-1}(\Delta W_j)^2 \to b-a, \quad \sum_{j=0}^{n-1}(\Delta W_j)\Delta t_j \to 0, \quad \text{and} \quad \sum_{j=0}^{n-1}(\Delta t_j)^2 \to 0$$

as $\|\mathcal{P}\| \to 0$. Using the table, we have

$$(dX)^2 = (F\,dW + G\,dt)^2 = F^2\,dt,$$

which leads to the second equality. ◊

11.4.7 Example. Let $h(t)$ be a differentiable function and X an Ito process. We calculate $d(hX)$ by applying the above formula to $f(t,x) = h(t)x$. From $f_t(t,x) = h'(t)x$, $f_x(t,x) = h(t)$, and $f_{xx}(t,x) = 0$ we have

$$d(hX) = h\,dX + h'X\,dt = h\,dX + X\,dh,$$

which conforms to the product rule in classical calculus. ◊

Ito-Doeblin Formula, Version 4

The general version of the Ito-Doeblin Formula allows functions of finitely many Ito processes $X_1, \ldots, X_n$. We state the formula for the case $n = 2$.

11.4.8 Theorem. *Suppose $f(t,x,y)$ is continuous with continuous partial derivatives f_t, f_x, f_y, f_{xx}, f_{xy}, and f_{yy}. Let X_j be an Ito process with differential*

$$dX_j = F_j\,dW_j + G_j\,dt, \quad j = 1, 2,$$

where W_1 and W_2 are Brownian motions. Then

$$\begin{aligned} df(t,X_1,X_2) = f_t\,(t,X_1,X_2)\,dt + f_x\,(t,X_1,X_2)\,dX_1 + f_y\,(t,X_1,X_2)\,dX_2 \\ + \tfrac{1}{2} f_{xx}\,(t,X_1,X_2)\,(dX_1)^2 + \tfrac{1}{2} f_{yy}\,(t,X_1,X_2)\,(dX_2)^2 \\ + f_{xy}\,(t,X_1,X_2)\,dX_1 \cdot dX_2. \end{aligned}$$

11.4.9 Remark. The differential $df\,(t,X_1,X_2)$ may be described in terms of dt, dW_1, and dW_2 by substituting $F_j\,dW_j + G_j\,dt$ for dX_j and using the formal multiplication rules given in Table 11.2. From the table, we see that

TABLE 11.2: Symbol Table Two.

$\cdot$	dt	dW_1	dW_2
dt	0	0	0
dW_1	0	dt	$dW_1 \cdot dW_2$
dW_2	0	$dW_1 \cdot dW_2$	dt

$$(dX_1)^2 = F_1^2\,dt, \quad (dX_2)^2 = F_2^2\,dt \;\text{ and }\; dX_1 \cdot dX_2 = F_1F_2\,dW_1 \cdot dW_2,$$

hence

$$\begin{aligned} df\,(t,X_1(t),X_2(t)) &= f_t\,dt + f_x \cdot (F_1\,dW_1 + G_1\,dt) + f_y \cdot (F_2\,dW_2 + G_2\,dt) \\ &\quad + \tfrac{1}{2} f_{xx}F_1^2\,dt + \tfrac{1}{2} f_{yy}F_2^2\,dt + f_{xy}F_1F_2\,dW_1 \cdot dW_2 \\ &= f_xF_1\,dW_1 + f_yF_2\,dW_2 + f_{xy}F_1F_2\,dW_1 \cdot dW_2 \\ &\quad + \left[f_t + f_xG_1 + f_yG_2 + \tfrac{1}{2} f_{xx}F_1^2 + \tfrac{1}{2} f_{yy}F_2^2\right] dt, \end{aligned}$$

where the partial derivatives of f are evaluated at $(t, X_1(t), X_2(t))$. The evaluation of the term $dW_1 \cdot dW_2$ depends on how dW_1 and dW_2 are related. For example, if W_1 and W_2 are independent, then $dW_1 \cdot dW_2 = 0$. On the other hand, if W_1 and W_2 are *correlated*, that is,

$$W_1 = \varrho W_2 + \sqrt{1 - \varrho^2}\, W_3,$$

where W_2 and W_3 are independent Brownian motions and $0 < |\varrho| \leq 1$, then

$$dW_1 \cdot dW_2 = \varrho\, dt.$$

(See, for example, [19].) ◊

11.4.10 Example. We use Theorem 11.4.8 to obtain Ito's product rule for the differentials $dX_j = F_j\, dW_j + G_j\, dt$, $j = 1, 2$. Taking $f(x, y) = xy$ we have $f_x = y$, $f_y = x$, $f_{xy} = 1$, and $f_t = f_{xx} = f_{yy} = 0$, hence

$$d(X_1 X_2) = \begin{cases} X_2\, dX_1 + X_1\, dX_2 & \text{if } W_1 \text{ and } W_2 \text{ are independent} \\ X_2\, dX_1 + X_1\, dX_2 + \varrho F_1 F_2\, dt & \text{if } W_1 \text{ and } W_2 \text{ are correlated.} \end{cases}$$

Thus in the independent case we obtain the familiar product rule of classical calculus. ◊

11.5 Stochastic Differential Equations

To put the theory in a suitable perspective, we begin with a brief discussion of classical differential equations.

Classical Differential Equations

An *ordinary differential equation* (ODE) is an equation involving an unknown function and its derivatives. A *first-order* ODE *with initial condition* is of the form

$$x' = f(t, x), \quad x(0) = x_0, \tag{11.10}$$

where f is a continuous function of (t, x) and x_0 is a given value. The variable t may be thought of as the *time* variable and x the *space* variable. A *solution* of (11.10) is a differentiable function $x = x(t)$ satisfying (11.10) on some open interval I containing 0. Equation (11.10) is frequently written in differential form as

$$dx = f(t, x)\, dt, \quad x(0) = x_0.$$

Explicit solutions of (11.10) are possible only in special cases. For example, if

$$f(x, t) = \frac{h(t)}{g(x)},$$

then (11.10) may be written

$$g(x(t))x'(t) = h(t), \quad x(0) = x_0$$

hence a solution $x(t)$ must satisfy $G(x(t)) = H(t) + c$, where $G(x)$ and $H(t)$ are antiderivatives of $g(x)$ and $h(t)$, respectively, and c is an arbitrary constant. The initial condition $x(0) = x_0$ may then be used to determine c. The result may be obtained formally by writing the differential equation in separated form as $g(x)dx = h(t)dt$ and then integrating.

11.5.1 Examples. **(a)** The differential equation $x'(t) = 2t \sec x(t)$, $x(0) = x_0$, has separated form $\cos x\, dx = 2t\, dt$, which integrates to $\sin x = t^2 + c$, $c = \sin x_0$. The solution may be written

$$x(t) = \sin^{-1}(t^2 + \sin x_0),$$

which is valid for $x_0 \in (-\pi/2, \pi/2)$ and for t sufficiently near 0.

(b) Let $x(t)$ denote the value at time t of an account earning interest at the time-dependent rate $r(t)$. For small Δt, the amount $\Delta x = x(t + \Delta t) - x(t)$ earned over the time interval $[t, t+\Delta t]$ is approximately $x(t)r(t)\Delta t$. This leads to the initial value problem

$$dx = x(t)r(t)dt, \quad x(0) = x_0,$$

where x_0 is the initial deposit. The solution is

$$x(t) = x_0 e^{R(t)}, \quad \text{where} \quad R(t) = \int_0^t r(\tau)\, d\tau. \qquad \diamond$$

Analogous to an ODE, a *partial differential equation* (PDE) is an equation involving an unknown function of *several* variables and its partial derivatives. As we shall see in Chapter 12, a stochastic differential equation can give rise to a PDE whose solution may be used to construct the solution of the original SDE. This method will be used in Chapter 12 to obtain the Black-Scholes option pricing formula.

Definition of Stochastic Differential Equation

An ODE is inherently *deterministic* in the sense that the initial condition x_0 and the rate $f(t,x)$ uniquely determine the solution $x(t)$ for all t sufficiently near 0. There are circumstances, however, where $f(t,x)$ is not completely determined but rather is subject to random fluctuations. For example, let $x(t)$ be the size of an account at time t. A model that incorporates random fluctuation of the interest rate is given by

$$x'(t) = [r(t) + \xi(t)]x(t), \tag{11.11}$$

where ξ is a random process. The same differential equation arises if $x(t)$ is the size of a population whose relative growth rate is subject to random fluctuations from environmental changes. Because (11.11) has a random component, one would expect its solution to be a random variable. Equations like this are called stochastic differential equations, formally defined as follows:

A *stochastic differential equation* (SDE) is an equation of the form

$$dX(t) = \alpha(t, F(t), X(t))\, dW(t) + \beta(t, G(t), X(t))\, dt,$$

where $\alpha(t, u, x)$ and $\beta(t, u, x)$ are continuous functions and F, G are stochastic processes satisfying the usual conditions. A *solution* of the SDE is a stochastic process X adapted to $(\mathcal{F}_t^W)$ and satisfying

$$X(t) = X(0) + \int_0^t \alpha(s, F(s), X(s))\, dW(s) + \int_0^t \beta(s, G(s), X(s))\, ds, \quad (11.12)$$

where $X(0)$ is a specified random variable.

In certain cases the Ito-Doeblin formula, which generates SDEs from Ito processes, may be used to find an explicit form of the solution (11.12). We illustrate with two general procedures, each based on an Ito process

$$Y(t) = Y(0) + \int_0^t F(s)\, dW(s) + \int_0^t G(s)\, ds. \quad (11.13)$$

Exponential Processes

Applying Version 3 of the Ito formula to the process $X(t) = e^{Y(t)}$ with $f(t, y) = e^y$ leads to

$$\begin{aligned} dX &= df(t, Y) = f_t(t, Y)\, dt + f_y(t, Y)\, dY + \tfrac{1}{2} f_{yy}(t, Y)\, (dY)^2 \\ &= X\, dY + \tfrac{1}{2} X\, (dY)^2. \end{aligned}$$

Since $dY = F\, dW + G\, dt$ and $(dY)^2 = F^2\, dt$ we obtain

$$dX = FX\, dW + \left(G + \tfrac{1}{2}F^2\right) X\, dt. \quad (11.14)$$

Equation 11.14 therefore provides a class of SDEs with solutions

$$X(t) = X(0) \exp\left\{\int_0^t F(s)\, dW(s) + \int_0^t G(s)\, ds\right\}. \quad (11.15)$$

11.5.2 Example. Let σ and μ be continuous stochastic processes. Taking $F = \sigma$ and $G = \mu - \frac{1}{2}\sigma^2$ in (11.14) yields the SDE

$$dX = \sigma X\, dW + \mu X\, dt,$$

which, by (11.15), has solution

$$X(t) = X(0) \exp\left\{\int_0^t \sigma(s)\, dW(s) + \int_0^t \left[\mu(s) - \tfrac{1}{2}\sigma^2(s)\right] ds\right\}.$$

In case μ and σ are constant, the solution reduces to

$$X(t) = X(0) \exp\left[\sigma W(t) + (\mu - \tfrac{1}{2}\sigma^2)t\right], \quad (11.16)$$

a process known as *geometric Brownian motion*. This example will form the basis of discussion in the next chapter. ◊

Product Processes

The second procedure applies Version 3 of the Ito-Doeblin formula to the process $X(t) = h(t)Y(t)$, where $h(t)$ is a nonzero differentiable function and Y is given by (11.13). By Example 11.4.7,

$$dX = h\,dY + h'Y\,dt = h(F\,dW + G\,dt) + h'Y\,dt.$$

Rearranging, we obtain the SDE

$$dX = hF\,dW + \left(hG + \frac{h'}{h}X\right) dt.$$

The solution $X = hY$ may be written

$$X(t) = h(t)\left(\frac{X(0)}{h(0)} + \int_0^t F(s)\,dW(s) + \int_0^t G(s)\,ds\right).$$

Taking $F = f/h$ and $G = g/h$, where f and g are continuous functions, we obtain the SDE

$$dX = f\,dW + \left(g + \frac{h'}{h}X\right) dt \tag{11.17}$$

with solution

$$X(t) = h(t)\left(\frac{X(0)}{h(0)} + \int_0^t \frac{f(s)}{h(s)}\,dW(s) + \int_0^t \frac{g(s)}{h(s)}\,ds\right). \tag{11.18}$$

11.5.3 Example. Let α, β, and σ be constants with $\beta > 0$ and take

$$f = \sigma, \quad g = \alpha, \quad \text{and,} \quad h(t) = \exp(-\beta t)$$

in (11.17). Then

$$dX = \sigma\,dW + (\alpha - \beta X)\,dt, \tag{11.19}$$

which, by (11.18), has solution

$$X(t) = e^{-\beta t}\left(X(0) + \frac{\alpha}{\beta}(e^{\beta t} - 1) + \sigma\int_0^t e^{\beta s}\,dW(s)\right). \tag{11.20}$$

The process X is called an *Ornstein-Uhlenbeck process.* In finance, the SDE in (11.19) is known as the *Vasicek equation* and is used to describe the evolution of stochastic interest rates (see, for example, [19]). For $\alpha = 0$ and $\sigma > 0$ (11.19) is called a *Langevin equation*, which plays a central role in statistical mechanics. ◊

11.6 Exercises

1. Find the solution of each of the following ODEs and the largest open interval on which it is defined.

   (a) $x' = x^2 \sin t,\ x(0) = 1/3$;

   (b) $x' = x^2 \sin t,\ x(0) = 2$;

   (c) $x' = \dfrac{2t + \cos t}{2x},\ x(0) = 1$;

   (d) $x' = \dfrac{x+1}{\tan t},\ x(\pi/6) = 1/2,\ 0 < t < \pi/2$.

2. The *variation of order* m of a (deterministic) function f on an interval $[a, b]$ is defined as the limit

$$\lim_{||\mathcal{P}||\to 0} V_{\mathcal{P}}^{(m)}(f).$$

   Prove the following:

   (a) If f is a bounded function with zero variation of order m on $[a, b]$, then f has zero variation of order $m + 1$ on $[a, b]$.

   (b) If f is continuous with bounded mth variation on $[a, b]$, then f has zero variation of order $k > m$ on $[a, b]$.

   (c) If f has a bounded first derivative on $[a, b]$, then it has bounded first variation and zero variation of order $m \geq 2$ on $[a, b]$.

   (d) The function $f(t) = t^{1+\varepsilon} \sin(1/t)$, $f(0) = 0$, has bounded first variation and zero variation of order $m \geq 2$ on $[0, 1]$.

3. Show that the Riemann-Stieltjes integral $\int_0^{2/\pi} 1\, dw$ does not exist for the function w defined in Example 11.2.1.

4. Show that for any nonzero constant c, $W_1(t) := cW(t/c^2)$ defines a Brownian motion.

5. Show that for $r \leq s \leq t$,

$$W(s) + W(t) \sim N(0, t + 3s) \quad \text{and} \quad W(r) + W(s) + W(t) \sim N(0, t + 3s + 5r).$$

   Generalize.

6. Show that for $a > 1/2$, $\lim_{s\to 0} s^a W(1/s) = 0$, where the limit is taken in the mean square sense.

7. Use Theorem 11.3.3 to find $\mathbb{V} X_t$ if $X_t =$

   (a) $\int_0^t \sqrt{s}\, W(s)\, dW(s)$. (b) $\int_0^t \exp\left(W^2(s)\right) dW(s)$. (c) $\int_0^t \sqrt{|W(s)|}\, dW(s)$.

8. Show that

$$\int_a^b [W(t) - W(b)]\, dt$$

   is normal with mean 0 and variance $(b - a)^3/3$. (See Example 11.3.1.)

9. Use the Ito-Doeblin formulas to show that

   (a) $\displaystyle\int_a^t e^{W(s)}\,dW(s) = e^{W(t)} - e^{W(a)} - \frac{1}{2}\int_a^t e^{W(s)}\,ds.$

   (b) $\displaystyle 2\int_0^t sW(s)\,dW(s) = tW^2(t) - \frac{t^2}{2} - \int_0^t W^2(s)\,ds.$

   (c) $\displaystyle d\left(\frac{X}{Y}\right) = \frac{X}{Y}\left[\frac{dX}{X} - \frac{dY}{Y} + \left(\frac{dY}{Y}\right)^2 - \frac{dX}{X}\frac{dY}{Y}\right]$, where X and Y are Ito processes.

10. Let X be an Ito process with $dX = FdW + Gdt$. Find the differentials in terms of dW and dt of

    (a) X^2. (b) $\ln X$. (c) tX^2.

11. Show that, for the process given in Equation (11.20), $\lim_{t\to\infty} \mathbb{E}\, X_t = \alpha/\beta$.

Chapter 12

The Black-Scholes-Merton Model

With the methods of Chapter 11 at our disposal, we are now able to derive the celebrated Black-Scholes formula for the price of a call option. The formula is based on the solution of a partial differential equation arising from a stochastic differential equation that governs the price of the underlying stock $\mathcal{S}$. We assume throughout that the market is arbitrage-free.

12.1 The Stock Price SDE

Let W be a Brownian motion on a probability space $(\Omega, \mathcal{F}, \mathbb{P})$. The price S of a single share of $\mathcal{S}$ is assumed to satisfy the SDE

$$\frac{dS}{S} = \sigma\, dW + \mu\, dt, \tag{12.1}$$

where μ and σ are constants called, respectively, the *drift* and *volatility* of the stock. Equation (12.1) asserts that the relative change in the stock price has two components: a deterministic part $\mu\, dt$, which accounts for the general trend of the stock, and a random component $\sigma\, dW$, which reflects the unpredictable nature of $\mathcal{S}$. The volatility is a measure of the riskiness of the stock and its sensitivity to changes in the market. If $\sigma = 0$, then (12.1) is an ODE with solution $S_t = S_0 e^{\mu t}$.

Equation (12.1) may be written in standard form as

$$dS = \sigma S\, dW + \mu S\, dt, \tag{12.2}$$

which is the SDE of Example 11.5.2. The solution there was found to be

$$S_t = S_0 \exp\left[\sigma W_t + \left(\mu - \tfrac{1}{2}\sigma^2\right) t\right]. \tag{12.3}$$

The integral version of (12.2) is

$$S_t = S_0 + \int_0^t \sigma S(s)\, dW(s) + \int_0^t \mu S(s)\, ds. \tag{12.4}$$

Taking expectations in (12.4) and using Theorem 11.3.3, we see that

$$\mathbb{E}\, S_t = S_0 + \mu \int_0^t \mathbb{E}\, S(s)\, ds.$$

The function $x(t) := \mathbb{E}\, S_t$ therefore satisfies the ODE $x' = \mu x$, hence $\mathbb{E}\, S_t = S_0 e^{\mu t}$. This is the solution of (12.1) for the case $\sigma = 0$ and represents the return on a risk-free investment. Thus taking expectations in (12.4) "removes" the random component of (12.1).

Both the drift μ and the volatility σ may be stochastic processes, in which case the solution to (12.1) is given by

$$S_t = S_0 \exp\left\{\int_0^t \sigma(s)\, dW(s) + \int_0^t \left[\mu(s) - \tfrac{1}{2}\sigma^2(s)\right] ds\right\},$$

as was shown in Example 11.5.2. However, we shall not consider such a general setting in what follows.

12.2 Continuous-Time Portfolios

As in the binomial model, the basic construct in determining the value of a claim is a self-financing, replicating portfolio based on $\mathcal{S}$ and a risk-free bond $\mathcal{B}$. The bond account is assumed to earn interest at a continuously compounded annual rate r. The value of the bond at time t is denoted by B_t, where we take the initial value B_0 to be one dollar. Thus $B_t = e^{rt}$, which is the solution of the ODE

$$dB = rB\, dt, \quad B_0 = 1.$$

We assume the market allows unlimited trading in shares of $\mathcal{S}$ and bonds $\mathcal{B}$.

We shall need the notion of *continuous-time portfolio* or *trading strategy*, defined as a two-dimensional stochastic process $(\phi, \theta) = (\phi_t, \theta_t)_{0 \le t \le T}$ adapted to the price filtration $(\mathcal{F}_t^S)_{0 \le t \le T}$. The random variables ϕ_t and θ_t are, respectively, the number of bonds $\mathcal{B}$ and shares of $\mathcal{S}$ held at time t. The *value* of the portfolio at time t is the random variable

$$V_t = \phi_t B_t + \theta_t S_t, \quad 0 \le t \le T.$$

The process $V = (V_t)_{0 \le t \le T}$ is the *value* or *wealth process* of the trading strategy and V_0 is the *initial investment* or *initial wealth.* A trading strategy (ϕ, θ) is *self-financing* if

$$dV = \phi\, dB + \theta\, dS. \tag{12.5}$$

To understand the implication of (12.5), consider a discrete version of the portfolio process defined at times $t_0 = 0 < t_1 < t_2 < \cdots < t_n = T$. At time t_j, the value of the portfolio before the price S_j is known is

$$\phi_j B_{j-1} + \theta_j S_{j-1},$$

where, for ease of notation, we have written S_j for S_{t_j}, etc. After S_j becomes known and the new bond value B_j is noted, the portfolio has value

$$V_j = \phi_j B_j + \theta_j S_j.$$

At this time, the number of stocks and bonds may be adjusted (based on the information provided by $\mathcal{F}_j$), but for the portfolio to be self-financing, this restructuring must be accomplished without changing the current value of the portfolio. Thus the new values ϕ_{j+1} and θ_{j+1} must satisfy

$$\phi_{j+1} B_j + \theta_{j+1} S_j = \phi_j B_j + \theta_j S_j.$$

It follows that

$$\begin{aligned}\Delta V_j &= \phi_{j+1} B_{j+1} + \theta_{j+1} S_{j+1} - (\phi_{j+1} B_j + \theta_{j+1} S_j)\\ &= \phi_{j+1} \Delta B_j + \theta_{j+1} \Delta S_j,\end{aligned}$$

which is the discrete version of (12.5), in agreement with Theorem 5.4.1(b).

As in the discrete case, a portfolio may be used as a hedging strategy, that is, an investment devised to cover the obligation of the writer of the claim at maturity T. In this case, the portfolio is said to *replicate* the claim, the latter formally defined as a $\mathcal{F}_T^S$-random variable. A market is *complete* if every claim can be replicated.

The importance of continuous time portfolios derives from the law of one price, which implies that in an arbitrage-free market the value of a claim is the same as that of a replicating, self-financing trading strategy. We use this observation in the next section to obtain a formula for the value of a claim with underlying $\mathcal{S}$.

12.3 The Black-Scholes Formula

To derive the formula, we begin by assuming the existence of a self-financing portfolio whose value V_T at time T is the payoff $f(S_T)$ of a European claim, where f is a continuous function with suitable growth conditions. For such a portfolio, the value of the claim at any time $t \in [0, T]$ is V_t. We seek a function $v(t, s)$ such that

$$v(t, S_t) = V_t \quad (0 \le t \le T) \quad \text{and} \quad v(T, S_T) = f(S_T).$$

Note that if $S_0 = 0$ then (12.3) implies that the process S is identically zero. In this case, the claim is worthless and $V_t = 0$ for all t. Therefore, v must satisfy the boundary conditions

$$v(T, s) = f(s), \ s \ge 0, \quad \text{and} \quad v(t, 0) = 0, \ 0 \le t \le T. \tag{12.6}$$

To find v, we begin by applying Version 3 of the Ito-Doeblin formula (Theorem 11.4.5) to the process $V_t = v(t, S_t)$. Using (12.2), we have

$$\begin{aligned} dV &= v_t\,dt + v_s\,dS + \tfrac{1}{2}v_{ss}\,(dS)^2 \\ &= \sigma v_s S\,dW + \left(v_t + \mu v_s S + \tfrac{1}{2}\sigma^2 v_{ss} S^2\right)\,dt, \end{aligned} \tag{12.7}$$

where the partial derivatives of v are evaluated at (t, S_t). Additionally, we have from (12.5)

$$\begin{aligned} dV &= \theta\,dS + \phi\,dB \\ &= \theta S(\mu\,dt + \sigma\,dW) + r\phi B\,dt \\ &= \sigma\theta S\,dW + [\mu\theta S + r(V - \theta)S]\,dt. \end{aligned} \tag{12.8}$$

Equating the respective coefficients of dt and dW in (12.7) and (12.8) leads to the equations

$$\mu\theta S + r(V - \theta S) = v_t + \mu v_s S + \tfrac{1}{2}\sigma^2 v_{ss} S^2 \quad \text{and} \quad \theta = v_s.$$

Substituting the second equation into the first and simplifying yields the partial differential equation

$$v_t + rsv_s + \tfrac{1}{2}\sigma^2 s^2 v_{ss} - rv = 0, \quad s > 0, \quad 0 \le t < T. \tag{12.9}$$

Equation (12.9) together with the boundary conditions in (12.6) is called the *Black-Scholes-Merton* (BSM) PDE.

The following theorem gives the solution $v(t, s)$ to (12.9). The assertion of the theorem may be verified directly, but it is instructive to see how the solution may be obtained from that of a simpler PDE. The latter approach is described in Appendix B.

12.3.1 Theorem (General Solution of the BSM PDE). *The solution of* (12.9) *with the boundary conditions* (12.6) *is given by*

$$\begin{aligned} v(t, s) &= e^{-r(T-t)}G(t, s), \quad 0 \le t < T, \quad \textit{where} \\ G(t, s) &:= \int_{-\infty}^{\infty} f\left(s\exp\left\{\sigma\sqrt{T-t}\,y + (r - \tfrac{1}{2}\sigma^2)(T - t)\right\}\right)\varphi(y)\,dy. \end{aligned}$$

Having obtained the solution v of the BSM PDE, to complete the circle of ideas we must show that $v(\,\cdot\,, S)$ is indeed the value process of a self-financing, replicating trading strategy (ϕ, θ). This is carried out in the following theorem, whose proof uses $v(\,\cdot\,, S)$ to construct the strategy.

12.3.2 Theorem. *Given a European claim with payoff* $f(S_T)$*, there exists a self-financing replicating strategy for the claim with value process*

$$V_t = v(t, S_t), \quad 0 \le t \le T, \tag{12.10}$$

where $v(t, s)$ *is the solution of the BSM PDE.*

Proof. Define V by (12.10) and define adapted processes θ and ϕ by

$$\theta(t) = v_s(t, S_t) \quad \text{and} \quad \phi(t) = B^{-1}(t)(V(t) - \theta(t)S(t)).$$

Then V is the value process of the strategy (ϕ, θ), and from (12.7) and (12.9)

$$\begin{aligned} dV &= \sigma\theta S\,dW + [r(V - \theta S) + \mu\theta S]\,dt \\ &= \theta S(\mu\,dt + \sigma\,dW) + r(V - \theta S)\,dt \\ &= \theta\,dS + \phi\,dB. \end{aligned}$$

Therefore, (ϕ, θ) is self-financing. Since $v(T, s) = f(s)$, the strategy replicates the claim. □

From Theorem 12.3.2 we obtain the celebrated *Black-Scholes option pricing formula*:

12.3.3 Corollary. *The value at time $t \in [0, T)$ of a standard call option with strike price K and maturity T is given by*

$$C_t = S_t\Phi\big(d_1(T - t, S_t, K, \sigma, r)\big) - Ke^{-r(T-t)}\Phi\big(d_2(T - t, S_t, K, \sigma, r)\big),$$

where the functions d_1 and d_2 are defined by

$$\begin{aligned} d_1 &= d_1(\tau, s, K, \sigma, r) = \frac{\ln(s/K) + (r + \frac{1}{2}\sigma^2)\tau}{\sigma\sqrt{\tau}} \quad \text{and} \\ d_2 &= d_2(\tau, s, K, \sigma, r) = \frac{\ln(s/K) + (r - \frac{1}{2}\sigma^2)\tau}{\sigma\sqrt{\tau}} = d_1 - \sigma\sqrt{\tau}. \end{aligned} \tag{12.11}$$

In particular, the cost of the option is

$$C_0 = S_0\Phi\big(d_1(T, S_0, K, \sigma, r)\big) - Ke^{-rT}\Phi\big(d_2(T, S_0, K, \sigma, r)\big). \tag{12.12}$$

Proof. Taking $f(s) = (s - K)^+$ in Theorem 12.3.1 and applying Theorem 12.3.2, we see that the value of the call option at time $t \in [0, T)$ is $C_t = e^{-r(T-t)}G(t, S_t)$, where

$$G(t, s) = \int_{-\infty}^{\infty} \left(s\exp\left\{\sigma\sqrt{\tau}\,y + (r - \tfrac{1}{2}\sigma^2)\tau\right\} - K\right)^+ \varphi(y)\,dy, \quad \tau := T - t.$$

To evaluate the integral, note that the integrand equals zero when $y < -d_2$. Therefore,

$$\begin{aligned} G(t, s) &= s\int_{-d_2}^{\infty} \exp\left\{\sigma\sqrt{\tau}\,y + \left(r - \tfrac{1}{2}\sigma^2\right)\tau\right\}\varphi(y)\,dy - K\int_{-d_2}^{\infty} \varphi(y)\,dy \\ &= \frac{se^{(r-\sigma^2/2)\tau}}{\sqrt{2\pi}} \int_{-d_2}^{\infty} \exp\left\{-\tfrac{1}{2}y^2 + \sigma\sqrt{\tau}\,y\right\}dy - K\left[1 - \Phi(-d_2)\right] \\ &= se^{r\tau}\Phi(d_1) - K\Phi(d_2), \end{aligned} \tag{12.13}$$

the last equality by Exercise 12. □

The module **CallPutPrice–BSM** uses 12.3.3 and the put-call parity formula to generate BSM prices of calls and puts. The companion module **GraphOfCallPutPrice–BSM** generated the following figures, which show graphically the dependence of call and put prices on the parameters. The analytical justification for the graphical properties will be discussed in the next section.

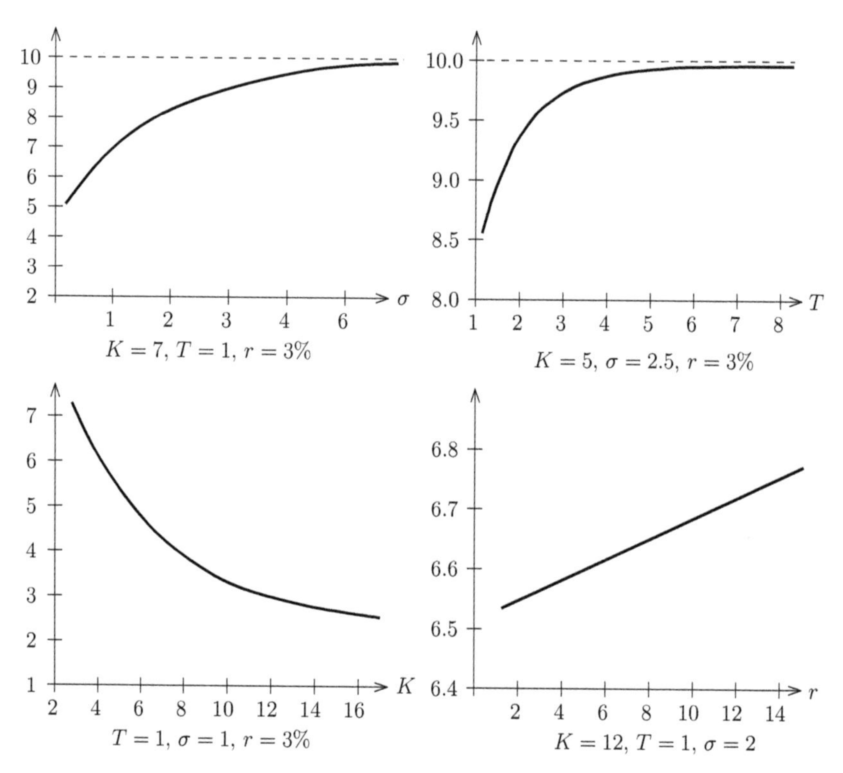

FIGURE 12.1: BSM call prices.

12.4 Properties of the Black-Scholes Call Function

The *Black-Scholes(B-S) call function* is defined by

$$C = C(\tau, s, K, \sigma, r) = s\Phi(d_1) - Ke^{-r\tau}\Phi(d_2), \quad \tau, s, K, \sigma, r > 0,$$

where d_1 and d_2 are given by (12.11). For the sake of brevity, we shall occasionally suppress one or more arguments in the function C. By (12.12), $C(T, S_0, K, \sigma, r)$ is the price C_0 of a call option with strike price K, maturity

T, and underlying stock price S_0. More generally, $C(T-t, S_t, K, \sigma, r)$ is the value of the call at time t.

The analogous *B-S put function* is defined as

$$P = P(\tau, s, K, \sigma, r) = C(\tau, s, K, \sigma, r) - s + Ke^{-r\tau}, \quad \tau, s, K, \sigma, r > 0. \quad (12.14)$$

By the put-call parity relation, $P(T, S_0, K, \sigma, r)$ is the price of the corresponding put option and $P(T-t, S_t, K, \sigma, r)$ is its value at time t.

We state below two theorems that summarize the analytical properties of the B-S call function. The first gives various measures of sensitivity (in the form of rates of change) of an option price to market parameters. The second describes the limiting behaviors of the price with respect to these parameters. The proofs are given in Appendix C. Analogous properties of the B-S put function may be derived from these theorems using (12.14).

12.4.1 Theorem (Growth Rates of C).

(i) $\dfrac{\partial C}{\partial s} = \Phi(d_1)$

(ii) $\dfrac{\partial^2 C}{\partial s^2} = \dfrac{1}{s\sigma\sqrt{\tau}}\varphi(d_1)$

(iii) $\dfrac{\partial C}{\partial \tau} = \dfrac{\sigma s}{2\sqrt{\tau}}\varphi(d_1) + Kre^{-r\tau}\Phi(d_2)$

(iv) $\dfrac{\partial C}{\partial \sigma} = s\sqrt{\tau}\,\varphi(d_1)$

(v) $\dfrac{\partial C}{\partial r} = K\tau e^{-r\tau}\Phi(d_2)$

(vi) $\dfrac{\partial C}{\partial K} = -e^{-r\tau}\Phi(d_2)$

Remarks. (a) The quantities

$$\frac{\partial C}{\partial s}, \quad \frac{\partial^2 C}{\partial s^2}, \quad -\frac{\partial C}{\partial \tau}, \quad \frac{\partial C}{\partial \sigma}, \quad \text{and} \quad \frac{\partial C}{\partial r}$$

are called, respectively, *delta*, *gamma*, *theta*, *vega*, and *rho*, and are known collectively as the *Greeks*. A detailed analysis with concrete examples may be found in [7].

(b) Theorem 12.4.1 shows that C is increasing in each of the variables s, τ, σ, and r, and decreasing in K. These analytical facts have simple financial interpretations. For example, an increase in S_0 and/or decrease in K will likely increase the payoff $(S_T - K)^+$ and therefore should have a higher call price. An increase in either T or r decreases the discounted strike price Ke^{-rT}, reducing the initial cash needed to cover the strike price at maturity, making the option more attractive.

(c) Since $v(t, s) = C(T-t, s, K, \sigma, r)$, property (i) implies that

$$v_s(t, s) = \Phi\left(d_1(T-t, s, K, \sigma, r)\right).$$

Recalling that $v_s(t, S_t) = \theta_t$ represents the stock holdings in the replicating portfolio at time t, we see that the product

$$S_t\Phi\big(d_1(T-t, S_t, K, \sigma, r)\big)$$

in the Black-Scholes formula gives the time-t value of the stock holdings, and the difference

$$v(t, S_t) - S_t\Phi\big(d_1(T-t, S_t, K, \sigma, r)\big) = -Ke^{-r(T-t)}\Phi\big(d_2(T-t, S_t, K, \sigma, r)\big)$$

is the time-t value of the bond holdings. In other words, the portfolio should always be long $\Phi(d_1)$ shares of the stock and short the cash amount $Ke^{-r(T-t)}\Phi(d_2)$. □

12.4.2 Theorem (Limiting Behavior of C).

(i) $\lim_{s\to\infty}(C - s) = -Ke^{-r\tau}$

(ii) $\lim_{s\to 0+} C = 0$

(iii) $\lim_{\tau\to\infty} C = s$

(iv) $\lim_{\tau\to 0+} C = (s-K)^+$

(v) $\lim_{K\to\infty} C = 0$

(vi) $\lim_{K\to 0+} C = s$

(vii) $\lim_{\sigma\to\infty} C = s$

(viii) $\lim_{\sigma\to 0+} C = (s - e^{-r\tau}K)^+$

(ix) $\lim_{r\to\infty} C = s$

Remarks. As with Theorem 12.4.1, the analytical assertions of Theorem 12.4.2 have simple financial interpretations. For example, part (i) implies that for large S_0, $C_0 \approx S_0 - Ke^{-rT}$, which is the initial value of a portfolio with payoff $S_T - K$. This is to be expected, as a larger S_0 makes it more likely that the option will be exercised, resulting in precisely this payoff.

Part (iii) asserts that for large T the cost of the option is roughly the same as the initial value of the stock. This can be understood by noting that if T is large the discounted strike price Ke^{-rT} is negligible. If the option finishes in the money, a portfolio consisting of cash in the (small) amount Ke^{-rT} and a long call will have the same maturity value as a portfolio that is long in the stock. The law of one price then dictates that the two portfolios have the same start-up cost, which is roughly that of the call. A similar explanation may be given for (ix).

Part (iv) implies that for $S_0 > K$ and small T the price of the option is the difference between the initial value of the stock and the strike price. This is to be expected, as the holder would likely receive an immediate payoff of $S_0 - K$.

Part (vi) confirms the following argument: For a small strike price (in comparison to S_0) the option will almost certainly finish in the money. Therefore, a portfolio long in the option will have about the same payoff as one long in the stock. By the law of one price, the portfolios should have the same initial cost.

Part (viii) asserts that if σ is small and $S_0 > e^{-rT}K$, then the option price is roughly the cost of a bond with face value $S_0e^{rT} - K$. As the stock has little volatility, this is also the expected payoff of the option. Therefore, the option and the bond should have the same price. □

*12.5 The BS Formula as a Limit of CRR Formulas

In this section we show that under suitable conditions the Black-Scholes formula for the price of a call option,

$$C_0 = S_0\Phi(\delta_1) - Ke^{-rT}\Phi(\delta_2), \quad \delta_{1,2} = \frac{\ln(S_0/K) + (r \pm \sigma^2/2)T}{\sigma\sqrt{T}},$$

is the limit of a sequence of CRR formulas based on n-step binomial models. For this we assume that all random variables in the following discussion are defined on a single probability space $(\Omega, \mathcal{F}, \mathbb{P})$. (This may be technically justified.) In addition for each $n = 1, 2, \ldots$ we make the following assumptions for the nth binomial model:

- Trading occurs only at the discrete times

$$t_{n,j} = jh_n,\ j = 0, 1, \ldots, n, \text{ where } h_n = T/n.$$

- The risk-free rate per period is $i_n = e^{rh_n} - 1$, so that the bond process is given by $(1 + i_n)^j$, $j = 0, 1, \ldots, n$.

- The up and down factors u_n and d_n are defined by

$$u_n = e^{\sigma\sqrt{h_n}} \text{ and } d_n = e^{-\sigma\sqrt{h_n}} = 1/u_n.$$

  Note that $d_n < 1 + i_n$, and for sufficiently large n, $1 + i_n < u_n$ since $h_n \to 0$.

- The risk-neutral probability is

$$p_n^* = \frac{1 + i_n - d_n}{u_n - d_n}.$$

- The nth binomial price process $(S_{n,j})_{j=0}^n$ satisfies $S_{n,0} = S_0$ for all n.

- The CRR call price for the n-binomial model is

$$C_{0,n} = S_0\Psi\big(m_n, n, \widehat{p}_n\big) - (1 + i_n)^{-n}K\Psi\big(m_n, n, p_n^*\big),$$

  where

$$m_n = \left\lfloor \frac{\ln K - \ln(d_n^n S_0)}{\ln u_n - \ln d_n} \right\rfloor + 1 \text{ and } \widehat{p}_n = \frac{p_n^* u_n}{1 + i_n}$$

With these assumptions we prove

12.5.1 Theorem. $\lim_{n\to+\infty} C_{n,0} = C_0$.

For the proof we need the following lemma.

12.5.2 Lemma. *For each n and $0 < p < 1$ define*

$$f_n(p) = \frac{n(1-p)}{\sqrt{np(1-p)}} \quad \textit{and} \quad g_n(p) = \frac{m_n - np}{\sqrt{np(1-p)}}.$$

Then

(a) $\lim_{n\to+\infty} \widehat{p}_n = \lim_{n\to+\infty} p_n^* = \dfrac{1}{2}$.

(b) $\lim_{n\to+\infty} f_n(\widehat{p}_n) = \lim_{n\to+\infty} f_n(p_n^*) = +\infty$.

(c) $\lim_{n\to+\infty} n(1-2p_n^*)\sqrt{h_n} = \left(\dfrac{\sigma}{2} - \dfrac{r}{\sigma}\right)T$.

(d) $\lim_{n\to+\infty} n(1-2\widehat{p}_n)\sqrt{h_n} = -\left(\dfrac{\sigma}{2} + \dfrac{r}{\sigma}\right)T$.

(e) $\lim_{n\to+\infty} g_n(\widehat{p}_n) = \delta_1$.

(f) $\lim_{n\to+\infty} g_n(p_n^*) = -\delta_2$.

Proof. For some of the limits it is convenient to replace the discrete variable $\sqrt{h_n}$ by x and consider the corresponding limits as $x \to 0$. Thus, since

$$p_n^* = \frac{e^{rh_n} - e^{-\sigma\sqrt{h_n}}}{e^{\sigma\sqrt{h_n}} - e^{-\sigma\sqrt{h_n}}} = \frac{e^{rh_n+\sigma\sqrt{h_n}} - 1}{e^{2\sigma\sqrt{h_n}} - 1},$$

we have, by L'Hôpital's rule,

$$\lim_{n\to+\infty} p_n^* = \lim_{x\to 0} \frac{e^{\sigma x + rx^2} - 1}{e^{2\sigma x} - 1} = \frac{1}{2}.$$

Since $u_n \to 1$ and $1 + i_n \to 1$, it follows that also $\widehat{p}_n \to 1/2$. This verifies (a), which in turn implies (b).

For (c) we have

$$n(1-2p_n^*)\sqrt{h_n} = \frac{T}{\sqrt{h_n}}\left(1 - 2\,\frac{e^{rh_n+\sigma\sqrt{h_n}} - 1}{e^{2\sigma\sqrt{h_n}} - 1}\right),$$

hence, by a double application L'Hôpital's rule,

$$\lim_{n\to+\infty} n(1-2p_n^*)\sqrt{h_n} = T\lim_{x\to 0} \frac{e^{2\sigma x} - 2e^{\sigma x + rx^2} + 1}{x(e^{2\sigma x} - 1)} = T\left(\frac{\sigma}{2} - \frac{r}{\sigma}\right).$$

A similar calculation yields (d).

For (e), set

$$\alpha_n := \frac{\ln K - \ln(d_n^n S_0)}{\ln u_n - \ln d_n} = \frac{\ln(K/S_0) - n\ln d_n}{\ln u_n - \ln d_n} = \frac{\ln(S_0/K) - n\sigma\sqrt{h_n}}{2\sigma\sqrt{h_n}}$$

and note that

$$\frac{\alpha_n - np}{\sqrt{np(1-p)}} \le g_n(p) \le \frac{\alpha_n - np}{\sqrt{np(1-p)}} + \frac{1}{\sqrt{np(1-p)}}.$$

Thus it suffices to show that

$$\lim_{n\to+\infty} \frac{\alpha_n - n\widehat{p}_n}{\sqrt{n\widehat{p}_n(1-\widehat{p}_n)}} = \delta_1.$$

This follows from

$$\begin{aligned}\frac{\alpha_n - n\widehat{p}_n}{\sqrt{n\widehat{p}_n(1-\widehat{p}_n)}} &= \frac{\dfrac{\ln(S_0/K) - n\sigma\sqrt{h_n}}{2\sigma\sqrt{h_n}} - n\widehat{p}_n}{\sqrt{n\widehat{p}_n(1-\widehat{p}_n)}} \\ &= \frac{\ln(S_0/K) - n\sigma\sqrt{h_n}(1-2\widehat{p}_n)}{2\sigma\sqrt{n\widehat{p}_n(1-\widehat{p}_n)h_n}} \\ &= \frac{\ln(S_0/K) - n\sigma\sqrt{h_n}(1-2\widehat{p}_n)}{2\sigma\sqrt{T\widehat{p}_n(1-\widehat{p}_n)}} \\ &\to \frac{\ln(S_0/K) + \sigma T(r/\sigma + \sigma/2)}{\sigma\sqrt{T}} = \delta_1,\end{aligned}$$

where in calculating the limit we have used (a). An entirely analogous argument proves (f). □

An examination of the formulas for C_0 and $C_{0,n}$ shows that Theorem 12.5.1 will follow if we show that

$$\text{(i) } \Phi(\delta_1) = \lim_{n\to+\infty} \Psi\big(m_n, \widehat{p}_n\big) \text{ and (ii) } e^{-rT}\Phi(\delta_2) = \lim_{n\to+\infty} (1+i_n)^{-n}\Psi\big(m_n, n, p_n^*\big).$$

To prove (i), for each n let $X_{n,1}, X_{n,2}, \ldots, X_{n,n}$, be independent Bernoulli random variables with parameter $\widehat{p}_n$ and set

$$Y_n = \sum_{j=1}^{n} X_{n,j}.$$

Then $Y_n \sim B(n, \widehat{p}_n)$ so by the version of the DeMoivre-Laplace theorem described in Remark 6.7.2 and by (b) and (d) of the lemma,

$$\lim_{n\to\infty} \mathbb{P}\left(g_n(\widehat{p}_n) \le \frac{Y_n - n\widehat{p}_n}{\sqrt{n\widehat{p}_n(1-\widehat{p}_n)}} \le f_n(\widehat{p}_n)\right) = \Phi(+\infty) - \Phi(\delta_1) = \Phi(-\delta_1).$$

But

$$\begin{aligned}\mathbb{P}\left(g_n(\widehat{p}_n) \le \frac{Y_n - n\widehat{p}_n}{\sqrt{n\widehat{p}_n(1-\widehat{p}_n)}} \le f_n(\widehat{p}_n)\right) &= \mathbb{P}\left(m_n \le Y_n \le n\right)) \\ &= \Psi\big(m_n, n, \widehat{p}_n\big).\end{aligned}$$

This verifies (i). Part (ii) may be proved in a similar manner, using the analogous assertions of the lemma for p_n^*.

12.6 Exercises

In these exercises, all options are assumed to have underlying $\mathcal{S}$ and maturity T. The price process of $\mathcal{S}$ is given by (12.3).

1. Use Theorem 12.3.2 with $f(s) = (s - K)^+$ and $(K - s)^+$ to obtain the put-call parity relation.

2. Show that the function $v(t, s) = \alpha s + \beta e^{rt}$ satisfies the BSM PDE (12.9), where α and β are constants. What portfolio does the function represent?

3. Show that the BSM put function P is decreasing in s. Calculate
$$\lim_{s\to\infty} P \quad \text{and} \quad \lim_{s\to 0+} P.$$

4. Show that
$$P_t(s) = Ke^{-r(T-t)}\Phi\big(-d_2(T-t, s)\big) - s\Phi\big(-d_1(T-t, s)\big).$$

5. A *cash-or-nothing call option* pays a constant amount A if $S_T > K$ and pays nothing otherwise. Use Theorem 12.3.2 to show that the value of the option at time t is
$$V_t = Ae^{-r(T-t)}\Phi\big(d_1(T-t, S_t, K, \sigma, r)\big).$$

6. An *asset-or-nothing call option* pays the amount S_T if $S_T > K$ and pays nothing otherwise. Use Theorem 12.3.2 to show that the value of the option at time t is
$$V_t = S_t\Phi\big(d_1(T-t, S_t, K, \sigma, r)\big).$$
Use this result together with that of Exercise 5 to show that in the BSM model a portfolio long in an asset-or-nothing call and short in a cash-or-nothing call with cash K has the same time-t value as a standard call option.

7. Use Exercise 6 to find the cost V_0 of a claim with payoff $V_T = S_T\mathbf{1}_{(K_1,K_2)}(S_T)$, where $0 < K_1 < K_2$.

8. A *collar option* has payoff
$$V_T = \min\big(\max(S_T, K_1), K_2\big), \quad \text{where} \quad 0 < K_1 < K_2.$$
Show that the value of the option at time t is
$$V_t = K_1e^{-r(T-t)} + C(T-t, S_t, K_1) - C(T-t, S_t, K_2).$$

9. A *break forward* is a derivative with payoff
$$V_T = \max(S_T, F) - K = (S_T - F)^+ + F - K,$$
where $F = S_0e^{rT}$ is the forward value of the stock and K is initially set at a value that makes the cost of the contract zero. Determine the value V_t of the derivative at time t and find K.

10. Find dS^k in terms of dW and dt.

11. Find the probability that a call option with underlying S finishes in the money.

12. Let p and q be constants with $p > 0$ and let x_1 and x_2 be extended real numbers. Verify that

$$\int_{x_1}^{x_2} e^{-px^2+qx}\,dx = e^{q^2/4p}\sqrt{\frac{\pi}{p}}\left[\Phi\left(\frac{q-2px_1}{\sqrt{2p}}\right) - \Phi\left(\frac{q-2px_2}{\sqrt{2p}}\right)\right],$$

where $\Phi(\infty) := 1$ and $\Phi(-\infty) := 0$.

13. The *elasticity of the call price* $C_0 = C(T, s, K, \sigma, r)$ *with respect to the stock price* s is defined as

$$E_C = \frac{s}{C_0}\frac{\partial C_0}{\partial s},$$

which is the percent increase in C_0 due to a 1% increase in s. Show that

$$E_C = \frac{s\Phi(d_1)}{s\Phi(d_1) - Ke^{-rT}\Phi(d_2)}, \qquad d_{1,2} := d_{1,2}(T, s, K, \sigma, r).$$

Conclude that $E_C > 1$, implying that the option is more sensitive to change than the underlying stock. Show also that

(a) $\lim_{s\to+\infty} E_C = 1$. (b) $\lim_{s\to 0+} E_C = +\infty$.

Interpret financially.

14. The *elasticity of the put price* $P_0 = P(T, s, K, \sigma, r)$ *with respect to the stock price* s is defined as

$$E_P = -\frac{s}{P_0}\frac{\partial P_0}{\partial s},$$

which is the percent decrease in P_0 due to a 1% increase in s. Show that

$$E_P = \frac{s\Phi(-d_1)}{Ke^{-rT}\Phi(-d_2) - s\Phi(-d_1)}$$

and that

(a) $\lim_{s\to+\infty} E_P = +\infty$. (b) $\lim_{s\to 0+} E_P = 0$.

Interpret financially and compare with Exercise 13.

15. Referring to Theorem 12.3.1 show that

$$G(t,s) = \frac{1}{\sigma\sqrt{2\pi(T-t)}}\int_0^\infty f(z)\exp\left(-d_2^2(T-t)/2\right)\frac{dz}{z}.$$

Chapter 13

Martingales in the Black-Scholes-Merton Model

In Chapter 9 we described option valuation in the binomial model in terms of discrete-time martingales. In this chapter, we carry out a similar program for the Black-Scholes-Merton model using continuous-time martingales. The first two sections develop the tools needed to implement this program. The central result here is Girsanov's Theorem, which guarantees the existence of risk-neutral probability measures. Throughout the chapter, $(\Omega, \mathcal{F}, \mathbb{P})$ denotes a fixed probability space with expectation operator $\mathbb{E}$, and W is a Brownian motion on $(\Omega, \mathcal{F}, \mathbb{P})$.

13.1 Continuous-Time Martingales

Definition and Examples

A continuous-time stochastic process $(M_t)_{t\geq 0}$ on $(\Omega, \mathcal{F}, \mathbb{P})$ adapted to a filtration $(\mathcal{F}_t)_{t\geq 0}$ is said to be a *martingale* (with respect to the given probability measure and filtration) if

$$\mathbb{E}(M_t \mid \mathcal{F}_s) = M_s, \quad 0 \leq s \leq t. \tag{13.1}$$

A martingale with respect to the Brownian filtration and with continuous paths is called a *Brownian martingale.* Note that, by the factor property, (13.1) is equivalent to

$$\mathbb{E}(M_t - M_s \mid \mathcal{F}_s) = 0, \quad 0 \leq s \leq t.$$

The following processes are examples of Brownian martingales.

13.1.1 Examples. (a) $\left(W_t\right)_{t\geq 0}$: The independent increment property of Brownian motion implies that $W_t - W_s$ is independent of $\mathcal{F}_s^W$ for all $s \leq t$. Therefore, by Theorem 8.3.6 and the mean zero property of Brownian motion,

$$\mathbb{E}(W_t - W_s \mid \mathcal{F}_s^W) = \mathbb{E}(W_t - W_s) = 0.$$

(b) $\left(W_t^2 - t\right)_{t\geq 0}$: For $0 \leq s \leq t$,

$$W_t^2 = [(W_t - W_s) + W_s]^2 = (W_t - W_s)^2 + 2W_s(W_t - W_s) + W_s^2.$$

Taking conditional expectations and using linearity and the factor and independence properties yields

$$\begin{aligned}\mathbb{E}(W_t^2 \mid \mathcal{F}_s^W) &= \mathbb{E}(W_t - W_s)^2 + 2W_s\mathbb{E}(W_t - W_s) + W_s^2 \\ &= t - s + W_s^2,\end{aligned}$$

hence $\mathbb{E}(W_t^2 - t \mid \mathcal{F}_s^W) = W_s^2 - s$.

(c) $\left(e^{W_t - t/2}\right)_{t\geq 0}$: For $0 \leq s \leq t$, the factor and independence properties imply that

$$\begin{aligned}\mathbb{E}(e^{W_t} \mid \mathcal{F}_s^W) &= e^{W_s}\mathbb{E}(e^{W_t - W_s} \mid \mathcal{F}_s^W) = e^{W_s}\mathbb{E}(e^{W_t - W_s}) \\ &= e^{W_s + (t-s)/2},\end{aligned}$$

the last equality by Exercise 6.14 and the normality of $W_t - W_s$. Therefore,

$$\mathbb{E}\left(e^{W_t - t/2} \mid \mathcal{F}_s^W\right) = e^{W_s - s/2}. \qquad \diamond$$

These examples are special cases of the following theorem.

13.1.2 Theorem. *Every Ito process of the form*

$$X_t = X_0 + \int_0^t F(s)\, dW(s)$$

is a Brownian martingale.

Proof. Let $\mathcal{P} = \{s = t_0 < t_1 < \cdots < t_n = t\}$ be a partition of $[s, t]$ and set

$$I(F) = \int_s^t F(u)\, dW(u) \quad \text{and} \quad I_{\mathcal{P}}(F) = \sum_{j=0}^{n-1} F(t_j)\Delta W_j.$$

Then

$$X_t - X_s = I(F) = \lim_{\|\mathcal{P}\|\to 0} I_{\mathcal{P}}(F),$$

where convergence is in the mean square sense:

$$\lim_{\|\mathcal{P}\|\to 0} \mathbb{E}\big(I_{\mathcal{P}}(F) - I(F)\big)^2 = 0.$$

Now let $A \in \mathcal{F}_s^W$. Since

$$\big|\mathbb{E}\left(\mathbf{1}_A I_{\mathcal{P}}(F)\right) - \mathbb{E}\left(\mathbf{1}_A I(F)\right)\big| \leq \mathbb{E}\big|\mathbf{1}_A(I_{\mathcal{P}}(F) - I(F))\big| \leq \mathbb{E}\big|I_{\mathcal{P}}(F) - I(F)\big|$$

and

$$\mathbb{E}^2|I_{\mathcal{P}}(F) - I(F)| \leq \mathbb{E}|I_{\mathcal{P}}(F) - I(F)|^2$$

(Exercise 6.17), we see that $\big|\mathbb{E}\left(\mathbf{1}_A I_{\mathcal{P}}(F)\right) - \mathbb{E}\left(\mathbf{1}_A I(F)\right)\big| \to 0$ as $\|\mathcal{P}\| \to 0$, that is,

$$\mathbb{E}\left[\mathbf{1}_A I(F)\right] = \lim_{\|\mathcal{P}\|\to 0} \mathbb{E}\left[\mathbf{1}_A I_{\mathcal{P}}(F)\right]. \tag{13.2}$$

Furthermore, since $s \le t_j$ for all j,

$$\begin{aligned}
\mathbb{E}\left[I_{\mathcal{P}}(F) \mid \mathcal{F}_s^W\right] &= \sum_{j=0}^{n-1} \mathbb{E}\left[F(t_j)\Delta W_j \mid \mathcal{F}_s^W\right] && \text{(linearity)} \\
&= \sum_{j=0}^{n-1} \mathbb{E}\left[\mathbb{E}\left(F(t_j)\Delta W_j \mid \mathcal{F}_{t_j}^W\right) \mid \mathcal{F}_s^W\right] && \text{(iterated conditioning)} \\
&= \sum_{j=0}^{n-1} \mathbb{E}\left[F(t_j)\mathbb{E}\left(\Delta W_j \mid \mathcal{F}_{t_j}^W\right) \mid \mathcal{F}_s^W\right] && \text{(independence)} \\
&= \sum_{j=0}^{n-1} \mathbb{E}\left[F(t_j)\mathbb{E}(\Delta W_j) \mid \mathcal{F}_s^W\right] && \text{(independence)} \\
&= 0 && \text{(since } \mathbb{E}(\Delta W_j) = 0\text{).}
\end{aligned}$$

Therefore, $\mathbb{E}\left[\mathbf{1}_A \mathbb{E}(I_{\mathcal{P}}(F) \mid \mathcal{F}_s^W)\right] = 0$. But by the conditional expectation property

$$\mathbb{E}\left[\mathbf{1}_A I_{\mathcal{P}}(F)\right] = \mathbb{E}\left[\mathbf{1}_A \mathbb{E}(I_{\mathcal{P}}(F) \mid \mathcal{F}_s^W)\right].$$

Therefore, $\mathbb{E}\left[\mathbf{1}_A I_{\mathcal{P}}(F)\right] = 0$, which, by (13.2), implies that $\mathbb{E}\left[\mathbf{1}_A I(F)\right] = 0$. Since $A \in \mathcal{F}_s^W$ was arbitrary, $\mathbb{E}(X_t - X_s \mid \mathcal{F}_s^W) = 0$, which is the desired martingale property.

The proof that X_t has continuous paths is more delicate. For this, the reader is referred to [11]. □

Theorem 13.1.2 asserts that Ito processes X with differential $dX = F\,dW$ are Brownian martingales. This is *not* generally true for Ito processes given by $dX = F\,dW + G\,dt$, as the reader may readily verify.

13.1.3 Example. Let Y_t be the Ito process

$$Y_t = Y_0 + \int_0^t F(s)\,dW(s) - \frac{1}{2}\int_0^t [F(s)]^2\,ds.$$

By (11.14), the process $X_t = e^{Y_t}$ satisfies the SDE

$$dX = FX\,dW.$$

Since there is no dt term, Theorem 13.1.2 implies that (X_t) is a Brownian martingale. In particular, taking F to be a constant α, we see that the process

$$\exp\left(\alpha W_t - \tfrac{1}{2}\alpha^2 t\right), \quad t \ge 0,$$

is a Brownian martingale. Example 13.1.1(c) is the special case $\alpha = 1$. ◊

Martingale Representation Theorem

A martingale of the form

$$X_t = X_0 + \int_0^t F(s)\, dW(s), \quad X_0 \text{ constant},$$

is *square integrable*, that is, $\mathbb{E}\, X_t^2 < \infty$ for all t. This may be seen by applying Theorem 11.3.3 to the integral terms in the equation

$$X_t^2 = X_0^2 + 2X_0 \int_0^t F(s)\, dW(s) + \left(\int_0^t F(s)\, dW(s)\right)^2.$$

It is a remarkable fact that all square integrable Brownian martingales are Ito processes of the above form. We state this result formally in the following theorem. For a proof, see, for example, [20].

13.1.4 Theorem (Martingale Representation Theorem). *If $(M_t)_{t\geq 0}$ is a square-integrable Brownian martingale, then there exists a square-integrable process (ψ_t) adapted to $(\mathcal{F}_t^W)$ such that*

$$M_t = M_0 + \int_0^t \psi(s)\, dW(s), \quad t \geq 0.$$

13.1.5 Example. Let H be an $\mathcal{F}_T^W$-random variable with $\mathbb{E}\, H^2 < \infty$. Define

$$M_t = \mathbb{E}(H \mid \mathcal{F}_t^W), \quad 0 \leq t \leq T.$$

The iterated conditioning property shows that (M_t) is a martingale. To see that (M_t) is square-integrable, note first that $H^2 \geq M_t^2 + 2M_t(H - M_t)$, as may be seen by expanding the nonnegative quantity $(H - M_t)^2$. For each positive integer n define $A_n = \{M_t \leq n\}$ and note that $A_n \in \mathcal{F}_t^W$. Since $\mathbf{1}_{A_n} M_t^2$ is bounded it has finite expectation, hence we may condition on the inequality

$$\mathbf{1}_{A_n} H^2 \geq \mathbf{1}_{A_n} M_t^2 + 2\mathbf{1}_{A_n} M_t(H - M_t)$$

to obtain

$$\mathbf{1}_{A_n}\mathbb{E}(H^2 \mid \mathcal{F}_t^W) \geq \mathbf{1}_{A_n} M_t^2 + 2\mathbf{1}_{A_n} M_t \mathbb{E}(H - M_t \mid \mathcal{F}_t^W) = \mathbf{1}_{A_n} M_t^2.$$

Letting $n \to \infty$ and noting that for each ω the sequence $\mathbf{1}_{A_n}(\omega)$ eventually equals 1, we see that

$$\mathbb{E}(H^2 \mid \mathcal{F}_t^W) \geq M_t^2.$$

Taking expectations yields $\mathbb{E}(H^2) \geq \mathbb{E}(M_t^2)$, hence (M_t) is square-integrable.

It may be shown that each M_t can be modified on a set of probability zero so that the resulting process has continuous paths (see [20]). Thus (M_t) has the representation described in Theorem 13.1.4. ◊

13.2 Change of Probability and Girsanov's Theorem

In Remark 9.5.5 we observed that if Ω is finite then, given a nonnegative random variable Z with $\mathbb{E}(Z) = 1$, the equation

$$\mathbb{P}^*(A) = \mathbb{E}(\mathbf{1}_A Z), \quad A \in \mathcal{F}, \tag{13.3}$$

defines a probability measure $\mathbb{P}^*$ on $(\Omega, \mathcal{F})$ such that $\mathbb{P}^*(\omega) > 0$ iff $\mathbb{P}(\omega) > 0$. Conversely, any probability measure $\mathbb{P}^*$ with this positivity property satisfies (13.3) for suitable Z. The positivity property is a special case of *equivalence of probability measures*, that is, measures having precisely the same sets of probability zero. These ideas carry over to the general setting as follows:

13.2.1 Theorem (Change of Probability). *Let Z be a positive random variable on $(\Omega, \mathcal{F})$ with $\mathbb{E}\, Z = 1$. Then* (13.3) *defines a probability measure $\mathbb{P}^*$ on $(\Omega, \mathcal{F})$ equivalent to $\mathbb{P}$. Moreover, all probability measures $\mathbb{P}^*$ equivalent to $\mathbb{P}$ arise in this manner and satisfy*

$$\mathbb{E}^*(X) = \mathbb{E}(XZ) \tag{13.4}$$

for all $\mathcal{F}$-random variables X for which $\mathbb{E}(XZ)$ is defined, where $\mathbb{E}^*$ is the expectation operator corresponding to $\mathbb{P}^*$.

The proof of Theorem 13.2.1 may be found in advanced texts on probability theory. The random variable Z is called the *Radon-Nikodym derivative* of $\mathbb{P}^*$ with respect to $\mathbb{P}$ and is denoted by $\frac{d\mathbb{P}^*}{d\mathbb{P}}$. The connection between $\mathbb{P}$ and $\mathbb{P}^*$ described in (13.3) and (13.4) is frequently expressed as $d\mathbb{P}^* = Zd\mathbb{P}$. Replacing X in (13.4) by XZ^{-1}, we obtain the companion formulas

$$\mathbb{E}(X) = \mathbb{E}^*(XZ^{-1}) \quad \text{and} \quad \mathbb{P}(A) = \mathbb{E}^*(\mathbf{1}_A Z^{-1}),$$

that is, $d\mathbb{P} = Z^{-1} d\mathbb{P}^*$.

For an illuminating example, consider the translation $Y := X + \alpha$ of a standard normal random variable X by a real number α. The random variable Y is normal so one might ask if there is a probability measure $\mathbb{P}^*$ equivalent to $\mathbb{P}$ under which Y is *standard* normal. To answer the question, suppose first that such a probability measure exists and set $Z = \frac{d\mathbb{P}^*}{d\mathbb{P}}$. Since Z should somehow depend on X, we assume that $Z = g(X)$ for some function $g(x)$ to be determined. By (13.3),

$$\mathbb{P}^*(Y \le y) = \mathbb{E}\left[\mathbf{1}_{\{Y \le y\}} Z\right] = \mathbb{E}\left[\mathbf{1}_{(-\infty, y]}(X + \alpha) g(X)\right],$$

and, since $X \sim N(0,1)$ under $\mathbb{P}$, the law of the unconscious statistician implies that

$$\mathbb{P}^*(Y \le y) = \int_{-\infty}^{\infty} \mathbf{1}_{(\infty, y]}(x + \alpha) g(x) \varphi(x)\, dx = \int_{-\infty}^{y-\alpha} g(x) \varphi(x)\, dx.$$

If Y is to be standard normal with respect to $\mathbb{P}^*$, we must therefore have

$$\int_{-\infty}^{y-\alpha} g(x)\varphi(x)\,dx = \Phi(y).$$

Differentiating yields $g(y-\alpha)\varphi(y-\alpha) = \varphi(y)$, hence

$$g(y) = \frac{\varphi(y+\alpha)}{\varphi(y)} = e^{-\alpha y - \alpha^2/2}.$$

Thus we are led to the probability measure $\mathbb{P}^*$ defined by

$$d\mathbb{P}^* = g(X)\,d\mathbb{P} = \exp\left(-\alpha X - \tfrac{1}{2}\alpha^2\right) d\mathbb{P}.$$

One easily checks that $X + \alpha$ is indeed standard normal under $\mathbb{P}^*$.

Girsanov's Theorem generalizes this result from a single random variable to an entire process. The proof uses moment generating functions discussed in §6.5.

13.2.2 Theorem (Girsanov's Theorem). *Let $(W_t)_{0\le t\le T}$ be a Brownian motion on the probability space $(\Omega, \mathcal{F}, \mathbb{P})$ and let α be a constant. Define*

$$Z_t = \exp\left(-\alpha W_t - \tfrac{1}{2}\alpha^2 t\right), \quad 0 \le t \le T.$$

Then the process W^* defined by

$$W_t^* := W_t + \alpha t, \quad 0 \le t \le T,$$

is a Brownian motion on the probability space $(\Omega, \mathcal{F}, \mathbb{P}^*)$, where $d\mathbb{P}^* = Z_T d\mathbb{P}$.

Proof. Note first that $(Z_t)_{0\le t\le T}$ is a martingale with respect to the law $\mathbb{P}$ (Example 13.1.3). In particular, $\mathbb{E}\, Z_T = \mathbb{E}\, Z_0 = 1$, hence $\mathbb{P}^*$ is well-defined. It is clear that W^* starts at 0 and has continuous paths, so it remains to show that, under $\mathbb{P}^*$, W^* has independent increments and $W_t^* - W_s^*$ is normally distributed with mean 0 and variance $t-s$, $0 \le s < t$.

Let $0 \le t_0 < t_1 < \cdots < t_n \le T$ and define random vectors

$$\boldsymbol{X} = (X_1, \ldots, X_n) \text{ and } \boldsymbol{X}^* = (X_1^*, \ldots, X_n^*),$$

where

$$X_j := W_{t_j} - W_{t_{j-1}} \text{ and } X_j^* := W_{t_j}^* - W_{t_{j-1}}^*, \quad j = 1, 2, \ldots, n.$$

The core of the proof is determining the mgf of X^* with respect to $\mathbb{P}^*$. To this end, let $\boldsymbol{\lambda} = (\lambda_1, \ldots, \lambda_n)$. By the factor and independence properties,

$$\begin{aligned}
\mathbb{E}^* e^{\boldsymbol{\lambda}\cdot\boldsymbol{X}} &= \mathbb{E}\left(e^{\boldsymbol{\lambda}\cdot\boldsymbol{X}} Z_T\right) = \mathbb{E}\left[e^{\boldsymbol{\lambda}\cdot\boldsymbol{X}}\mathbb{E}\left(Z_T \mid \mathcal{F}_{t_n}\right)\right] = \mathbb{E}\left(e^{\boldsymbol{\lambda}\cdot\boldsymbol{X}} Z_{t_n}\right)\\
&= \mathbb{E}\exp\left(\boldsymbol{\lambda}\cdot\boldsymbol{X} - \alpha W_{t_n} - \tfrac{1}{2}\alpha^2 t_n\right)\\
&= \mathbb{E}\exp\left(\sum_{j=1}^n (\lambda_j - \alpha)X_j - \alpha W_{t_0} - \tfrac{1}{2}\alpha^2 t_n\right)\\
&= \mathbb{E}\exp\left(-\alpha W_{t_0} - \tfrac{1}{2}\alpha^2 t_n\right)\prod_{j=1}^n \mathbb{E}\exp\left((\lambda_j - \alpha)X_j\right).
\end{aligned}$$

The factors in the last expression are mgfs of normal random variables hence, by Example 6.5.1,

$$\mathbb{E}^* e^{\boldsymbol{\lambda}\cdot\boldsymbol{X}} = \exp\left(\frac{1}{2}\alpha^2(t_0 - t_n) + \frac{1}{2}\sum_{j=1}^{n}(\lambda_j - \alpha)^2\Delta t_j\right), \quad \Delta t_j := (t_j - t_{j-1}).$$

Since $\boldsymbol{\lambda}\cdot\boldsymbol{X}^* = \boldsymbol{\lambda}\cdot\boldsymbol{X} + \alpha\sum_{j=1}^{n}\lambda_j\Delta t_j)$, we see from the above equation that $\mathbb{E}^* e^{\boldsymbol{\lambda}\cdot\boldsymbol{X}^*} = e^{\frac{1}{2}A}$, where

$$A = \alpha^2(t_0 - t_n) + \sum_{j=1}^{n}(\lambda_j - \alpha)^2\Delta t_j + 2\alpha\sum_{j=1}^{n}\lambda_j\Delta t_j = \sum_{j=1}^{n}\lambda_j^2\Delta t_j.$$

Thus

$$\mathbb{E}^* e^{\boldsymbol{\lambda}\cdot\boldsymbol{X}^*} = \exp\left(\frac{1}{2}\sum_{j=1}^{n}\lambda_j^2\Delta t_j\right).$$

In particular,

$$\mathbb{E}^* e^{\lambda_j X_j^*} = \exp\left(\Delta t_j\lambda_j^2/2\right),$$

so $W_{t_j}^* - W_{t_{j-1}}^*$ is normally distributed with mean zero and variance $t_j - t_{j-1}$ (Example 6.5.1 and Theorem 6.5.2). Since

$$\mathbb{E}^* e^{\boldsymbol{\lambda}\cdot\boldsymbol{X}^*} = \prod_{j=1}^{n}\mathbb{E}^* e^{\lambda_j X_j^*},$$

the increments $W_{t_j}^* - W_{t_{j-1}}^*$ are independent (Corollary 6.5.5). Therefore, W^* is a Brownian motion under $\mathbb{P}^*$. □

13.2.3 Remark. The general version of Girsanov's Theorem allows more than one Brownian motion. It asserts that for *independent* Brownian motions W_j on $(\Omega, \mathcal{F}, \mathbb{P})$ and constants α_j $(j = 1, \ldots, d)$ there exists a *single* probability measure $\mathbb{P}^*$ relative to which the processes $W_j^*(t) := W_j(t) + \alpha_j t$ $(0 \le t \le T)$, are Brownian motions with respect to filtration generated by the processes W_j. The α_j's may even be stochastic processes provided they satisfy the *Novikov condition*

$$\mathbb{E}\left[\exp\left(\frac{1}{2}\int_0^T \beta(s)\,ds\right)\right] < \infty, \quad \beta := \sum_{j=1}^{d}\alpha_j^2.$$

In this case, $W_j^*(t)$ is defined as

$$W_j(t) + \int_0^t \alpha_j(s)\,ds.$$

(See, for example, [19].) ◊

13.3 Risk-Neutral Valuation of a Derivative

In this section we show how continuous-time martingales provide an alternative method of determining the fair price of a derivative in the Black-Scholes-Merton model. As in Chapter 12, our market consists of a risk-free bond $\mathcal{B}$ with price process governed by the ODE

$$dB = rB\,dt, \quad B_0 = 1$$

and a stock $\mathcal{S}$ with price process S following the SDE

$$dS = \sigma S\,dW + \mu S\,dt,$$

where μ and σ are constants. As shown in Chapter 11, the solution to the SDE is

$$S_t = S_0 \exp\left(\sigma W_t + (\mu - \tfrac{1}{2}\sigma^2)t\right), \quad 0 \le t \le T.$$

All martingales in this and the remaining sections of the chapter are relative to the filtration $(\mathcal{F}_t^S)_{t=0}^T$. Note that, because S_t and W_t may each be expressed in terms of the other by a continuous function, $\mathcal{F}_t^S = \mathcal{F}_t^W$ for all t. The probability measure $\mathbb{P}^*$ on $(\Omega, \mathcal{F})$ defined by

$$d\mathbb{P}^* = Z_T d\mathbb{P}, \quad Z_T := e^{-\alpha W_T - \frac{1}{2}\alpha^2 T}, \quad \alpha := \frac{\mu - r}{\sigma},$$

is called the *risk-neutral probability measure* for the price process S. The corresponding expectation operator is denoted by $\mathbb{E}^*$.

The following theorem and its corollary are the main results of the chapter. Martingale proofs are given in the next section. Note that the conclusion of the corollary is in agreement with Theorem 12.3.2, obtained by PDE methods.

13.3.1 Theorem. *Let H be a European claim, that is, an $\mathcal{F}_T$-random variable, with $\mathbb{E}\,H^2 < \infty$. Then there exists a unique self-financing, replicating strategy for H with value process V such that*

$$V_t = e^{-r(T-t)}\mathbb{E}^*\left(H \mid \mathcal{F}_t\right), \quad 0 \le t \le T. \tag{13.5}$$

13.3.2 Corollary. *If H is a European claim of the form $H = f(S_T)$, where f is continuous and $\mathbb{E}\,H^2 < \infty$, then*

$$V_t = e^{-r(T-t)}\mathbb{E}^*\left(f(S_T) \mid \mathcal{F}_t\right) = e^{-r(T-t)}G(t, S_t), \quad 0 \le t \le T, \tag{13.6}$$

where

$$G(t,s) := \int_{-\infty}^{\infty} f\left(s \exp\left\{\sigma\sqrt{T-t}\,y + (r - \tfrac{1}{2}\sigma^2)(T-t)\right\}\right)\varphi(y)\,dy. \tag{13.7}$$

As noted in Chapter 12, under the assumption that the market is arbitrage free, V_t must be the time-t value of the claim.

13.3.3 Example. By Corollary 13.3.2, the time-t value of a forward contract with forward price K is

$$F_t = e^{-r(T-t)}\mathbb{E}^*\left(S_T - K \mid \mathcal{F}_t\right). \tag{13.8}$$

Because there is no cost in entering a forward contract, $F_0 = 0$, and therefore

$$K = \mathbb{E}^* S_T = e^{rT}\mathbb{E}^* \widetilde{S}_T = e^{rT} S_0. \tag{13.9}$$

Here we have used the fact that $\widetilde{S}$ is a martingale with respect to $\mathbb{P}^*$ (Lemma 13.4.2, below). (Recall that Equation (13.9) was obtained in Section 4.4 using a general arbitrage argument.) By the martingale property again,

$$\mathbb{E}^*(S_T \mid \mathcal{F}_t) = e^{rT}\mathbb{E}^*(\widetilde{S}_T \mid \mathcal{F}_t) = e^{rT}\widetilde{S}_t = e^{r(T-t)}S_t. \tag{13.10}$$

Substituting (13.9) and (13.10) into (13.8), we see that

$$F_t = e^{-r(T-t)}\left(e^{r(T-t)}S_t - e^{rT}S_0\right) = S_t - e^{rt}S_0,$$

which is in agreement with Equation (4.4) of Section 4.4. ◊

13.3.4 Example. The time-t value of a call option with strike price K and maturity T is

$$e^{-r(T-t)}\mathbb{E}^*\left[(S_T - K)^+ \mid \mathcal{F}_t\right] = e^{-r(T-t)}G(t, S_t), \quad 0 \le t \le T,$$

where $G(t, s)$ is given by (13.7) with $f(x) = (x - K)^+$. Evaluating (13.7) for this f produces the Black-Scholes formula, exactly as in Corollary 12.3.3. ◊

13.4 Proofs of the Valuation Formulas

Some Preparatory Lemmas

13.4.1 Lemma. *Under the risk-neutral probability $\mathbb{P}^*$, the process*

$$W_t^* := W_t + \alpha t, \quad \alpha := \frac{\mu - r}{\sigma}, \qquad 0 \le t \le T,$$

is a Brownian motion with Brownian filtration $(\mathcal{F}_t^S)$.

Proof. Since W_t and W_t^* differ by a constant, $\mathcal{F}_t^{W^*} = \mathcal{F}_t^W = \mathcal{F}_t^S$. The conclusion now follows from Girsanov's Theorem. □

13.4.2 Lemma. *The price process S is given in terms of W^* as*

$$S_t = S_0 \exp\left(\sigma W_t^* - \tfrac{1}{2}\sigma^2 t + rt\right), \quad 0 \le t \le T,$$

and the discounted price process $\widetilde{S}$ is given by

$$\widetilde{S}_t := e^{-rt} S_t = S_0 \exp\left(\sigma W_t^* - \tfrac{1}{2}\sigma^2 t\right), \quad 0 \le t \le T.$$

Moreover, $\widetilde{S}$ is a martingale under $\mathbb{P}^*$ satisfying $d\widetilde{S} = \sigma \widetilde{S} dW^*$.

Proof. One readily verifies that the first assertion of the lemma holds, from which it is seen that $\widetilde{S}$ has the desired form. That $\widetilde{S}$ is a martingale follows from Example 13.1.3. The last assertion of the lemma is clear. □

13.4.3 Lemma. *Let (ϕ, θ) be a self-financing portfolio adapted to $(\mathcal{F}_t^S)$ with value process*

$$V_t = \phi_t B_t + \theta_t S_t, \quad 0 \le t \le T.$$

Then the discounted value process $\widetilde{V}$, given by $\widetilde{V}_t := e^{-rt} V_t$, is a martingale with respect to $\mathbb{P}^*$.

Proof. By Ito's product rule and the self-financing condition $dV = \phi\, dB + \theta\, dS$,

$$\begin{aligned} d\widetilde{V}_t &= -re^{-rt} V_t\, dt + e^{-rt} dV_t \\ &= -re^{-rt}\left[\phi_t B_t + \theta_t S_t\right] dt + e^{-rt}\left[r\phi_t B_t\, dt + \theta_t\, dS_t\right] \\ &= -re^{-rt}\theta_t S_t\, dt + e^{-rt}\theta_t\, dS_t \\ &= \theta_t\, d\widetilde{S}_t \\ &= \sigma\theta_t \widetilde{S}_t\, dW_t^*, \end{aligned}$$

the last equality from the Ito-Doeblin formula and Lemma 13.4.2. It follows from Theorem 13.1.2 that $\widetilde{V}$ is a martingale under $\mathbb{P}^*$. □

13.4.4 Lemma. *Let $\mathcal{G}$ be a σ-field contained in $\mathcal{F}$, let X be a $\mathcal{G}$-random variable and Y an $\mathcal{F}$-random variable independent of $\mathcal{G}$. If $g(x, y)$ is a continuous function with $\mathbb{E}^* |g(X, Y)| < \infty$, then*

$$\mathbb{E}^*\left[g(X, Y) \mid \mathcal{G}\right] = G(X), \tag{13.11}$$

where $G(x) := \mathbb{E}^* g(x, Y)$.

Proof. We give an outline of the proof. Let $\mathcal{R}$ denote the smallest σ-field of subsets of $\mathbf{R}^2$ containing all rectangles $R = J \times K$, where J and K are intervals. If $g = \mathbf{1}_R$, then

$$\begin{aligned} \mathbb{E}^*\left[g(X, Y) \mid \mathcal{G}\right] = \mathbb{E}^*\left[\mathbf{1}_J(X)\mathbf{1}_K(Y) \mid \mathcal{G}\right] &= \mathbf{1}_J(X)\mathbb{E}^*\left[\mathbf{1}_K(Y) \mid \mathcal{G}\right] \\ &= \mathbf{1}_J(X)\mathbb{E}^*\, \mathbf{1}_K(Y), \end{aligned}$$

where we have used the $\mathcal{G}$-measurability of $\mathbf{1}_J(X)$ and the independence of

$\mathbf{1}_K(Y)$ and $\mathcal{G}$. Since $G(x) = \mathbb{E}^* [\mathbf{1}_J(x)\mathbf{1}_K(Y)] = \mathbf{1}_J(x)\mathbb{E}^* \mathbf{1}_K(Y)$, we see that (13.11) holds for indicator functions of rectangles. From this it may be shown that (13.11) holds for the indicator functions of *all* members of $\mathcal{R}$ and hence for linear combinations of these indicator functions. Because a continuous function is the limit of a sequence of such linear combinations, (13.11) holds for any function g satisfying the conditions of the lemma. □

13.4.5 Remark. For future reference, we note that Lemma 13.4.4 extends (with a similar proof) to more than one $\mathcal{G}$-measurable random variable X. For example, if X_1 and X_2 are $\mathcal{G}$-measurable and if $g(x_1, x_2, y)$ is continuous with $\mathbb{E}^* |g(X_1, X_2, Y)| < \infty$, then

$$\mathbb{E}^* [g(X_1, X_2, Y) \mid \mathcal{G}] = G(X_1, X_1),$$

where $G(x_1, x_2) = \mathbb{E}^* g(x_1, x_2, Y)$. ◊

<u>Proof of Theorem 13.3.1</u>

Define a process V by Equation (13.5). Then $\widetilde{V}_T = e^{-rT}H$ and

$$\widetilde{V}_t = e^{-rt}V_t = \mathbb{E}^* \left(\widetilde{V}_T \mid \mathcal{F}_t\right), \quad 0 \le t \le T.$$

By Example 13.1.5, $\widetilde{V}$ is a square-integrable martingale with respect to $\mathbb{P}^*$ so, by the Martingale Representation Theorem (Theorem 13.1.4),

$$\widetilde{V}_t = \widetilde{V}_0 + \int_0^t \psi(s)\, dW^*(s), \quad 0 \le t \le T,$$

for some process ψ adapted to $(\mathcal{F}_t^S)$. Now set

$$\theta = \frac{\psi}{\sigma \widetilde{S}} \quad \text{and} \quad \phi = B^{-1}(V - \theta S).$$

Then (ϕ, θ) is adapted to $(\mathcal{F}_t)$ and $V = \theta S + \phi B$. Furthermore,

$$\begin{aligned}
dV_t &= e^{rt}\, d\widetilde{V}_t + rV_t\, dt \\
&= e^{rt}\psi_t\, dW_t^* + rV_t\, dt \\
&= \frac{e^{rt}\psi_t}{\sigma \widetilde{S}_t}\, d\widetilde{S}_t + rV_t\, dt \qquad \text{(by Lemma 13.4.2)} \\
&= e^{rt}\theta_t \left[-re^{-rt}S_t\, dt + e^{-rt}dS_t\right] + rV_t\, dt \\
&= \theta_t\, dS_t + r\left[V_t - \theta_t S_t\right] dt \\
&= \theta_t\, dS_t + \phi_t\, dB_t.
\end{aligned}$$

Therefore, (ϕ, θ) is a self-financing, replicating trading strategy for H with value process V.

To show uniqueness, suppose that (ϕ', θ') is a self-financing, replicating trading strategy for H based on $\mathcal{S}$ and $\mathcal{B}$. By Lemma 13.4.3, the value process V' of the strategy is a martingale, hence

$$\widetilde{V}_t' = \mathbb{E}^*(\widetilde{V}_T' \mid \mathcal{F}_t^S) = \mathbb{E}^*(e^{-rT}H \mid \mathcal{F}_t^S) = \widetilde{V}_t. \qquad \square$$

Proof of Corollary 13.3.2

By Lemma 13.4.2, we may write

$$S_T = S_t \exp\left(\sigma\sqrt{T-t}\,Y_t + (r - \tfrac{1}{2}\sigma^2)(T-t)\right), \tag{13.12}$$

where

$$Y_t := \frac{W_T^* - W_t^*}{\sqrt{T-t}}.$$

Now define

$$g(t,x,y) = f\left(x\exp\left\{\sigma\sqrt{T-t}\,y + (r - \tfrac{1}{2}\sigma^2)(T-t)\right\}\right).$$

Since $Y_t \sim N(0,1)$ under $\mathbb{P}^*$, the law of the unconscious statistician implies that

$$\begin{aligned}\mathbb{E}^*\, g(t,x,Y_t) &= \int_{-\infty}^{\infty} f\left(s\exp\left\{\sigma\sqrt{T-t}\,y + (r - \tfrac{1}{2}\sigma^2)(T-t)\right\}\right)\varphi(y)\,dy \\ &= G(t,x).\end{aligned}$$

Moreover, from (13.12), $g(t, S_t, Y_t) = f(S_T)$. Therefore, by Lemma 13.4.4,

$$\mathbb{E}^*[f(S_T) \mid \mathcal{F}_t^S] = \mathbb{E}^*[g(t,S_t,Y_t) \mid \mathcal{F}_t^S] = G(t,S_t). \qquad \Diamond$$

---

*13.5 Valuation under $\mathbb{P}$

The following theorem expresses the value process (V_t) in terms of the original probability measure $\mathbb{P}$. It is the continuous-time analog of Theorem 9.5.1.

13.5.1 Theorem. *The time-t value of a European claim H with $\mathbb{E}\,H^2 < \infty$ is given by*

$$\begin{aligned}V_t &= e^{-r(T-t)}\frac{\mathbb{E}(HZ_T \mid \mathcal{F}_t^S)}{\mathbb{E}(Z_T \mid \mathcal{F}_t^S)} \\ &= e^{-(r+\alpha^2/2)(T-t)}\mathbb{E}(e^{-\alpha(W_T - W_t)}H \mid \mathcal{F}_t^S),\end{aligned} \tag{13.13}$$

where

$$Z_T := e^{-\alpha W_T - \frac{1}{2}\alpha^2 T} \quad \text{and} \quad \alpha := \frac{\mu - r}{\sigma}.$$

Proof. Since $\mathbb{E}\,|HZ_T| = \mathbb{E}^*|H|$ is finite, $\mathbb{E}(HZ_T \mid \mathcal{F}_t^S)$ is defined. Since $V_t = e^{-r(T-t)}\mathbb{E}^*\left(H \mid \mathcal{F}_t^S\right)$, the first equality in (13.13) is equivalent to

$$\mathbb{E}(HZ_T \mid \mathcal{F}_t^S) = \mathbb{E}^*(H \mid \mathcal{F}_t^S)\mathbb{E}(Z_T \mid \mathcal{F}_t^S). \tag{13.14}$$

To verify (13.14), let $A \in \mathcal{F}_t^S$ and set $X_t = \mathbb{E}^*(\mathbf{1}_A H \mid \mathcal{F}_t^S)$. Then,

$$\begin{aligned}\mathbb{E}(\mathbf{1}_A H Z_T) &= \mathbb{E}^*(\mathbf{1}_A H) = \mathbb{E}^* X_t = \mathbb{E}\left[X_t Z_T\right] = \mathbb{E}\left[\mathbb{E}(X_t Z_T \mid \mathcal{F}_t^S)\right] \\ &= \mathbb{E}\left[\mathbf{1}_A \mathbb{E}^*(H \mid \mathcal{F}_t^S)\mathbb{E}(Z_T \mid \mathcal{F}_t^S)\right],\end{aligned}$$

establishing (13.14) and hence the first equality in (13.13).

For the second equality we have, by definition of Z_T,

$$\frac{\mathbb{E}(HZ_T \mid \mathcal{F}_t^S)}{\mathbb{E}(Z_T \mid \mathcal{F}_t^S)} = \frac{\mathbb{E}(e^{-\alpha W_T} H \mid \mathcal{F}_t^S)}{\mathbb{E}(e^{-\alpha W_T} \mid \mathcal{F}_t^S)}.$$

By the factor and independence properties, the denominator in the last expression may be written

$$\begin{aligned}\mathbb{E}(e^{-\alpha W_T} \mid \mathcal{F}_t^S) &= \mathbb{E}(e^{-\alpha(W_T - W_t)} e^{-\alpha W_t} \mid \mathcal{F}_t^S) = e^{-\alpha W_t}\mathbb{E}\left(e^{-\alpha(W_T - W_t)}\right) \\ &= e^{-\alpha W_t} e^{\alpha^2(T-t)/2},\end{aligned}$$

the last equality by Exercise 6.14. Therefore

$$\frac{\mathbb{E}(HZ_T \mid \mathcal{F}_t^S)}{\mathbb{E}(Z_T \mid \mathcal{F}_t^S)} = e^{-\alpha^2(T-t)/2}\mathbb{E}\left(e^{-\alpha(W_T - W_t)} H \mid \mathcal{F}_t^S\right). \qquad \square$$

---

*13.6 The Feynman-Kac Representation Theorem

We now have two ways of deriving the Black-Scholes pricing formula, one using PDE techniques and the other using martingale methods. The connection between the two methods is made in the Feynman-Kac Representation Theorem, which gives a probabilistic solution to a class of PDEs. The following version of the theorem is sufficient for our purposes. It expresses the conditional expectation of a solution of an SDE in terms of the solution to an associated PDE.

13.6.1 Theorem. *Let $\mu(t,x)$, $\sigma(t,x)$ and $f(x)$ be continuous functions. Suppose that, for $0 \le t \le T$, X_t is the solution of the SDE*

$$dX_t = \mu(t, X_t)\,dt + \sigma(t, X_t)\,dW_t \tag{13.15}$$

and $w(t,x)$ is the solution of the boundary value problem

$$w_t(t,x) + \mu(t,x)w_x(t,x) + \tfrac{1}{2}\sigma^2(t,x)w_{xx}(t,x) = 0, \quad w(T,x) = f(x). \tag{13.16}$$

If

$$\int_0^T \mathbb{E}\left[\sigma(t, X_t)w_x(t, X_t)\right]^2 dt < \infty, \tag{13.17}$$

then

$$w(t, X_t) = \mathbb{E}\left[f(X_T) \mid \mathcal{F}_t^W\right], \quad 0 \le t \le T.$$

Proof. Since $f(X_T) = w(T, X_T)$, the conclusion of the theorem may be written

$$w(t, X_t) = \mathbb{E}\left[w(T, X_T) \mid \mathcal{F}_t^W\right], \quad 0 \le t \le T.$$

Thus we must show that $(w(t, X_t))_{t=0}^T$ is a martingale. By Version 3 of the Ito-Doeblin formula,

$$\begin{aligned} dw(t, X) &= w_t(t, X)\,dt + w_x(t, X)\,dX + \tfrac{1}{2}w_{xx}(t, X)(dX)^2 \\ &= w_t(t, X)\,dt + w_x(t, X)\left[\mu(t, X)\,dt + \sigma(t, X)\,dW\right] \\ &\qquad + \tfrac{1}{2}\sigma^2(t, X)w_{xx}(t, X)\,dt \\ &= \left[w_t(t, X) + \mu(t, X)w_x(t, X) + \tfrac{1}{2}\sigma^2(t, X)w_{xx}(t, X)\right]\,dt \\ &\qquad + \sigma(t, X)w_x(t, X)\,dW \\ &= \sigma(t, X)w_x(t, X)\,dW, \end{aligned}$$

the last equality by (13.16). It follows from (13.17) and Theorem 11.3.3(vi) that $w(t, X_t)$ is an Ito process. An application of Theorem 13.1.2 completes the proof. □

13.6.2 Corollary. *Suppose that X_t satisfies* (13.15) *and that $v(t, x)$ is the solution of the boundary value problem*

$$v_t(t, x) + \mu(t, x)v_x(t, x) + \tfrac{1}{2}\sigma^2(t, x)v_{xx}(t, x) - rv(t, x) = 0, \quad v(T, x) = f(x),$$

where r is a constant and

$$\int_0^T \mathbb{E}\left[\sigma(t, X_t)v_x(t, X_t)\right]^2 dt < \infty.$$

Then

$$v(t, X_t) = e^{-r(T-t)}\mathbb{E}\left[f(X_T) \mid \mathcal{F}_t^W\right], \quad 0 \le t \le T.$$

Proof. One easily checks that

$$w(t, x) := e^{r(T-t)}v(t, x)$$

satisfies (13.16) and (13.17). □

Remark. In the derivation of the Black-Scholes formula in Chapter 12, financial considerations led to the PDE

$$v_t + rxv_x + \tfrac{1}{2}\sigma^2x^2v_{xx} - rv = 0, \quad 0 \le t < T.$$

The corollary therefore provides the desired connection between PDE and martingale methods of option valuation. ◊

13.7 Exercises

1. Show that, for $0 \le s \le t$, $\mathbb{E}(W_s W_t) = s$

2. Show that $\mathbb{E}(W_t \mid W_s) = W_s$ $(0 \le s \le t)$.

3. Show that, for $0 < s < t$, the joint density $f_{t,s}$ of (W_t, W_s) is given by

$$f_{t,s}(x,y) := \frac{1}{\sqrt{s(t-s)}}\varphi\left(\frac{x-y}{\sqrt{t-s}}\right)\varphi\left(\frac{y}{\sqrt{s}}\right).$$

4. Use (8.4) and Exercise 3 to find $\mathbb{E}(W_s \mid W_t)$ for $0 < s < t$.

5. Show that $M := \left(e^{\alpha W_t + h(t)}\right)_{t \ge 0}$ is a martingale iff $h(t) = -\alpha^2 t/2 + h(0)$.

6. Show that $\mathbb{E}\left(W_t^3 \mid \mathcal{F}_s^W\right) = W_s^3 + 3(t-s)W_s$, $0 \le s \le t$. Conclude that $\left(W_t^3 - 3tW_t\right)$ is a martingale.
   Hint. Expand $[(W_t - W_s) + W_s]^3$.

7. Find $\mathbb{E}(W_t^2 \mid W_s)$ and $\mathbb{E}(W_t^3 \mid W_s)$ for $0 < s < t$.

8. The *Hermite polynomials* $H_n(x,t)$ are defined by $H_n(x,t) = f_{x,t}^{(n)}(0)$, where

$$f_{x,t}(\lambda) := f(\lambda, x, t) = \exp\left(\lambda x - \tfrac{1}{2}\lambda^2 t\right).$$

   (a) Show that $\exp\left(\lambda x - \frac{1}{2}\lambda^2 t\right) = \sum_{n=0}^{\infty} H_n(x,t)\frac{\lambda^n}{n!}$.

   (b) Use (a) and the fact that $\left(\exp\left(\lambda W_t - \frac{1}{2}\lambda^2 t\right)\right)_{t \ge 0}$ is a Brownian martingale to show that the process $(H_n(W_t,t))_{t \ge 0}$ is a Brownian martingale for each n. (Example 13.1.1(b) and Exercise 6 are the special cases $H_2(W_t,t)$ and $H_3(W_t,t)$, respectively.)

   (c) Show that $f_{x,t}^{(n+1)}(\lambda) = (x - \lambda t) f_{x,t}^{(n)}(\lambda) - nt f_{x,t}^{(n-1)}(\lambda)$ and hence that

$$H_{n+1}(x,t) = xH_n(x,t) - ntH_{n-1}(x,t).$$

   (d) Use (c) to find explicit representations of the martingales $H_4(W_t,t)$ and $H_5(W_t,t)$.

   (e) Use the Ito-Doeblin formula to show that

$$H_n(W_t,t) = \int_0^t nH_{n-1}(W_s,s)\,dW_s.$$

   This gives another proof that $(H_n(W_t,t))_{t \ge 0}$ is a martingale.

9. Let $(W_t)_{0\le t\le T}$ be a Brownian motion on the probability space $(\Omega, \mathcal{F}, \mathbb{P})$ and let α be a constant. Suppose that X is a random variable independent of W_T. Show that X has the same cdf under $\mathbb{P}$ as under $\mathbb{P}^*$, where

$$d\mathbb{P}^* = \exp\left(-\alpha W_T - \tfrac{1}{2}\alpha^2 T\right) d\mathbb{P}.$$

10. Let W_1 and W_2 be independent Brownian motions and $0 < |\varrho| < 1$. Show that the process $W = \varrho W_1 + \sqrt{1-\varrho^2}\,W_2$ is a Brownian motion.

Chapter 14

Path-Independent Options

In Chapter 13 the price of a standard European call option in the BSM model was determined using the risk-neutral probability measure given by Girsanov's Theorem. There are a variety of other European options that may be similarly valued. In this chapter we consider the most common of these. We also give a brief analysis of American claims in the BSM model.

Throughout, $(\Omega, \mathcal{F}, \mathbb{P})$ denotes a fixed probability space with expectation operator $\mathbb{E}$, where $\mathbb{P}$ is the law governing the behavior of the price of an asset $\mathcal{S}$. Risk-neutral measures on Ω will be denoted, as usual, by the generic notation $\mathbb{P}^*$, with $\mathbb{E}^*$ the corresponding expectation operator. As before, $(\mathcal{F}_t^S)$ denotes the natural filtration for the price process S of the asset $\mathcal{S}$. We assume that the markets under consideration are arbitrage-free, so that the value of a claim is that of a self-financing, replicating portfolio.

14.1 Currency Options

In this section, we consider derivatives whose underlying asset is a euro bond. For this, let $D_t = e^{r_d t}$ and $E_t = e^{r_e t}$ denote the price processes of a US dollar bond and a euro bond, respectively, where r_d and r_e are the dollar and euro interest rates. Further, let Q denote the exchange rate process in dollars per euro (see Section 4.5). To model the volatility of the exchange rate, we take Q to be a geometric Brownian motion on $(\Omega, \mathcal{F}, \mathbb{P})$ given by

$$Q_t = Q_0 \exp\left[\sigma W_t + (\mu - \tfrac{1}{2}\sigma^2)t\right], \quad 0 \le t \le T, \tag{14.1}$$

where σ and μ are constants. Define

$$S_t = Q_t E_t = S_0 \exp[\sigma W_t + (\mu + r_e - \tfrac{1}{2}\sigma^2)t], \tag{14.2}$$

which is the dollar value of the euro bond at time t. Because of the volatility of the exchange rate, from the point of view of the domestic investor, the euro bond is a risky asset. The form of the price process S clearly reflects this view.

Given a European claim H with $\mathbb{E}\, H^2 < \infty$, we can apply the methods of Chapter 13 to construct a self-financing replicating trading strategy (ϕ, θ)

for H, where ϕ_t and θ_t are, respectively, the number of dollar bonds and euro bonds held at time t. Set

$$r := r_d - r_e, \quad \alpha := \frac{\mu - r}{\sigma}, \quad \text{and} \quad W_t^* := W_t + \alpha t. \tag{14.3}$$

By Girsanov's Theorem, W^* is a Brownian measure under $\mathbb{P}^*$, where

$$d\,\mathbb{P}^* := e^{-\alpha W_T - \frac{1}{2}\alpha^2 T} d\,\mathbb{P}.$$

Let V denote the value process of (ϕ, θ), and set

$$\widetilde{S} := D^{-1}S = D^{-1}QE \quad \text{and} \quad \widetilde{V} := D^{-1}V.$$

Since

$$S_t = S_0 \exp[\sigma W_t^* + (r_d - \sigma^2/2)t],$$

the process $\widetilde{S}$ and hence $\widetilde{V}$ are martingales with respect to $\mathbb{P}^*$. Risk-neutral pricing and the no-arbitrage assumption therefore imply that the time-t dollar value of H is given by

$$V_t = e^{-r_d(T-t)}\mathbb{E}^*\left(H \mid \mathcal{F}_t\right), \quad 0 \le t \le T.$$

Now suppose that $H = f(Q_T) = f(E_T^{-1}S_T)$, where $f(x)$ is continuous. Set $f_1(x) = f\left(e^{-r_e T}x\right) = f\left(E_T^{-1}x\right)$ so that $H = f_1(S_T)$. From (14.2), S is the price process of the stock in Chapter 13 with μ replaced by $\mu + r_e$. By Theorem 12.3.2 or Corollary 13.3.2, $V_t = e^{-r_d(T-t)}G_1(t, S_t)$, where

$$\begin{aligned} G_1(t,s) &:= \int_{-\infty}^{\infty} f_1\left(s \exp\left\{\sigma\sqrt{T-t}\,y + \left(r_d - \tfrac{1}{2}\sigma^2\right)(T-t)\right\}\right)\varphi(y)\,dy \\ &= \int_{-\infty}^{\infty} f\left(s \exp\left\{-r_e T + \sigma\sqrt{T-t}\,y + \left(r_d - \tfrac{1}{2}\sigma^2\right)(T-t)\right\}\right)\varphi(y)\,dy. \end{aligned}$$

Now define $G(t,s) = G_1\left(t, e^{r_e t}s\right)$, so that $G_1(t, S_t) = G(t, Q_t)$ and

$$G(t,s) = f\left(s \exp\left\{r_e t - r_e T + \sigma\sqrt{T-t}\,y + \left(r_d - \tfrac{1}{2}\sigma^2\right)(T-t)\right\}\right)\varphi(y)\,dy.$$

Thus we have the following result:

14.1.1 Theorem. *Let $H = f(Q_T)$, where f is continuous and $\mathbb{E}\,H^2 < \infty$. Then the value of H at time t is*

$$V_t = e^{-r_d(T-t)}G(t, Q_t),$$

where

$$G(t,s) := \int_{-\infty}^{\infty} f\left(s \exp\left\{\sigma\sqrt{T-t}\,y + (r_d - r_e - \tfrac{1}{2}\sigma^2)(T-t)\right\}\right)\varphi(y)\,dy.$$

14.1.2 Example. (Currency Forward). Consider a forward contract for the purchase in dollars of one euro at time T. Let K denote the forward price of the euro. At time T, the euro costs Q_T dollars, hence the dollar value of the forward at time t is

$$V_t = e^{-r_d(T-t)}\mathbb{E}^*\left(Q_T - K \mid \mathcal{F}_t\right). \tag{14.4}$$

Because there is no cost to enter a forward contract, $V_0 = 0$ and therefore $K = \mathbb{E}^* Q_T$. Since $Q = D\widetilde{S}E^{-1}$, we have

$$\begin{aligned}\mathbb{E}^*(Q_T \mid \mathcal{F}_t) &= D_T E_T^{-1}\mathbb{E}^*(\widetilde{S}_T \mid \mathcal{F}_t) = D_T E_T^{-1}\widetilde{S}_t \\ &= e^{r(T-t)}Q_t \end{aligned} \tag{14.5}$$

and in particular

$$K = \mathbb{E}^* Q_T = e^{rT}Q_0. \tag{14.6}$$

Substituting (14.5) and (14.6) into (14.4), we obtain

$$\begin{aligned}V_t &= e^{-r_d(T-t)}\left(e^{r(T-t)}Q_t - e^{rT}Q_0\right) \\ &= e^{-r_e T}\left(e^{r_e t}Q_t - e^{r_d t}Q_0\right). \end{aligned}$$

◊

14.1.3 Example. (Currency Call Option). Taking $f(x) = (x - K)^+$ in Theorem 14.1.1, we see that the time-t dollar value of an option to buy one euro for K dollars at time T is $C_t = e^{-r_d(T-t)}G(t, Q_t)$, where, as in the proof of Corollary 12.3.3 with r replaced by $r_d - r_e$,

$$G(t,s) = se^{(r_d-r_e)(T-t)}\Phi\left(d_1(s, T-t)\right) - K\Phi\left(d_2(s, T-t)\right),$$

$$d_{1,2}(s,\tau) := \frac{\ln(s/K) + (r_d - r_e \pm \sigma^2/2)\tau}{\sigma\sqrt{\tau}}.$$

Thus,

$$C_t = e^{-r_e(T-t)}Q_t\Phi\big(d_1(Q_t, T-t)\big) - e^{-r_d(T-t)}K\Phi\big(d_2(Q_t, T-t)\big).$$

This may also be expressed in terms of the forward price $K_t = e^{(r_d-r_e)(T-t)}Q_t$ of a euro (see Section 4.5). Indeed, since $Q_t = e^{(r_e-r_d)(T-t)}K_t$ we have

$$C_t = e^{-r_d(T-t)}K_t\Phi\big(\widehat{d}_1(K_t, T-t)\big) - e^{-r_d(T-t)}K\Phi\big(\widehat{d}_2(K_t, T-t)\big),$$

where

$$\widehat{d}_{1,2}(s,\tau) = \frac{\ln(s/K) \pm \tau\sigma^2/2}{\sigma\sqrt{\tau}}.$$

◊

14.2 Forward Start Options

A *forward start option* is a contract that gives the holder at time T_0, for no extra cost, an option with maturity $T > T_0$ and strike price $K = S_{T_0}$. Consider, for example, a *forward start call option*, whose underlying call has payoff $(S_T - S_{T_0})^+$. Let V_t denote the value of the option at time t. At time T_0 the strike price S_{T_0} is known, hence the forward start option has the value of the underlying call at that time. Thus, in the notation of Section 12.4,

$$V_{T_0} = C(T - T_0, S_{T_0}, S_{T_0}, \sigma, r) = C(T - T_0, 1, 1, \sigma, r)S_{T_0},$$

which is the value at time T_0 of a portfolio consisting of $C(T - T_0, 1, 1, \sigma, r)$ units of the underlying security. Since the values of the forward start option and the portfolio agree at time T_0, by the law of one price they must agree at all times $t \le T_0$. In particular, the initial cost of the forward start call option is

$$V_0 = C(T - T_0, 1, 1, \sigma, r)S_0 = \left[\Phi(d_1) - e^{r(T-T_0)}\Phi(d_2)\right] S_0,$$

where

$$d_{1,2} = d_{1,2}(T - T_0, 1, 1, \sigma, r) = \left(\frac{r}{\sigma} \pm \frac{\sigma}{2}\right)\sqrt{T - T_0}.$$

14.3 Chooser Options

A *chooser option* gives the holder the right to select at some future date T_0 whether the option is to be a call or a put with common exercise price K, maturity $T > T_0$, and underlying $\mathcal{S}$. Let V_t, C_t, and P_t denote, respectively, the time-t values of the chooser option, the call, and the put. In the notation of Section 12.4,

$$C_t = C(T - t, S_t, K, \sigma, r) \quad \text{and} \quad P_t = P(T - t, S_t, K, \sigma, r).$$

Since at time T_0 the holder will choose the option with the higher value,

$$\begin{aligned} V_{T_0} &= \max(C_{T_0}, P_{T_0}) \\ &= \max\left(C_{T_0}, C_{T_0} - S_{T_0} + Ke^{-r(T-T_0)}\right) \\ &= C_{T_0} + \left(Ke^{-r(T-T_0)} - S_{T_0}\right)^+, \end{aligned}$$

where we have used the put-call parity relation in the second equality. The last expression is the value at time T_0 of a portfolio consisting of a long call option with strike price K and maturity T and a long put option with strike price $K_1 := Ke^{-r(T-T_0)}$ and maturity T_0. Since the values of the chooser option and the portfolio are the same at time T_0, they must be the same for all times $t \le T_0$. Thus, using put-call parity again and noting that $K_1 e^{-r(T_0-t)} = Ke^{-r(T-t)}$, we have for $0 \le t \le T_0$

$$\begin{aligned}V_t &= C(T-t, S_t, K, \sigma, r) + P(T_0 - t, S_t, K_1, \sigma, r)\\ &= C(T-t, S_t, K, \sigma, r) + C(T_0 - t, S_t, K_1, \sigma, r) - S_t + Ke^{-r(T-t)}.\end{aligned}$$

In particular,

$$V_0 = C(T, S_0, K, \sigma, r) + C(T_0, S_0, K_1, \sigma, r) - S_0 + Ke^{-rT}. \tag{14.7}$$

Now recall the Black-Scholes valuation formulas

$$\begin{aligned}C(T, S_0, K, \sigma, r) &= S_0\Phi(d_1) - Ke^{-rT}\Phi(d_2) \quad \text{and}\\ C(T_0, S_0, K_1, \sigma, r) &= S_0\Phi(\widehat{d}_1) - K_1 e^{-rT_0}\Phi(\widehat{d}_2),\end{aligned}$$

where

$$\begin{aligned}d_{1,2} = d_{1,2}(T, S_0, K, \sigma, r) &= \frac{\ln(S_0/K) + (r \pm \sigma^2/2)T}{\sigma\sqrt{T}} \quad \text{and}\\ \widehat{d}_{1,2} = d_{1,2}(T_0, S_0, K_1, \sigma, r) &= \frac{\ln(S_0/K_1) + (r \pm \sigma^2/2)T_0}{\sigma\sqrt{T_0}}\\ &= \frac{\ln(S_0/K) + rT \pm \sigma^2 T_0/2}{\sigma\sqrt{T_0}}.\end{aligned}$$

Substituting into (14.7), we obtain the formula

$$\begin{aligned}V_0 &= S_0\Phi(d_1) - Ke^{-rT}\Phi(d_2) + S_0\Phi(\widehat{d}_1) - Ke^{-rT}\Phi(\widehat{d}_2) - S_0 + Ke^{-rT}\\ &= S_0\big[\Phi(d_1) + \Phi(\widehat{d}_1) - 1\big] - Ke^{-rT}\big[\Phi(d_2) + \Phi(\widehat{d}_2) - 1\big]\\ &= S_0\big[\Phi(d_1) - \Phi(-\widehat{d}_1)\big] - Ke^{-rT}\big[\Phi(d_2) - \Phi(-\widehat{d}_2)\big].\end{aligned}$$

The value of the option for $T_0 \le t \le T$ is either C_t or P_t, depending on whether the call or put was chosen at time T_0. To distinguish between the two scenarios let

$$A = \{C_{T_0} > P_{T_0}\}.$$

Since $\mathbf{1}_A = 1$ iff the call was chosen and $\mathbf{1}_{A'} = 1$ iff the put was chosen, we have

$$V_t = C_t\mathbf{1}_A + P_t\mathbf{1}_{A'}, \quad T_0 \le t \le T.$$

In particular, the payoff of the chooser option may be written

$$V_T = \big(S_T - K\big)^+\mathbf{1}_A + \big(K - S_T\big)^+\mathbf{1}_{A'}.$$

14.4 Compound Options

A *compound option* is a call or put option whose underlying is another call or put option, the latter with underlying $\mathcal{S}$. Consider the case of a *call-on-call option.* Suppose that the underlying call has strike price K and maturity T, and that the compound option has strike price K_0 and maturity $T_0 < T$. We seek the fair price V_0^{cc} of the compound option at time 0.

The value of the underlying call at time T_0 is $C(S_{T_0})$, where, by the Black-Scholes formula,

$$C(s) = C(T - T_0, s, K, \sigma, r) = s\Phi\big(d_1(s)\big) - Ke^{-r(T-T_0)}\Phi\big(d_2(s)\big), \quad (14.8)$$

$$d_{1,2}(s) = \frac{\ln(s/K) + (r \pm \sigma^2/2)(T - T_0)}{\sigma\sqrt{T - T_0}}.$$

The payoff of the compound option at T_0 is therefore $\big[C(S(T_0)) - K_0\big]^+$. By Corollary 13.3.2 with $f(s) = \big[C(s) - K_0\big]^+$, the cost of the compound option is

$$V_0^{cc} = e^{-rT_0}\int_{-\infty}^{\infty}\big[C(g(y)) - K_0\big]^+\varphi(y)\,dy, \quad (14.9)$$

where

$$g(y) = S_0\exp\Big\{\sigma\sqrt{T_0}\,y + \big(r - \tfrac{1}{2}\sigma^2\big)T_0\Big\}.$$

By (14.8),

$$\begin{aligned}C(g(y)) &= g(y)\Phi\big(d_1(g(y))\big) - Ke^{-r(T-T_0)}\Phi\big(d_2(g(y))\big)\\ &= g(y)\Phi\big(\widehat{d}_1(y)\big) - Ke^{-r(T-T_0)}\Phi\big(\widehat{d}_2(y)\big),\end{aligned}$$

where

$$\widehat{d}_1(y) := d_1\big(g(y)\big) = \frac{\ln(S_0/K) + \sigma\sqrt{T_0}\,y + rT + \sigma^2(T - 2T_0)/2}{\sigma\sqrt{T - T_0}} \quad \text{and}$$

$$\widehat{d}_2(y) := d_2\big(g(y)\big) = \frac{\ln(S_0/K) + \sigma\sqrt{T_0}\,y + (r - \sigma^2/2)T}{\sigma\sqrt{T - T_0}}.$$

Also, since $\big[C(g(y)) - K_0\big]^+$ is increasing in y,

$$\big[C(g(y)) - K_0\big]^+ = \big[C(g(y)) - K_0\big]\mathbf{1}_{(y_0,\infty)},$$

where

$$y_0 := \inf\{y \mid C(g(y)) > K_0\}.$$

Therefore,

$$\begin{aligned}V_0^{cc} = e^{-rT_0}\int_{y_0}^{\infty}\big[g(y)\Phi\big(\widehat{d}_1(y)\big) - Ke^{-r(T-T_0)}\Phi\big(\widehat{d}_2(y)\big)\big]\varphi(y)\,dy\\ - e^{-rT_0}K_0\Phi(-y_0).\end{aligned}$$

14.5 Quantos

A *quanto* is a derivative with underlying asset denominated in one currency and payoff denominated in another. Consider the case of a foreign stock with price process S^e denominated in euros. Let Q denote the exchange rate in dollars per euro, so that the stock has dollar price process $S := S^eQ$. A standard call option on the foreign stock has payoff

$$(S_T^e - K)^+ \text{ euros} = (S_T - KQ_T)^+ \text{ dollars},$$

but this value might be adversely affected by the exchange rate. Instead, a domestic investor could buy a quanto call option with payoff $(S_T^e - K)^+$ dollars. Here, the strike price K is in dollars and S_T^e is interpreted as a dollar value (e.g., $S_T^e = 5$ is interpreted as 5 dollars).

More generally, consider a claim with dollar value $H = f(S_T^e)$, where $f(x)$ is continuous and $\mathbb{E}\, H^2 < \infty$. The methods of Chapter 13 may be easily adapted to the current setting to find the fair price of such a claim. Our point of view is that of a domestic investor, that is, one who constantly computes investment transactions in dollars.

To begin, assume that S and Q are geometric Brownian motion processes, say

$$S_t = S_0 e^{\sigma_1 W_1(t) + (\mu_1 - \sigma_1^2/2)t} \quad \text{and} \quad Q_t = Q_0 e^{\sigma_2 W_2(t) + (\mu_2 - \sigma_2^2/2)t},$$

where σ_1, σ_2, μ_1, and μ_2 are constants and W_1 and W_2 are Brownian motions on $(\Omega, \mathcal{F}, \mathbb{P})$. For ease of exposition, we take W_1 and W_2 to be independent processes.[1] As in Section 14.1 we let $D_t = e^{r_d t}$ and $E_t = e^{r_e t}$ denote the price processes of a dollar bond and a euro bond, respectively, where r_d and r_e are the dollar and euro interest rates. Set

$$\alpha_1 := \frac{\mu_1 - r_d}{\sigma_1} \quad \text{and} \quad \alpha_2 := \frac{\mu_2 + r_e - r_d}{\sigma_2}.$$

By the general Girsanov's Theorem (Remark 13.2.3), there exists a single probability measure $\mathbb{P}^*$ on Ω (the so-called *domestic risk-neutral probability measure*) relative to which the processes

$$W_1^*(t) := W_1(t) + \alpha_1 t \quad \text{and} \quad W_2^*(t) := W_2(t) + \alpha_2 t, \quad 0 \le t \le T,$$

are independent Brownian motions with respect to the filtration $(\mathcal{G}_t)$ generated by W_1 and W_2. In terms of W_1^* and W_2^*, the process $S^e = SQ^{-1}$ may be

[1] A more realistic model assumes that the processes are *correlated*, that is, $W_2 = \varrho W_1 + \sqrt{1-\varrho^2}\, W_3$, where W_1 and W_3 are independent Brownian motions and $0 < |\varrho| < 1$. See, for example, [19].

written

$$\begin{aligned} S_t^e &= S_0^e \exp\left\{\sigma_1 W_1(t) + \left(\mu_1 - \frac{\sigma_1^2}{2}\right)t - \sigma_2 W_2(t) - \left(\mu_2 - \frac{\sigma_2^2}{2}\right)t\right\} \\ &= S_0^e \exp\left\{\sigma_1\left[W_1(t) + \alpha_1 t\right] - \sigma_2\left[W_2(t) + \alpha_2 t\right] + \left(r_e - \frac{\sigma_1^2}{2} + \frac{\sigma_2^2}{2}\right)t\right\} \\ &= S_0^e \exp\left\{\sigma_1 W_1^*(t) - \sigma_2 W_2^*(t) + \left(r_e - \frac{\sigma_1^2}{2} + \frac{\sigma_2^2}{2}\right)t\right\} \\ &= S_0^e \exp\left\{\sigma W^*(t) + \left(r_e + \sigma_2^2 - \frac{\sigma^2}{2}\right)t\right\}, \end{aligned} \tag{14.10}$$

where

$$\sigma := (\sigma_1^2 + \sigma_2^2)^{1/2} \quad \text{and} \quad W^* := \sigma^{-1}\left(\sigma_1 W_1^* - \sigma_2 W_2^*\right).$$

W^* is easily seen to be a Brownian motion with respect to $(\mathcal{G}_t)$. By the continuous parameter version of Theorem 9.1.4, W^* is also a Brownian motion with respect to the Brownian filtration $(\mathcal{F}_t^{W^*})$.

Now let H be a claim of the form $f(S_T^e)$ and let (ϕ, θ) be a self-financing, replicating trading strategy for H based on the dollar bond and a risky asset with dollar price process X given by

$$X_t := e^{\zeta t} S_t^e = S_0^e \exp\left\{\sigma W^*(t) + \left(r_d - \frac{\sigma^2}{2}\right)t\right\}, \qquad \zeta := r_d - r_e - \sigma_2^2.$$

Let $V = \phi D + \theta X$ denote the value process of the portfolio and set $f_1(x) = f\left(e^{-\zeta T}x\right)$, so that $H = f_1(X_T)$. Note that the processes $D^{-1}X$ and hence $D^{-1}V$ are martingales with respect to $\mathbb{P}^*$, which is the precise reason for the choice of ζ. By Corollary 13.3.2, the value of the claim at time t is

$$V_t = e^{-r_d(T-t)}\mathbb{E}^*(H \mid \mathcal{F}_t^{W^*}) = e^{-r_d(T-t)} G_1(t, X_t),$$

where

$$\begin{aligned} G_1(t,s) &:= \int_{-\infty}^{\infty} f_1\left(s \exp\left\{\sigma\sqrt{T-t}\,y + (r_d - \sigma^2/2)(T-t)\right\}\right)\varphi(y)\,dy \\ &= \int_{-\infty}^{\infty} f\left(s \exp\left\{-\zeta T + \sigma\sqrt{T-t}\,y + (r_d - \sigma^2/2)(T-t)\right\}\right)\varphi(y)\,dy \end{aligned}$$

Now define $G(t,s) := G_1\left(t, e^{\zeta t}s\right)$, so that $G_1(t, X_t) = G(t, S_t^e)$ and

$$G(t,s) = \int_{-\infty}^{\infty} f\left(s \exp\left\{\sigma\sqrt{T-t}\,y + (-\zeta + r_d - \sigma^2/2)(T-t)\right\}\right)\varphi(y)\,dy.$$

Recalling the definition of ζ we arrive at the following result:

14.5.1 Theorem. *Let $H = f(S_T^e)$, where f is continuous and $\mathbb{E}\,H^2 < \infty$. Then the dollar value of H at time t is*

$$V_t = e^{-r_d(T-t)} G(t, S_t^e),$$

where S_t^e is interpreted as a dollar value and

$$G(t,s) := \int_{-\infty}^{\infty} f\left(s\exp\left\{\sigma\sqrt{T-t}\,y + (r_e + \sigma_2^2 - \sigma^2/2)(T-t)\right\}\right)\varphi(y)\,dy.$$

By taking $f(x) = x - K$ one obtains a quanto forward, and an analysis similar to that in 13.3.3 may be carried out for the current setting. Here is another example:

14.5.2 Example. (Quanto Call Option). Taking $f(x) = (x-K)^+$ in Theorem 14.5.1 and using (12.13) with r replaced by $r_e + \sigma_2^2$, we obtain

$$G(t,s) = se^{(r_e+\sigma_2^2)(T-t)}\Phi\big(d_1(s,T-t)\big) - K\Phi\big(d_2(s,T-t)\big),$$

where

$$d_{1,2}(s,\tau) = \frac{\ln(s/K) + (r_e + \sigma_2^2 \pm \sigma^2/2)\tau}{\sigma\sqrt{\tau}}.$$

The dollar value of a quanto call option at time t is therefore

$$V_t = e^{-(r_d - r_e - \sigma_2^2)(T-t)}S_t^e\Phi\big(d_1(S_t^e, T-t)\big) - e^{-r_d(T-t)}K\Phi\big(d_2(S_t^e, T-t)\big).$$

14.6 Options on Dividend-Paying Stocks

In this section we determine the price of a claim based on a dividend-paying stock. We consider two cases: continuous dividends and periodic dividends. We begin with the former, which is somewhat easier to model.

Continuous Dividend Stream

Assume our stock pays a dividend of $\delta S_t\,dt$ in the time interval from t to $t+dt$, where δ is a constant between 0 and 1. Since dividends reduce the value of the stock, the price process must be adjusted to reflect this reduction. Therefore, we have

$$dS_t = \sigma S_t\,dW_t + \mu S_t\,dt - \delta S_t\,dt = \sigma S_t\,dW_t + (\mu - \delta)S_t\,dt. \tag{14.11}$$

Now let (ϕ,θ) be a self-financing trading strategy with value process $V = \phi B + \theta S$. The definition of self-financing must be modified to take into account the dividend stream, as the change in V depends not only on changes in the stock and bond values but also on the dividend increment. Thus we require that

$$dV_t = \phi_t\,dB_t + \theta_t\,dS_t + \delta\theta_t S_t\,dt.$$

From (14.11),

$$\begin{aligned} dV &= \phi\, dB + \theta\big[\sigma S\, dW + (\mu-\delta)S\, dt\big] + \delta\theta S\, dt \\ &= \phi\, dB + \theta S\big(\sigma\, dW + \mu\, dt\big). \end{aligned} \tag{14.12}$$

Now set

$$\widehat{\theta}_t = e^{-\delta t}\theta_t \quad \text{and} \quad \widehat{S}_t = e^{\delta t} S_t = S_0 \exp\left(\sigma W_t + (\mu - \tfrac{1}{2}\sigma^2)t\right).$$

Note that $\widehat{S}_t$ is the price process of the stock without dividends (or with dividends reinvested) and $V = \phi B + \widehat{\theta}\widehat{S}$ is the value process of a trading strategy $(\phi, \widehat{\theta})$ based on the bond and the non-dividend-paying version of our stock. Since $d\widehat{S} = \widehat{S}(\sigma\, dW + \mu\, dt)$, we see from (14.12) that

$$dV = \phi\, dB + \widehat{\theta}\, d\widehat{S},$$

that is, the trading strategy $(\phi, \widehat{\theta})$ is self-financing. The results of Chapter 13 therefore apply to $\widehat{S}$ and we see that the value of a claim H is the same as that for a stock without dividends, namely, $e^{-r(T-t)}\mathbb{E}^*\left(H \mid \mathcal{F}_t\right)$, where $\mathbb{P}^*$ is the risk-neutral measure. The difference here is that the discounted process $\widetilde{S}$ is no longer a martingale under $\mathbb{P}^*$, as seen from the representation of S as

$$S_t = e^{-\delta t}\widehat{S}_t = S_0 e^{\sigma W_t^* + (r-\delta-\sigma^2/2)t}. \tag{14.13}$$

Now suppose that $H = f(S_T)$, where $f(x)$ is continuous and $\mathbb{E}^* H^2 < \infty$. Set $f_1(x) = f\left(e^{-\delta T}x\right)$, so that $H = f_1(\widehat{S}_T)$. By Corollary 13.3.2, the value of the claim at time t is $V_t = e^{-r(T-t)}G_1(t, \widehat{S}_t)$, where

$$\begin{aligned} G_1(t,s) &:= \int_{-\infty}^{\infty} f_1\left(s\exp\left\{\sigma\sqrt{T-t}\,y + (r-\sigma^2/2)(T-t)\right\}\right)\varphi(y)\,dy \\ &= \int_{-\infty}^{\infty} f\left(s\exp\left\{-\delta T + \sigma\sqrt{T-t}\,y + (r-\sigma^2/2)(T-t)\right\}\right)\varphi(y)\,dy. \end{aligned}$$

Now define $G(t,s) = G_1\left(t, e^{\delta t}s\right)$, so that $G_1(t, \widehat{S}_t) = G(t, S_t)$ and

$$G(t,s) = \int_{-\infty}^{\infty} f\left(s\exp\left\{\sigma\sqrt{T-t}\,y + (-\delta + r - \sigma^2/2)(T-t)\right\}\right)\varphi(y)\,dy.$$

Thus we have the following result:

14.6.1 Theorem. *Let $H = f(S_T)$, where f is continuous and $\mathbb{E}\, H^2 < \infty$. Then the value of H at time t is $V_t = e^{-r(T-t)}G(t, S_t)$, where*

$$G(t,s) = \int_{-\infty}^{\infty} f\left(s\exp\left\{\sigma\sqrt{T-t}\,y + (r-\delta-\sigma^2/2)(T-t)\right\}\right)\varphi(y)\,dy.$$

14.6.2 Example. (Call option on a dividend-paying stock). Taking $f(x) = (x-K)^+$ in Theorem 14.6.1 and using (12.13) with r replaced by $r-\delta$, we have

$$G(t,s) = se^{(r-\delta)(T-t)}\Phi\left(d_1(s,T-t)\right) - K\Phi\left(d_2(s,T-t)\right),$$

where

$$d_{1,2}(s,\tau) = \frac{\ln(s/K) + (r-\delta \pm \sigma^2/2)\tau}{\sigma\sqrt{\tau}}.$$

The time-t value of a call option on a dividend-paying stock is therefore

$$C_t = e^{-\delta(T-t)}S_t\Phi\big(d_1(S_t,T-t)\big) - e^{-r(T-t)}K\Phi\big(d_2(S_t,T-t)\big).$$

Discrete Dividend Stream

Now suppose that our stock pays a dividend only at the discrete times t_j, where $0 < t_1 < t_2 < \cdots < t_n < T$. Set $t_0 = 0$ and $t_{n+1} = T$. Between dividends, the stock price is assumed to follow the SDE $dS_t = \sigma S_t\,dW_t + \mu S_t\,dt$. At each dividend-payment time, the stock value is reduced by the amount of the dividend, which we again assume is a fraction $\delta \in (0,1)$ of the value of the stock. The price process is no longer continuous but has jumps at the dividend-payment times t_j. This may be modeled by the equations

$$\begin{aligned}
S_t &= S_{t_j}e^{\sigma\left(W_t - W_{t_j}\right) + (\mu-\sigma^2/2)(t-t_j)}, \quad t_j \le t < t_{j+1}, \ \ j = 0,1,\ldots,n,\\
S_{t_{j+1}} &= S_{t_j}e^{\sigma\left(W(t_{j+1}) - W_{t_j}\right) + (\mu-\sigma^2/2)(t_{j+1}-t_j)}(1-\delta), \ \ j = 0,1,\ldots,n-1.
\end{aligned}$$

Setting $\widehat{S}_t = S_0 e^{\sigma W_t + (\mu-\sigma^2/2)t}$ we may rewrite these equations as

$$\begin{aligned}
S_t &= \frac{S_{t_j}\widehat{S}_t}{\widehat{S}_{t_j}}, \quad t_j \le t < t_{j+1}, \quad j = 0,1,\ldots,n,\\
S_{t_{j+1}} &= \frac{S_{t_j}\widehat{S}_{t_{j+1}}}{\widehat{S}_{t_j}}(1-\delta), \quad j = 0,1,\ldots,n-1. \qquad (14.14)
\end{aligned}$$

If $j = n$, the first formula also holds for $t = t_{n+1} (= T)$. Let (ϕ,θ) be a trading strategy based on the stock and the bond. Between dividends the value process is given by

$$V_t = \phi_t B_t + \theta_t S_t = \phi_t B_t + \theta_t \frac{S_{t_j}\widehat{S}_t}{\widehat{S}_{t_j}}, \quad t_j \le t < t_{j+1}.$$

At time t_{j+1} the stock portion of the portfolio decreases in value by the amount

$$\delta\theta_{t_j}\frac{S_{t_j}\widehat{S}_{t_{j+1}}}{\widehat{S}_{t_j}}$$

but the cash portion increases by the same amount because of the dividend. Since the net change is zero, V is a continuous process. Moreover, assuming that (ϕ, θ) is self-financing between dividends, we have for $t_j \le t < t_{j+1}$

$$dV_t = \phi_t \, dB_t + \theta_t \big[\sigma S_t \, dW_t + \mu S_t \, dt\big] = \big[r\phi_t B_t + \mu \theta_t S_t\big] \, dt + \sigma \theta_t S_t \, dW_t.$$

Now, from (14.14), for $1 \le m \le n$ and $t_m \le t < t_{m+1}$, or $m = n$ and $t = T$, we have

$$\frac{S_t}{S_0} = \frac{S_t}{S_{t_m}} \prod_{j=1}^{m} \frac{S_{t_j}}{S_{t_{j-1}}} = (1-\delta)^m \frac{\widehat{S}_t}{\widehat{S}_{t_m}} \prod_{j=1}^{m} \frac{\widehat{S}_{t_j}}{\widehat{S}_{t_{j-1}}} = (1-\delta)^m \frac{\widehat{S}_t}{S_0},$$

that is,

$$S_t = (1-\delta)^m \widehat{S}_t, \quad t_m \le t < t_{m+1}, \quad \text{and} \quad S_T = (1-\delta)^n \widehat{S}_T.$$

In particular, S_T is the value at time T of a non-dividend-paying stock with the geometric Brownian motion price process $(1-\delta)^n \widehat{S}$. Therefore, we can replicate a claim $H = f(S_T)$ with a self-financing strategy based on a stock with this price process and the original bond. By Corollary 13.3.2, the value of the claim at time t is

$$V_t = e^{-r(T-t)} G\left(t, (1-\delta)^{n-m} \widehat{S}_t\right), \quad t_m \le t < t_{m+1},$$

where $G(t, s$ is given by (13.7).

14.7 American Claims

Many of the results concerning valuation of American claims in the binomial model carry over to the BSM setting. The verifications are more difficult, however, and require advanced techniques from martingale theory. In this section, we give a brief overview of the main ideas, without proofs. For details the reader is referred to [8, 14, 15].

Assume the payoff of a (path-independent) American claim at time t is of the form $g(t, S_t)$. The holder will try to choose an exercise time that optimizes his payoff. The exercise time, being a function of the price of the underlying, is a random variable, the determination of which cannot rely on future information. As in the discrete case, such a random variable is called a *stopping time*, formally defined as a function τ on Ω with values in $[0, T]$ such that

$$\{\tau \le t\} \in \mathcal{F}_t, \quad 0 \le t \le T.$$

Since $\{\tau > t - 1/n\} = \{\tau \le t - 1/n\}' \in \mathcal{F}_t$ for any positive integer n it follows that

$$\{\tau = t\} = \bigcap_{n=1}^{\infty} \{t - 1/n < \tau \le t\} \in \mathcal{F}_t, \quad 0 \le t \le T.$$

Now let $\mathcal{T}_{t,T}$ denote the set of all stopping times with values in the interval $[t,T]$, $0 \le t \le T$. If at time t the holder of the claim chooses the stopping time $\tau \in \mathcal{T}_{t,T}$, her payoff will be $g(\tau, S_\tau)$. The writer will therefore need a portfolio that covers this payoff. By risk-neutral pricing, the time-t value of the payoff is $\mathbb{E}^*\left(e^{-r(\tau-t)}g(\tau,S_\tau) \mid \mathcal{F}_t\right)$. Thus, if the writer is to cover the claim for any choice of stopping time, then the value of the portfolio at time t should be[2]

$$V_t = \max_{\tau \in \mathcal{T}_{t,T}} \mathbb{E}^*\left(e^{-r(\tau-t)}g(\tau,S_\tau) \,\middle|\, \mathcal{F}_t\right). \tag{14.15}$$

One can show that a trading strategy (ϕ, θ) exists with value process given by (14.15). With this trading strategy, the writer may hedge the claim, so it is natural to take the value of the claim at time t to be V_t. In particular, the fair price of the claim is

$$V_0 = \max_{\tau \in \mathcal{T}_{0,T}} \mathbb{E}^*\left(e^{-r\tau}g(\tau,S_\tau)\right). \tag{14.16}$$

It may be shown that price V_0 given by (14.16) guarantees that neither the holder nor the writer of the claim has an arbitrage opportunity. More precisely:

- It is not possible for the holder to initiate a self-financing trading strategy (ϕ', θ') with initial value
$$V_0' := \phi_0' + \theta_0' S_0 + V_0 = 0$$
such that the terminal value of the portfolio obtained by exercising the claim at some time τ and investing the proceeds in risk-free bonds, namely,
$$V_T' := e^{r(T-\tau)}\left[\phi_\tau' e^{r\tau} + \theta_\tau' S_\tau + g(\tau, S_\tau)\right],$$
is nonnegative and has a positive probability of being positive.
- For any writer-initiated self-financing trading (ϕ', θ') with initial value
$$V_0' := \phi_0' + \theta_0' S_0 - V_0 = 0,$$
there exists a stopping time τ for which it is *not* the case that the terminal value of the portfolio resulting from the holder exercising the claim at time τ, namely,
$$V_T' := e^{r(T-\tau)}\left(\phi_\tau' e^{r\tau} + \theta_\tau' S_\tau - g(\tau, S_\tau)\right),$$
is nonnegative and has a positive probability of being positive.

Finally, it may be shown that after time t the optimal time for the holder to exercise the claim is

$$\tau_t = \inf\{u \in [t,T] \mid g(u, S_u) = V_u\}.$$

Explicit formulas are available in the special case of an American put, for which $g(t,s) = (K-s)^+$.

[2]Since the conditional expectations in this equation are defined only up to a set of probability one, the maximum must be interpreted as the *essential supremum*, defined in standard texts on real analysis.

14.8 Exercises

1. Let Q be the exchange rate process in dollars per euro, as given by (14.1). Show that

$$Q_t = Q_0 \exp\left[\sigma W_t^* + (r_d - r_e - \sigma^2/2)t\right],$$

where W^* is defined in (14.3). Use this to derive the SDEs

$$\text{(a)}\ \frac{dQ}{Q} = \sigma\, dW^* + (r_d - r_e)\, dt \quad \text{and} \quad \text{(b)}\ \frac{dQ^{-1}}{Q^{-1}} = \sigma\, dW^* + (r_e - r_d + \sigma^2)\, dt.$$

Remark. Since Q^{-1} is the exchange rate process in euros per dollar, the term σ^2 in (b) is at first surprising, as it suggests an asymmetric relationship between the currencies. This phenomenon, known as *Siegel's paradox*, may be explained by observing that the probability measure $\mathbb{P}^*$ is risk neutral when the dollar bond is the numeraire. This is the appropriate measure when pricing in the domestic currency and for that reason $\mathbb{P}^*$ is called the *domestic risk-neutral probability measure.* Both (a) and (b) are derived in this context. When calculating prices in a foreign currency, the *foreign risk-neutral probability measure* must be used. This is the probability measure under which $W_t^* - \sigma t$ is a Brownian motion, and is risk-neutral when the euro bond is taken as the numeraire. With respect to this measure, the Ito-Doeblin formula gives the expected form of dQ^{-1}.

2. Referring to the subsection on discrete dividends, show that if dividend payments are made at the equally spaced times $t_j = jT/(n+1)$, $j = 1, 2, \ldots, n$, then for $0 \le t < T$,

$$S_t = (1-\delta)^{\lfloor (n+1)t/T \rfloor} \hat{S}_t \quad \text{and} \quad V_t = e^{-r(T-t)} G\left((1-\delta)^{n - \lfloor (n+1)t/T \rfloor} S(t)\right),$$

where $\lfloor x \rfloor$ denotes the greatest integer in x.

3. Referring to the subsection on continuous dividends, show that $\left(e^{(\delta - r)t} S_t\right)$ is a $\mathbb{P}^*$-martingale.

4. Referring to Section 14.3, find the probability under the risk-neutral measure $\mathbb{P}^*$ that the call is chosen at time T_0.

5. A *shout option* is a European option that allows the holder to "shout" to the writer, at some time τ before maturity, her wish to lock in the current price S_τ of the security. For a call option, the holder's payoff at maturity, assuming that a shout is made, is

$$V_T := \max(S_\tau - K, S_T - K),$$

where K is the strike price. (If no shout is made, then the payoff is the usual amount $(S_T - K)^+$.) Show that

$$V_T = S_\tau - K + (S_T - S_\tau)^+$$

and use this to find the value of the shout option at time $t \ge \tau$.

6. A *cliquet* (or *ratchet*) *option* is a derivative with payoff $(S_{T_0} - K)^+$ at time T_0 and payoff $(S_{T_j} - S_{T_{j-1}})^+$ at time T_j, $j = 1, \ldots, n$, where $0 < T_0 < T_1 < \cdots < T_n$. Thus the strike price of the option is initially set at K, but at times T_j, $0 \le j \le n-1$, it is reset to S_{T_j}. Find the cost of the option.

Chapter 15

Path-Dependent Options

A path-dependent derivative is a contract with payoff that depends not just on the value of the underlying asset at maturity but on the entire history of the asset over the duration of the contract. Because of this dependency, the valuation of path-dependent derivatives is more complex than that of path-independent derivatives. In this chapter we consider the most common path-dependent derivatives: barrier options, lookback options, and Asian options, which were introduced in §4.11. The notation and assumptions of the preceding chapter are still in force.

15.1 Barrier Options

Recall that the payoff for a barrier option depends on whether the value of the asset has crossed a predetermined level called a barrier. In this section, we show that the price C_0^{do} of a down-and-out call option is given by the formula

$$C_0^{do} = S_0\left[\Phi(d_1) - \left(\frac{c}{S_0}\right)^{\frac{2r}{\sigma^2}+1}\Phi(\delta_1)\right] - Ke^{-rT}\left[\Phi(d_2) - \left(\frac{c}{S_0}\right)^{\frac{2r}{\sigma^2}-1}\Phi(\delta_2)\right] \quad (15.1)$$

where, with $M := \max(K, c)$,

$$d_{1,2} = \frac{\ln(S_0/M) + (r \pm \frac{1}{2}\sigma^2)T}{\sigma\sqrt{T}}, \quad \delta_{1,2} = \frac{\ln(c^2/S_0M) + (r \pm \frac{1}{2}\sigma^2)T}{\sigma\sqrt{T}}. \quad (15.2)$$

To establish (15.1), note first that the payoff for a down-and-out call is

$$C_T^{do} = (S_T - K)^+\mathbf{1}_A, \quad \text{where} \quad A = \{m^S \geq c\}.$$

Since the option is out of the money if $S_T < K$, the payoff may be written somewhat more simply as

$$C_T^{do} = (S_T - K)\mathbf{1}_B, \quad \text{where} \quad B = \{S_T \geq K, m^S \geq c\}.$$

By risk-neutral pricing, the cost of the option is therefore

$$C_0^{do} = e^{-rT}\mathbb{E}^*\big[(S_T - K)\mathbf{1}_B\big]. \quad (15.3)$$

By Lemma 13.4.2, the value of the underlying at time t may be expressed as

$$S_t = S_0 e^{\sigma(W_t^* + \beta t)}, \quad \beta := r/\sigma - \sigma/2, \quad 0 \le t \le T, \tag{15.4}$$

where W_t^* is a Brownian motion under $\mathbb{P}^*$. By Girsanov's Theorem,

$$\widehat{W}_t := W_t^* + \beta t$$

is a Brownian motion under the probability measure $\widehat{\mathbb{P}}$ given by

$$d\widehat{\mathbb{P}} = Z_T d\mathbb{P}^*, \quad \text{where} \quad Z_T := e^{-\beta W_T^* - \frac{1}{2}\beta^2 T} = e^{-\beta \widehat{W}_T + \frac{1}{2}\beta^2 T}.$$

Since $S_T = S_0 e^{\sigma \widehat{W}_T}$ and $m^S = S_0 e^{\sigma m^{\widehat{W}}}$, we may express the event B as

$$B = \left\{\widehat{W}_T \ge a,\ m^{\widehat{W}} \ge b\right\}, \quad a := \sigma^{-1} \ln(K/S_0), \quad b := \sigma^{-1} \ln(c/S_0). \tag{15.5}$$

Note that the barrier is set so that $b < 0$. Since

$$S_T Z_T^{-1} = S_0 e^{\sigma \widehat{W}_T} e^{\beta \widehat{W}_T - \frac{1}{2}\beta^2 T} = S_0 e^{\gamma \widehat{W}_T - \frac{1}{2}\beta^2 T}, \quad \gamma := \sigma + \beta = r/\sigma + \sigma/2,$$

Equation (15.3) and the change of measure formula yield

$$\begin{aligned} C_0^{do} &= e^{-rT} \widehat{\mathbb{E}}\big[(S_T - K)\mathbf{1}_B Z_T^{-1}\big] \\ &= e^{-(r+\beta^2/2)T} \left[S_0 \widehat{\mathbb{E}}\big(e^{\gamma \widehat{W}_T} \mathbf{1}_B\big) - K \widehat{\mathbb{E}}\big(e^{\beta \widehat{W}_T} \mathbf{1}_B\big)\right]. \end{aligned} \tag{15.6}$$

It remains to evaluate $\widehat{\mathbb{E}}\big(e^{\lambda \widehat{W}_T} \mathbf{1}_B\big)$ for $\lambda = \gamma$ and β. For this we shall need the following lemma, which we state without proof.[1]

15.1.1 Lemma. *The joint density $\widehat{f}_m(x, y)$ of $(\widehat{W}_T, m^{\widehat{W}})$ under $\widehat{\mathbb{P}}$ is given by*

$$\begin{aligned} \widehat{f}_m(x, y) &= \widehat{g}_m(x, y) \mathbf{1}_E(x, y), \quad \textit{where} \\ \widehat{g}_m(x, y) &= \frac{2(x - 2y)}{T\sqrt{2\pi T}} \exp\left[\frac{-(x - 2y)^2}{2T}\right] \quad \textit{and} \\ E &= \{(x, y) \mid y \le 0, y \le x\}. \end{aligned} \tag{15.7}$$

By the lemma, Equation (15.5), and the law of the unconscious statistician we see that for any real number λ

$$\begin{aligned} \widehat{\mathbb{E}}\left(e^{\lambda \widehat{W}_T} \mathbf{1}_B\right) &= \widehat{\mathbb{E}}\left(e^{\lambda \widehat{W}_T} \mathbf{1}_{[a,\infty) \times [b,\infty)}(\widehat{W}_T, m^{\widehat{W}})\right) \\ &= \iint_D e^{\lambda x} \widehat{g}_m(x, y)\, dA, \end{aligned} \tag{15.8}$$

[1] The derivation of the density formula (15.7) is based on the *reflection principle* of Brownian motion, which states that the first time the process hits a specified nonzero level l it starts anew and its probabilistic behavior thereafter is invariant under reflection in the horizontal line through l. For a detailed account, the reader is referred to [3] or [19].

where

$$D = E \cap \big([a, \infty) \times [b, \infty)\big) = \{(x, y) \mid b \le y \le 0, x \ge a, x \ge y\}.$$

The integral in (15.8) depends on the relative values of K and c and also of K and S_0. To facilitate its evaluation, we prepare the following lemma.

15.1.2 Lemma. *For real numbers λ, x_j, y_j,*

$$\int_{x_1}^{x_2}\int_{y_1}^{y_2} e^{\lambda x}\widehat{g}_m(x, y)\,dy\,dx = I_\lambda(y_2; x_1, x_2) - I_\lambda(y_1; x_1, x_2) \quad \textit{and}$$

$$\int_{x_1}^{x_2}\int_{y_1}^{x} e^{\lambda x}\widehat{g}_m(x, y)\,dy\,dx = I_\lambda(0; x_1, x_2) - I_\lambda(y_1; x_1, x_2), \quad \textit{where}$$

$$I_\lambda(y; x_1, x_2) = e^{2y\lambda+\lambda^2T/2}\left\{\Phi\left(\frac{2y - x_1 + \lambda T}{\sqrt{T}}\right) - \Phi\left(\frac{2y - x_2 + \lambda T}{\sqrt{T}}\right)\right\}.$$

Proof. A simple substitution yields

$$\int_{y_1}^{z} \widehat{g}_m(x, y)\,dy = \frac{1}{\sqrt{2\pi T}}\left[e^{u(x,z)} - e^{u(x,y_1)}\right], \tag{15.9}$$

where

$$u(x, z) = -\left(\frac{x - 2z}{\sqrt{2T}}\right)^2.$$

Taking $z = y_2$ in (15.9) we have

$$\int_{x_1}^{x_2}\int_{y_1}^{y_2} e^{\lambda x}\widehat{g}_m(x, y)\,dy\,dx = J_\lambda(y_2; x_1, x_2) - J_\lambda(y_1; x_1, x_2),$$

where

$$J_\lambda(y; x_1, x_2) = \frac{1}{\sqrt{2\pi T}}\int_{x_1}^{x_2} e^{\lambda x+u(x,y)}\,dx.$$

Also, taking $z = x$ in (15.9) and using $u(x, x) = u(x, 0)$, we obtain

$$\int_{x_1}^{x_2}\int_{y_1}^{x} e^{\lambda x}\widehat{g}_m(x, y)\,dy\,dx = J_\lambda(0; x_1, x_2) - J_\lambda(y_1; x_1, x_2),$$

It remains to show that $J_\lambda(y; x_1, x_2) = I_\lambda(y; x_1, x_2)$. Since

$$\lambda x + u(x, y) = \lambda x - \frac{x^2 - 4xy + 4y^2}{2T} = -\frac{x^2}{2T} - \frac{y^2}{T} + \left(\lambda + \frac{2y}{T}\right)x,$$

we have

$$\begin{aligned} J_\lambda(y; x_1, x_2) &= \frac{e^{-2y^2/T}}{\sqrt{2\pi T}}\int_{x_1}^{x_2} e^{-x^2/(2T)+(\lambda+2y/T)x}\,dx \\ &= e^{2y\lambda+\lambda^2T/2}\left\{\Phi\left(\frac{2y - x_1 + \lambda T}{\sqrt{T}}\right) - \Phi\left(\frac{2y - x_2 + \lambda T}{\sqrt{T}}\right)\right\}, \end{aligned}$$

where for the last equality we used Exercise 12.12 with $p = 1/(2T)$ and $q = \lambda + 2y/T$. Thus, $J_\lambda = I_\lambda$, completing the proof. □

We may now evaluate (15.8):

15.1.3 Lemma. *If $K > c$, then*

$$\widehat{\mathbb{E}}\left(e^{\lambda \widehat{W}_T}\mathbf{1}_B\right) = e^{\lambda^2 T/2}\left\{\Phi\left(\frac{-a+\lambda T}{\sqrt{T}}\right) - e^{2b\lambda}\Phi\left(\frac{2b-a+\lambda T}{\sqrt{T}}\right)\right\}. \quad (15.10)$$

If $K \le c$, then

$$\widehat{\mathbb{E}}\left(e^{\lambda \widehat{W}_T}\mathbf{1}_B\right) = e^{\lambda^2 T/2}\left\{\Phi\left(\frac{-b+\lambda T}{\sqrt{T}}\right) - e^{2b\lambda}\Phi\left(\frac{b+\lambda T}{\sqrt{T}}\right)\right\}. \quad (15.11)$$

Proof. Suppose first that $K > c$, so that $a > b$. If $K \ge S_0$, then $a \ge 0$ and $D = [a,\infty) \times [b,0]$ hence, letting $x_2 \to +\infty$ in Lemma 15.1.2, we obtain

$$\widehat{\mathbb{E}}\left(e^{\lambda \widehat{W}_T}\mathbf{1}_B\right) = \int_a^\infty \int_b^0 e^{\lambda x}\widehat{g}_m(x,y)\,dy\,dx = I_\lambda(0;,a,\infty) - I_\lambda(b;a,\infty),$$

which is (15.10). On the other hand, if $K < S_0$ then $a < 0$ and

$$D = \{(x,y) \mid a \le x \le 0,\, b \le y \le x\} \cup \big([0,\infty) \times [b,0]\big),$$

so, again by Lemma 15.1.2,

$$\begin{aligned}\widehat{\mathbb{E}}\left(e^{\lambda \widehat{W}_T}\mathbf{1}_B\right) &= \int_a^0 \int_b^x e^{\lambda x}\widehat{g}_m(x,y)\,dy\,dx + \int_0^\infty \int_b^0 e^{\lambda x}\widehat{g}_m(x,y)\,dy\,dx \\ &= I_\lambda(0;a,0) - I_\lambda(b;a,0) + I_\lambda(0;0,\infty) - I_\lambda(b;0,\infty)\end{aligned}$$

By cancellations, the last expression reduces to (15.10) which therefore holds for all values of S_0 and K with $K > c$.

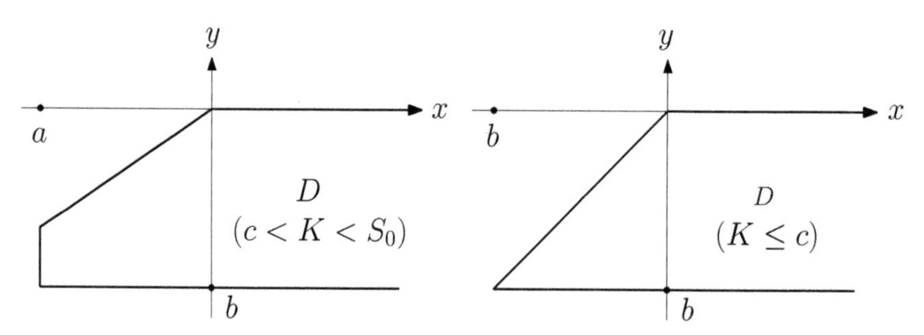

FIGURE 15.1: The region D.

Now suppose $K \le c$. Then $a \le b$ hence $D = \{(x,y) \mid b \le y \le 0, x \ge y\}$ and

$$\begin{aligned}\widehat{\mathbb{E}}\left(e^{\lambda \widehat{W}_T}\mathbf{1}_B\right) &= \int_b^0 \int_b^x e^{\lambda x}\widehat{g}_m(x,y)\,dy\,dx + \int_0^\infty \int_b^0 e^{\lambda x}\widehat{g}_m(x,y)\,dy\,dx \\ &= I_\lambda(0;b,0) - I_\lambda(b;b,0) + I_\lambda(0;0,\infty) - I_\lambda(b;0,\infty),\end{aligned}$$

which is (15.11). □

We are now ready to evaluate C_0^{do} for the case $K > c$. Taking $\lambda = \gamma$ and β in (15.10) we see from (15.6) that

$$C_0^{do} = e^{(\gamma^2-\beta^2-2r)T/2}S_0\left\{\Phi\left(\frac{-a+\gamma T}{\sqrt{T}}\right) - e^{2b\gamma}\Phi\left(\frac{2b-a+\gamma T}{\sqrt{T}}\right)\right\}$$
$$- Ke^{-rT}\left\{\Phi\left(\frac{-a+\beta T}{\sqrt{T}}\right) - e^{2b\beta}\Phi\left(\frac{2b-a+\beta T}{\sqrt{T}}\right)\right\}. \quad (15.12)$$

Recalling that

$$a = \sigma^{-1}\ln(K/S_0), \quad b = \sigma^{-1}\ln(c/S_0), \quad \beta = r/\sigma - \sigma/2, \quad \text{and} \quad \gamma = r/\sigma + \sigma/2,$$

we have

$$\gamma^2 - \beta^2 = 2r, \quad e^{2b\gamma} = \left(\frac{c}{S_0}\right)^{\frac{2r}{\sigma^2}+1}, \text{ and } e^{2b\beta} = \left(\frac{c}{S_0}\right)^{\frac{2r}{\sigma^2}-1}.$$

Furthermore, one readily checks that

$$\begin{aligned}
\frac{-a+\gamma T}{\sqrt{T}} &= \frac{\ln(S_0/K) + (r+\sigma^2/2)T}{\sigma\sqrt{T}} &&= d_1,\\
\frac{-a+\beta T}{\sqrt{T}} &= \frac{\ln(S_0/K) + (r-\sigma^2/2)T}{\sigma\sqrt{T}} &&= d_2,\\
\frac{2b-a+\gamma T}{\sqrt{T}} &= \frac{\ln(c^2/S_0K) + (r+\sigma^2/2)T}{\sigma\sqrt{T}} &&= \delta_1,\\
\frac{2b-a+\beta T}{\sqrt{T}} &= \frac{\ln(c^2/S_0K) + (r-\sigma^2/2)T}{\sigma\sqrt{T}} &&= \delta_2.
\end{aligned}$$

Inserting these expressions into (15.12) establishes (15.1) for the case $K > c$ $(M = K)$.

Now suppose $K \le c$. Then from (15.6) and (15.11),

$$C_0^{do} = e^{(\gamma^2-\beta^2-2r)T/2}S_0\left\{\Phi\left(\frac{-b+\gamma T}{\sqrt{T}}\right) - e^{2b\gamma}\Phi\left(\frac{b+\gamma T}{\sqrt{T}}\right)\right\}$$
$$-e^{-rT}K\left\{\Phi\left(\frac{-b+\beta T}{\sqrt{T}}\right) - e^{2b\beta}\Phi\left(\frac{b+\beta T}{\sqrt{T}}\right)\right\}. \quad (15.13)$$

Since

$$\begin{aligned}
\frac{-b+\gamma T}{\sqrt{T}} &= \frac{\ln(S_0/c) + (r+\sigma^2/2)T}{\sigma\sqrt{T}} &&= d_1,\\
\frac{b+\gamma T}{\sqrt{T}} &= \frac{\ln(c/S_0) + (r+\sigma^2/2)T}{\sigma\sqrt{T}} &&= \delta_1,\\
\frac{-b+\beta T}{\sqrt{T}} &= \frac{\ln(S_0/c) + (r-\sigma^2/2)T}{\sigma\sqrt{T}} &&= d_2,\\
\frac{b+\beta T}{\sqrt{T}} &= \frac{\ln(c/S_0) + (r-\sigma^2/2)T}{\sigma\sqrt{T}} &&= \delta_2,
\end{aligned}$$

we see that (15.1) holds for the case $K \le c$ $(M = c)$, as well. This completes the derivation of (15.1).

We remark that the barrier level c for a down-and-out call is usually set to a value less than K; otherwise, the option could be knocked out even if it expires in the money.

15.1.4 Example. Table 15.1 gives prices C_0^{do} of down-and-out call options based on a stock that sells for $S_0 = \$50.00$. The parameters are $T = .5$, $r = .10$ and $\sigma = .20$. The cost of the corresponding standard call is \$7.64 for $K = 45$ and \$1.87 for $K = 55$. Notice that the price of the barrier option decreases as

c	39	42	45	47	49	49.99
C_0^{do}	\$7.64	\$7.54	\$6.74	\$5.15	\$2.16	\$0.02
$K = 45$						
c	42	43	45	47	49	49.99
C_0^{do}	\$1.87	\$1.86	\$1.81	\$1.57	\$0.78	\$0.01
$K = 55$						

TABLE 15.1: Variation of C_0^{do} with the barrier level c.

the barrier level increases. This is to be expected, since the higher the barrier the more likely the option will be knocked out, hence the less attractive the option.

15.2 Lookback Options

The payoff of a lookback option depends on the maximum or minimum value of the asset over the contract period. The payoffs of lookback call and put options are, respectively, $S_T - m^S$ and $M^S - S_T$, where m^S and M^S are defined by

$$M^S := \max\{S_t \mid 0 \le t \le T\} \quad \text{and} \quad m^S := \min\{S_t \mid 0 \le t \le T\}.$$

In this section we determine the value V_t of a floating strike lookback call option.

By risk-neutral pricing,

$$V_t = e^{-r(T-t)}\mathbb{E}^*(S_T - m^S|\mathcal{F}_t) = S_t - e^{-r(T-t)}\mathbb{E}^*(m^S|\mathcal{F}_t), \quad 0 \le t \le T, \tag{15.14}$$

where we have used the fact that the discounted asset price is a martingale under $\mathbb{P}^*$. As in §15.1, the value of the underlying at time t may be expressed as

$$S_t = S_0 e^{\sigma \widehat{W}_t}, \quad \widehat{W}_t := W_t^* + \beta t, \quad \beta := r/\sigma - \sigma/2, \qquad 0 \le t \le T,$$

where W_t^* is a Brownian motion under the law $\mathbb{P}^*$. To evaluate $\mathbb{E}^*\left(m^S|\mathcal{F}_t\right)$, we introduce the following additional notation: For a process X and for $t \in [0, T]$, set

$$m_t^X := \min\{X_u \mid 0 \le u \le t\} \quad \text{and} \quad m_{t,T}^X := \min\{X_u \mid t \le u \le T\}.$$

Now let $t < T$ and set

$$Y_t = e^{\sigma \min\{\widehat{W}_u - \widehat{W}_t | t \le u \le T\}}.$$

Since $S_u = S_t e^{\sigma(\widehat{W}_u - \widehat{W}_t)}$,

$$m_{t,T}^S = \min\{S_t e^{\sigma(\widehat{W}_u - \widehat{W}_t)} \mid t \le u \le T\} = S_t Y_t$$

and therefore

$$m^S = \min(m_t^S, m_{t,T}^S) = \min(m_t^S, S_t Y_t).$$

Since m_t^S is $\mathcal{F}_t$-measurable and Y_t is independent of $\mathcal{F}_t$, we can apply Remark 13.4.5 with $g(x_1, x_2, y) = \min(x_1, x_2 y)$ to conclude that

$$\mathbb{E}^*(m^S|\mathcal{F}_t) = \mathbb{E}^*(g(m_t^S, S_t, Y_t)|\mathcal{F}_t) = G_t(m_t^S, S_t),$$

where

$$G_t(m, s) = \mathbb{E}^* g(m, s, Y_t) = \mathbb{E}^* \min(m, sY_t), \quad 0 < m \le s. \tag{15.15}$$

From (15.14) we see that V_t may now be expressed as

$$V_t = v_t(m_t^S, S_t), \text{ where } v_t(m, s) = s - e^{-r(T-t)} G_t(m, s). \tag{15.16}$$

The remainder of the section is devoted to evaluating G_t.

Fix t, m, and s with $0 < m \le s$ and set

$$A = \{m_\tau^{\widehat{W}} \le a\}, \quad \tau := T - t, \quad a := \frac{1}{\sigma} \ln\left(\frac{m}{s}\right).$$

Since $\widehat{W}_u - \widehat{W}_t = W_u^* - W_t^* + \beta(u - t)$ and $\widehat{W}_{u-t} = W_{u-t}^* + \beta(u - t)$, we see that $\widehat{W}_u - \widehat{W}_t$ and $\widehat{W}_{u-t}$ have the same distribution under $\mathbb{P}^*$. Therefore Y_t and $e^{\sigma m_\tau^{\widehat{W}}}$ have the same distribution under $\mathbb{P}^*$. It follows that

$$G_t(m, s) - m = \mathbb{E}^*\left[\min\left(m, s e^{\sigma m_\tau^{\widehat{W}}}\right) - m\right] = \mathbb{E}^*\left[\min\left(0, s e^{\sigma m_\tau^{\widehat{W}}} - m\right)\right].$$

Noting that $\min\left(0, s e^{\sigma m_\tau^{\widehat{W}}} - m\right) = s e^{\sigma m_\tau^{\widehat{W}}} - m$ iff $m_\tau^{\widehat{W}} \le a$, we see that

$$\begin{aligned} G_t(m, s) &= \mathbb{E}^*\left[(s e^{\sigma m_\tau^{\widehat{W}}} - m)\mathbf{1}_A\right] + m \\ &= s\mathbb{E}^*\left(e^{\sigma m_\tau^{\widehat{W}}} \mathbf{1}_A\right) + m\left(1 - \mathbb{E}^* \mathbf{1}_A\right). \end{aligned} \tag{15.17}$$

It remains then to evaluate $\mathbb{E}^*\left(e^{\sigma m_\tau^{\widehat{W}}} \mathbf{1}_A\right)$ and $\mathbb{E}^*\left(\mathbf{1}_A\right)$. For this, we shall need the following lemmas.

15.2.1 Lemma. *The joint density $f_m(x,y)$ of $(\widehat{W}_\tau, m_\tau^{\widehat{W}})$ under $\mathbb{P}^*$ is*

$$f_m(x,y) = e^{\beta x - \frac{1}{2}\beta^2\tau} g_m(x,y)\mathbf{1}_E(x,y), \quad \textit{where}$$

$$g_m(x,y) = \frac{2(x-2y)}{\tau\sqrt{2\pi\tau}} \exp\left\{-\frac{(x-2y)^2}{2\tau}\right\} \quad \textit{and}$$

$$E = \{(x,y) \mid y \le 0, y \le x\}.$$

Proof. By Girsanov's Theorem, $(\widehat{W}_u)_{0 \le u \le \tau}$ is a Brownian motion under the probability measure $\widehat{\mathbb{P}}$ given by

$$d\widehat{\mathbb{P}} = Z_\tau \, d\mathbb{P}^*, \quad Z_\tau = e^{-\beta\widehat{W}_\tau + \frac{1}{2}\beta^2\tau}.$$

By Lemma 15.1.1 with T replaced by τ, the joint density of $(\widehat{W}_\tau, m_\tau^{\widehat{W}})$ under $\widehat{\mathbb{P}}$ is $g_m(x,y)\mathbf{1}_E(x,y)$, where g_m and E are defined as above. The cdf of $(\widehat{W}_\tau, m_\tau^{\widehat{W}})$ under $\mathbb{P}^*$ is therefore

$$\begin{aligned}
\mathbb{P}^*\left(\widehat{W}_\tau \le x, m_\tau^{\widehat{W}} \le y\right) &= \mathbb{E}^*\left(\mathbf{1}_{\{\widehat{W}_\tau \le x, m_\tau^{\widehat{W}} \le y\}}\right) \\
&= \widehat{\mathbb{E}}\left(\mathbf{1}_{\{\widehat{W}_\tau \le x, m_\tau^{\widehat{W}} \le y\}} Z_\tau^{-1}\right) \\
&= \widehat{\mathbb{E}}\left(\mathbf{1}_{\{\widehat{W}_\tau \le x, m_\tau^{\widehat{W}} \le y\}} e^{\beta\widehat{W}_\tau - \frac{1}{2}\beta^2\tau}\right) \\
&= \int_{-\infty}^{x}\int_{-\infty}^{y} e^{\beta u - \frac{1}{2}\beta^2\tau} g_m(u,v)\mathbf{1}_E(u,v)\, dv\, du,
\end{aligned}$$

verifying the lemma. □

15.2.2 Lemma. *The density f_m of $m_\tau^{\widehat{W}}$ under $\mathbb{P}^*$ is given by*

$$f_m(z) = g_m(z)\mathbf{1}_{(-\infty,0]}(z), \quad \textit{where}$$

$$g_m(z) = \frac{2}{\sqrt{\tau}}\varphi\left(\frac{z-\beta\tau}{\sqrt{\tau}}\right) + 2\beta e^{2\beta z}\Phi\left(\frac{z+\beta\tau}{\sqrt{\tau}}\right).$$

Proof. By Lemma 15.2.1, the cdf of $m_\tau^{\widehat{W}}$ under $\mathbb{P}^*$ is

$$\begin{aligned}
\mathbb{P}^*\left(m_\tau^{\widehat{W}} \le z\right) &= \int_{-\infty}^{\infty}\int_{-\infty}^{z} f_m(x,y)\, dy\, dx \\
&= e^{-\beta^2\tau/2} \iint_D e^{\beta x} g_m(x,y)\, dA,
\end{aligned}$$

where $D = \{(x,y) \mid y \le \min(0,x,z)\}$. Suppose first that $z \le 0$. The integral over D may then be expressed as $I' + I''$, where

$$I' = \int_{z}^{\infty}\int_{-\infty}^{z} e^{\beta x} g_m(x,y)\, dy\, dx \quad \text{and} \quad I'' = \int_{-\infty}^{z}\int_{-\infty}^{x} e^{\beta x} g_m(x,y)\, dy\, dx$$

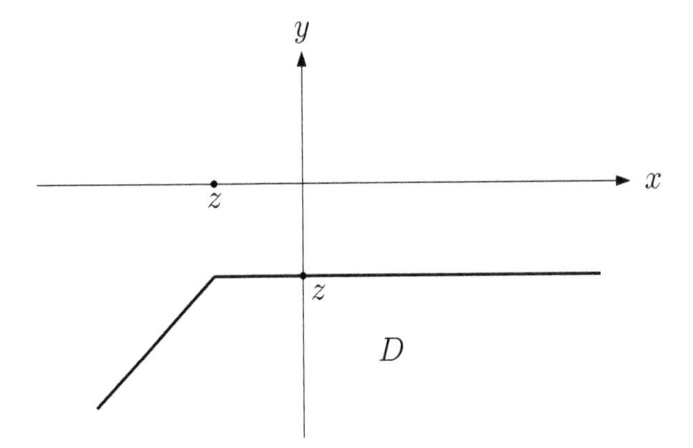

FIGURE 15.2: $D = \{(x, y) \mid y \le \min\{x, z\}\}$, $z \le 0$.

(Figure 15.2). By Lemma 15.1.2 (with T replaced by τ),

$$I' = I_\beta(z; z, \infty) - I_\beta(-\infty; z, \infty) = e^{2z\beta+\beta^2\tau/2}\Phi\left(\frac{z+\beta\tau}{\sqrt{\tau}}\right) \quad \text{and}$$

$$I'' = I_\beta(0; -\infty, z) - I_\beta(-\infty; -\infty, z) = e^{\beta^2\tau/2}\Phi\left(\frac{z-\beta\tau}{\sqrt{\tau}}\right).$$

Thus if $z \le 0$,

$$\mathbb{P}^*\left(m^{\widehat{W}} \le z\right) = e^{2\beta z}\Phi\left(\frac{z+\beta\tau}{\sqrt{\tau}}\right) + \Phi\left(\frac{z-\beta\tau}{\sqrt{\tau}}\right).$$

Differentiating the expression on the right with respect to z and using the identity

$$e^{2z\beta}\varphi\left(\frac{z+\beta\tau}{\sqrt{\tau}}\right) = \varphi\left(\frac{z-\beta\tau}{\sqrt{\tau}}\right)$$

produces $g_m(z)$. Since $\mathbb{P}^*\left(m^{\widehat{W}} \le z\right) = \mathbb{P}^*\left(m^{\widehat{W}} \le 0\right)$ for $z > 0$, the conclusion of the lemma follows. □

We are now in a position to evaluate $\mathbb{E}^*\left(e^{\lambda m_\tau^{\widehat{W}}}\mathbf{1}_A\right)$. By Lemma 15.2.2 (noting that $a \le 0$), we have

$$\mathbb{E}^*\left(e^{\lambda m_\tau^{\widehat{W}}}\mathbf{1}_A\right) = \int_{-\infty}^{a} e^{\lambda z} g_m(z)\,dz = \frac{2}{\sqrt{\tau}}J'_\lambda + 2\beta J''_\lambda, \tag{15.18}$$

where

$$J'_\lambda = \int_{-\infty}^{a} e^{\lambda z}\varphi\left(\frac{z-\beta\tau}{\sqrt{\tau}}\right)dz = \frac{e^{-\beta^2\tau/2}}{\sqrt{2\pi}}\int_{-\infty}^{a} e^{-z^2/(2\tau)+(\beta+\lambda)z}\,dz \quad \text{and}$$

$$J''_\lambda = \int_{-\infty}^{a} e^{(\lambda+2\beta)z}\Phi\left(\frac{z+\beta\tau}{\sqrt{\tau}}\right)dz = \frac{1}{\sqrt{2\pi}}\int_{-\infty}^{a} e^{(\lambda+2\beta)z}\int_{-\infty}^{\frac{z+\beta\tau}{\sqrt{\tau}}} e^{-x^2/2}\,dx\,dz.$$

By Exercise 12.12,

$$J'_\lambda = \sqrt{\tau}e^{\lambda\beta\tau+\lambda^2\tau/2}\Phi\left(\frac{a-(\lambda+\beta)\tau}{\sqrt{\tau}}\right). \tag{15.19}$$

To evaluate J''_λ, we reverse the order of integration: For $\lambda \neq -2\beta$,

$$\begin{aligned}
J''_\lambda &= \frac{1}{\sqrt{2\pi}}\int_{-\infty}^{b} e^{-x^2/2}\int_{\sqrt{\tau}x-\beta\tau}^{a} e^{(\lambda+2\beta)z}\,dz\,dx, \qquad b := \frac{a+\beta\tau}{\sqrt{\tau}} \\
&= \frac{1}{(\lambda+2\beta)\sqrt{2\pi}}\int_{-\infty}^{b} e^{-x^2/2}\left(e^{(\lambda+2\beta)a} - e^{(\lambda+2\beta)(\sqrt{\tau}x-\beta\tau)}\right)dx \\
&= \frac{1}{(\lambda+2\beta)}\left[e^{(\lambda+2\beta)a}\Phi(b) - e^{\lambda\beta\tau+\lambda^2\tau/2}\Phi\big(b-(\lambda+2\beta)\sqrt{\tau}\big)\right], \qquad (15.20)
\end{aligned}$$

where to obtain the last equality we used Exercise 12.12 again. From (15.19) and (15.20),

$$\begin{aligned}
J'_\sigma &= \sqrt{\tau}e^{\sigma\beta\tau+\sigma^2\tau/2}\Phi\left(\frac{a-(\sigma+\beta)\tau}{\sqrt{\tau}}\right) \quad \text{and} \\
J''_\sigma &= \frac{1}{(\sigma+2\beta)}\left[e^{(\sigma+2\beta)a}\Phi(b) - e^{\sigma\beta\tau+\sigma^2\tau/2}\Phi\big(b-(\sigma+2\beta)\sqrt{\tau}\big)\right].
\end{aligned}$$

Recalling that

$$a = \frac{1}{\sigma}\ln\left(\frac{m}{s}\right),\quad \beta = \frac{r}{\sigma}-\frac{\sigma}{2},\quad \text{and}\quad b = \frac{a+\beta\tau}{\sqrt{\tau}} = \frac{1}{\sigma\sqrt{\tau}}\ln\left(\frac{m}{s}\right)+\beta\sqrt{\tau},$$

we see that

$$\begin{aligned}
\sigma\beta\tau + \frac{\sigma^2\tau}{2} &= \tau r, \quad (\sigma+2\beta)a = \frac{2r}{\sigma^2}\ln\left(\frac{m}{s}\right) \quad \text{and} \\
\frac{a-(\sigma+\beta)\tau}{\sqrt{\tau}} &= b-(\sigma+2\beta)\sqrt{\tau} = \frac{1}{\sqrt{\tau}}\left[\frac{1}{\sigma}\ln\left(\frac{m}{s}\right) - \left(\frac{r}{\sigma}+\frac{\sigma}{2}\right)\tau\right].
\end{aligned}$$

Therefore, setting

$$\begin{aligned}
\delta_{1,2} = \delta_{1,2}(\tau,m,s) &= \frac{\ln(s/m)+(r\pm\sigma^2/2)\tau}{\sigma\sqrt{\tau}} \quad \text{and} \\
d = d(\tau,m,s) &= \frac{\ln(m/s)+(r-\sigma^2/2)\tau}{\sigma\sqrt{\tau}},
\end{aligned}$$

we have

$$J'_\sigma = \sqrt{\tau}e^{r\tau}\Phi(-\delta_1) \quad \text{and} \quad J''_\sigma = \frac{\sigma}{2r}\left[\left(\frac{m}{s}\right)^{2r/\sigma^2}\Phi(d) - e^{r\tau}\Phi(-\delta_1)\right].$$

From (15.18),

$$\begin{aligned}\mathbb{E}^*\left(e^{\sigma m_\tau^{\widehat{W}}}\mathbf{1}_A\right) &= \frac{2}{\sqrt{\tau}}J'_\sigma + 2\beta J''_\sigma \\ &= 2e^{r\tau}\Phi(-\delta_1) + \left(1-\frac{\sigma^2}{2r}\right)\left[\left(\frac{m}{s}\right)^{2r/\sigma^2}\Phi(d) - e^{r\tau}\Phi(-\delta_1)\right] \\ &= e^{r\tau}\left(1+\frac{\sigma^2}{2r}\right)\Phi(-\delta_1) + \left(1-\frac{\sigma^2}{2r}\right)\left(\frac{m}{s}\right)^{2r/\sigma^2}\Phi(d).\end{aligned} \tag{15.21}$$

Similarly, if $\beta \neq 0$,

$$\begin{aligned}J'_0 &= \sqrt{\tau}\Phi\left(\frac{a-\beta\tau}{\sqrt{\tau}}\right) = \sqrt{\tau}\Phi(-\delta_2) \quad \text{and} \\ J''_0 &= \frac{1}{2\beta}\left[e^{2\beta a}\Phi(b) - \Phi\left(b-2\beta\sqrt{\tau}\right)\right] = \frac{1}{2\beta}\left[\left(\frac{m}{s}\right)^{\frac{2r}{\sigma^2}-1}\Phi(d) - \Phi(-\delta_2)\right],\end{aligned}$$

hence

$$\begin{aligned}\mathbb{E}^*(\mathbf{1}_A) &= \frac{2}{\sqrt{\tau}}J'_0 + 2\beta J''_0 = 2\Phi(-\delta_2) + \left[\left(\frac{m}{s}\right)^{\frac{2r}{\sigma^2}-1}\Phi(d) - \Phi(-\delta_2)\right] \\ &= \Phi(-\delta_2) + \left(\frac{m}{s}\right)^{\frac{2r}{\sigma^2}-1}\Phi(d).\end{aligned} \tag{15.22}$$

The reader may verify that (15.22) also holds if $\beta = 0$. From (15.17), (15.21), and (15.22),

$$\begin{aligned}G_t(m,s) &= s\left\{e^{r\tau}\left(1+\frac{\sigma^2}{2r}\right)\Phi(-\delta_1) + \left(1-\frac{\sigma^2}{2r}\right)\left(\frac{m}{s}\right)^{\frac{2r}{\sigma^2}}\Phi(d)\right\} \\ &\qquad + m\left\{1-\Phi(-\delta_2) - \left(\frac{m}{s}\right)^{\frac{2r}{\sigma^2}-1}\Phi(d)\right\} \\ &= se^{r\tau}\left(1+\frac{\sigma^2}{2r}\right)\Phi\big(-\delta_1(\tau,m,s)\big) + m\Phi\big(\delta_2(\tau,m,s)\big) \\ &\qquad - \frac{s\sigma^2}{2r}\left(\frac{m}{s}\right)^{\frac{2r}{\sigma^2}}\Phi\big(d(\tau,m,s)\big).\end{aligned} \tag{15.23}$$

Finally, recalling (15.16), we obtain the formula

$$V_t = v(m_t^S, S_t),$$

where

$$\begin{aligned}v_t(m,s) &= s - e^{-r\tau}G_t(m,s) \\ &= s\Phi\big(\delta_1(m,s,\tau)\big) - me^{-r\tau}\Phi\big(\delta_2(m,s,\tau)\big) - \frac{s\sigma^2}{2r}\Phi\big(-\delta_1(m,s,\tau)\big) \\ &\qquad + e^{-r\tau}\frac{s\sigma^2}{2r}\left(\frac{m}{s}\right)^{\frac{2r}{\sigma^2}}\Phi\big(d(m,s,\tau)\big).\end{aligned}$$

15.3 Asian Options

Recall that an *Asian* or *average option* has payoff that depends on an average $A(S)$ of the price process S of the underlying asset. The reader is referred to §4.11 for details on these options.

The fixed strike geometric average option readily lends itself to Black-Scholes-Merton risk-neutral pricing, as the following theorems illustrate.

15.3.1 Theorem. *The cost V_0 of a fixed strike, continuous, geometric average call option is*

$$V_0 = S_0 e^{-rT/2-\sigma^2 T/12}\Phi\left(d_1\right) - Ke^{-rT}\Phi\left(d_2\right), \tag{15.24}$$

where

$$d_2 := \frac{\ln\left(\frac{S_0}{K}\right) + (r - \frac{1}{2}\sigma^2)\frac{T}{2}}{\sigma\sqrt{\frac{T}{3}}}, \qquad d_1 := d_2 + \sigma\sqrt{\frac{T}{3}}.$$

Proof. By risk-neutral pricing,

$$V_0 = e^{-rT}\mathbb{E}^*\left[A(S) - K\right]^+, \quad \text{where} \quad A(S) = \exp\left(\frac{1}{T}\int_0^T \ln S_t\, dt\right).$$

From Lemma 13.4.2,

$$\ln S_t = \ln S_0 + \sigma W_t^* + \left(r - \frac{\sigma^2}{2}\right)t$$

hence

$$\frac{1}{T}\int_0^T \ln S_t\, dt = \ln S_0 + \frac{\sigma}{T}\int_0^T W_t^*\, dt + \left(r - \frac{\sigma^2}{2}\right)\frac{T}{2}.$$

By Example 11.3.1, $\int_0^T W_t^*\, dt$ is normal under $\mathbb{P}^*$ with mean zero and variance $T^3/3$. Therefore

$$A(S) = S_0 \exp\left\{\sigma\sqrt{\frac{T}{3}}\,Z + \left(r - \frac{\sigma^2}{2}\right)\frac{T}{2}\right\},$$

where $Z \sim N(0,1)$. It follows that

$$e^{rT}V_0 = \int_{-\infty}^{\infty}\left(S_0 \exp\left\{\sigma\sqrt{\frac{T}{3}}\,z + \left(r - \frac{\sigma^2}{2}\right)\frac{T}{2}\right\} - K\right)^+ \varphi(z)\, dz.$$

Since the integrand is zero when $z < -d_2$,

$$e^{rT}V_0 = S_0 \int_{-d_2}^{\infty} \exp\left\{\sigma\sqrt{\frac{T}{3}}\,z + \left(r - \frac{\sigma^2}{2}\right)\frac{T}{2}\right\}\varphi(z)\,dz - K\int_{-d_2}^{\infty}\varphi(z)\,dz$$

$$= S_0 e^{(r-\sigma^2/2)T/2}\frac{1}{\sqrt{2\pi}}\int_{-d_2}^{\infty}\exp\left\{\sigma\sqrt{\frac{T}{3}}\,z - \frac{1}{2}z^2\right\}dz - K\Phi(d_2)$$

$$= S_0 e^{rT/2-\sigma^2T/12}\Phi(d_1) - K\Phi(d_2),$$

the last equality by Exercise 12.12. □

To price the discrete geometric average call option, we first establish the following lemma.

15.3.2 Lemma. *Let $t_0 := 0 < t_1 < t_2 < \cdots < t_n \le T$. The joint density of $(W^*_{t_1}, W^*_{t_2}, \ldots, W^*_{t_n})$ under $\mathbb{P}^*$ is given by*

$$f(x_1, x_2, \ldots, x_n) = \prod_{j=1}^{n} f_j(x_j - x_{j-1}),$$

where $x_0 = 0$ and f_j is the density of $W^*_{t_j} - W^*_{t_{j-1}}$:

$$f_j(x) = \frac{1}{\sqrt{2\pi(t_j - t_{j-1})}}\varphi\left(\frac{x}{\sqrt{t_j - t_{j-1}}}\right).$$

Proof. Set

$$A = (-\infty, z_1] \times (-\infty, z_2] \times \cdots \times (-\infty, z_n]$$

and

$$B = \{(y_1, y_2, \ldots, y_n) \mid (y_1, y_1 + y_2, \ldots, y_1 + y_2 + \cdots + y_n) \in A\}.$$

By independent increments,

$$\mathbb{P}^*\big((W^*_{t_1}, W^*_{t_2}, \ldots, W^*_{t_n}) \in A\big) = \mathbb{P}^*\big((W^*_{t_1}, W^*_{t_2} - W^*_{t_1}, \ldots, W^*_{t_n} - W^*_{t_{n-1}}) \in B\big)$$
$$= \int_B f_1(y_1)f_2(y_2)\cdots f_n(y_n)\,dy.$$

With the substitution $x_j = y_1 + y_2 + \cdots + y_j$, we obtain

$$\mathbb{P}^*\big((W^*_{t_1}, W^*_{t_2}, \ldots, W^*_{t_n}) \in A\big) = \int_A f_1(x_1)f_2(x_2 - x_1)\cdots f_n(x_n - x_{n-1})\,dx,$$

which establishes the lemma. □

15.3.3 Corollary. $W^*_{t_1} + W^*_{t_2} + \cdots + W^*_{t_n} \sim N(0, \sigma_n^2)$ *under $\mathbb{P}^*$, where*

$$\sigma_n^2 := 1^2\tau_n + 2^2\tau_{n-1} + \cdots + n^2\tau_1, \quad \tau_j := t_j - t_{j-1}.$$

Proof. By Lemma 15.3.2,

$$\mathbb{E}^*\left[e^{\lambda(W^*_{t_1}+W^*_{t_2}+\cdots+W^*_{t_n})}\right] = \int_{-\infty}^{\infty}\cdots\int_{-\infty}^{\infty} e^{\lambda(x_1+\cdots+x_n)} f(x_1,\ldots,x_n)\,dx$$

$$= \int_{-\infty}^{\infty}\cdots\int_{-\infty}^{\infty} e^{\lambda(x_1+x_2+\cdots+x_{n-1})} f_1(x_1)\cdots f_{n-1}(x_{n-1}-x_{n-2})$$

$$\int_{-\infty}^{\infty} e^{\lambda x_n} f_n(x_n - x_{n-1})\,dx_n\,dx_{n-1}\cdots dx_1.$$

The innermost integral evaluates to

$$\frac{1}{\sqrt{2\pi\tau_n}}\int_{-\infty}^{\infty} e^{\lambda x_n-(x_n-x_{n-1})^2/(2\tau_n)}\,dx_n = \frac{e^{\lambda x_{n-1}}}{\sqrt{2\pi\tau_n}}\int_{-\infty}^{\infty} e^{\lambda x_n - x_n^2/(2\tau_n)}\,dx_n$$

$$= e^{\lambda x_{n-1}+\tau_n\lambda^2/2}.$$

Therefore,

$$\mathbb{E}^*\left[e^{\lambda(W^*_{t_1}+W^*_{t_2}+\cdots+W^*_{t_n})}\right]$$

$$= e^{\tau_n\lambda^2/2}\int_{-\infty}^{\infty}\cdots\int_{-\infty}^{\infty} e^{\lambda(x_1+\cdots+x_{n-2})} f_1(x_1)\cdots f_{n-2}(x_{n-2}-x_{n-3})$$

$$\int_{-\infty}^{\infty} e^{2\lambda x_{n-1}} f_{n-1}(x_{n-1}-x_{n-2})\,dx_{n-1}\,dx_{n-2}\cdots dx_1.$$

Repeating the calculation we arrive at

$$\mathbb{E}^*\left[e^{\lambda(W^*_{t_1}+\cdots+W^*_{t_n})}\right] = e^{\sigma_n^2\lambda^2/2}.$$

The corollary now follows from Theorem 6.5.2 and Example 6.5.1. □

15.3.4 Theorem. *The cost of a fixed strike, discrete, geometric average call option with payoff* $(A(S)-K)^+$, *where*

$$A(S) := \left(\prod_0^n S_{t_j}\right)^{1/(n+1)}, \quad t_j := \frac{jT}{n}, \quad j = 0,1,\ldots,n,$$

is given by

$$V_0^{(n)} = S_0\exp\left\{-\left(r+\frac{\sigma^2}{2}\right)\frac{T}{2}+\frac{\sigma^2 T a_n}{2}\right\}\Phi\left(\sigma\sqrt{a_n T}+d_n\right) - K\Phi\left(d_n\right),$$

where

$$d_n := \frac{\ln\left(S_0/K\right)+\frac{1}{2}(r-\sigma^2/2)T}{\sigma\sqrt{a_n T}}, \quad a_n := \frac{2n+1}{6(n+1)}.$$

Proof. By Corollary 15.3.3,

$$\begin{aligned} A(S) &= S_0 \exp\left\{\sigma(n+1)^{-1}\sum_{j=1}^{n} W^*_{t_j} + (r - \tfrac{1}{2}\sigma^2)(n+1)^{-1}\sum_{j=1}^{n} t_j\right\} \\ &= S_0 \exp\left\{\sigma(n+1)^{-1}\sigma_n Z + \left(r - \tfrac{1}{2}\sigma^2\right)\bar{t}\right\}, \end{aligned}$$

where

$$\begin{aligned} \sigma_n^2 &= \frac{T}{n}\sum_{j=1}^{n} j^2 = \frac{(n+1)(2n+1)T}{6} = (n+1)^2 a_n T, \\ Z &= \frac{1}{\sigma_n}\sum_{j=1}^{n} W^*_{t_j} \sim N(0,1) \quad \text{and} \\ \bar{t} &= \frac{1}{n+1}\sum_{j=1}^{n} t_j = \frac{T}{2}. \end{aligned}$$

Since $\sigma_n(n+1)^{-1} = \sqrt{a_n T}$, risk-neutral pricing implies that

$$\begin{aligned} e^{rT}V_0^{(n)} &= \mathbb{E}^*\left(A(S) - K\right)^+ \\ &= \int_{-\infty}^{\infty}\left(S_0 \exp\left\{\sigma\sqrt{a_n T}z + \left(r - \frac{\sigma^2}{2}\right)\frac{T}{2}\right\} - K\right)^+ \varphi(z)\,dz \\ &= S_0 e^{\left(r-\frac{\sigma^2}{2}\right)\frac{T}{2}}\frac{1}{\sqrt{2\pi}}\int_{-d_n}^{\infty} e^{\sigma\sqrt{a_n T}z - \frac{1}{2}z^2}\,dz - K\int_{-d_n}^{\infty}\varphi(z)\,dz \\ &= S_0 e^{\left(r-\frac{\sigma^2}{2}\right)\frac{T}{2} + \frac{1}{2}\sigma^2 a_n T}\,\Phi\left(\sigma\sqrt{a_n T} + d_n\right) - K\Phi(d_n), \end{aligned}$$

the last equality by Exercise 12.12. □

Arguments similar to those in Theorems 15.3.1 and 15.3.4 establish formulas for the value of the options at arbitrary time $t \in [0, T]$ (see [12]).

In the arithmetic case, $A(S)$ does not have a lognormal distribution, hence arithmetic average options are more difficult to price. Indeed, as of this writing there is no known closed form pricing formula as in the geometric average case. Techniques used to price arithmetic average options typically involve approximation, Monte Carlo simulation, or partial differential equations. Accounts of these approaches along with further references may be found in [1, 12, 14, 19].

15.4 Other Options

Because of the limited scope of the text in this chapter and the last, we have described only a few of the many intricate options, path dependent or otherwise, available in the market. Omissions include

- *call or put options based on stocks with jumps*, thus incorporating into the model the realistic possibility of market shock.
- *stock index option*, where the underlying is a weighted average of stocks with interrelated price processes.
- *exchange option*, giving the holder the right to exchange one risky asset for another.
- *basket option*, where the payoff is a weighted average of a group of underlying assets.
- *Bermuda option*, similar to an American option but with a finite set of prescribed exercise dates.
- *Russian option*, where the payoff is the discounted maximum value of the underlying up to exercise time.
- *rainbow option*, where the payoff is usually based on the maximum or minimum value of a group of correlated assets.

The interested reader may find descriptions of these and other options, as well as expositions of related topics, in [1, 12, 14, 19].

15.5 Exercises

1. Let V_0^{cp} denote the price of a call-on-put option with strike price K_0 and maturity T_0, where the underlying put has strike price K and matures at time $T > T_0$. Show that

$$V_0^{cp} = e^{-rT_0} \int_{-\infty}^{y_1} \left[P_{T_0}\left(g(y)\right) - K_0\right] \varphi(y)\, dy,$$

where

$$y_1 := \sup\{y \mid P_{T_0}\left(g(y)\right) > K_0\},$$

$$P_{T_0}(s) = Ke^{-r(T-T_0)}\Phi\big(-d_2(T-T_0, s)\big) - s\Phi\big(-d_1(T-T_0, s)\big),$$

and g and $d_{1,2}$ are defined as in Section 14.4.

2. Let V_0^{pc} and V_0^{cc} denote, respectively, the prices of a put-on-call and a call-on-call option with strike price K_0 and maturity T_0, where the underlying call has strike price K and matures at time $T > T_0$. Prove the put-on-call, call-on-call parity relation

$$V_0^{pc} - V_0^{cc} + e^{-rT_0} \int_{-\infty}^{\infty} C\left(g(y)\right) \varphi(y)\, dy = K_0 e^{-rT_0},$$

where g and C are defined as in Section 14.4.

3. Let C_t^{di} and C_t^{do} denote, respectively, the time-t values of a down-and-in call option and a down-and-out call option, each with underlying $\mathcal{S}$, strike price K, and barrier level c. Show that $C_0^{do} + C_0^{di} = C_0$, where C_0 is the price of the corresponding standard call option.

4. (*Down-and-out forward*). Show that the price V_0 of a derivative with payoff $(S_T - K)\mathbf{1}_A$, where $A = \{m^S \geq c\}$, is given by (15.1) and (15.2) with $M = c$.

5. Let C_t^{do} and P_t^{do} denote, respectively, the time-t values of a down-and-out call option and a down-and-out put option, each with underlying $\mathcal{S}$, strike price K, and barrier level c. Let $A = \{m^S \geq c\}$. Show that $C_0^{do} - P_0^{do} = V_0$, where V_0 is as in Exercise 4.

6. (*Currency barrier option*). Referring to Section 14.1, show that the cost of a down-and-out option to buy one euro for K dollars at time T is

$$C_0^{do} = e^{-(r_d+\beta^2/2)T}\left[S_0\hat{\mathbb{E}}\left(e^{\gamma\hat{W}_T}\mathbf{1}_B\right) - K\hat{\mathbb{E}}\left(e^{\beta\hat{W}_T}\mathbf{1}_B\right)\right],$$

where

$$S = QE, \quad \beta = \frac{r}{\sigma} - \frac{\sigma}{2}, \quad r = r_d - r_e, \quad \text{and} \quad \gamma = \beta + \sigma,$$

Conclude that

$$C_0^{do} = S_0 e^{-r_e T}\left[\Phi(d_1) - \left(\frac{c}{S_0}\right)^{\frac{2r}{\sigma^2}+1}\Phi(\delta_1)\right] - Ke^{-r_d T}\left[\Phi(d_2) - \left(\frac{c}{S_0}\right)^{\frac{2r}{\sigma^2}-1}\Phi(\delta_2)\right],$$

where $d_{1,2}$ and $\delta_{1,2}$ are defined as in (15.2) with $r = r_d - r_e$.

7. A *fixed strike lookback put option* has payoff $(K - m^S)^+$. Show that the value of the option at time t is

$$V_t = e^{-r(T-t)}\mathbb{E}^*\left[(K - m^S)^+|\mathcal{F}_t^S\right] = e^{-r(T-t)}\left[K - G_t(\min(m_t^S, K), S_t)\right],$$

where $G_t(m, s)$ is given by (15.23).

8. Referring to Lemma 15.1.1, use the fact that $\hat{U} := -\hat{W}$ is a $\hat{\mathbb{P}}$-Brownian motion to show that the joint density $\hat{f}_M(x, y)$ of $(\hat{W}_T, M^{\hat{W}})$ under $\hat{\mathbb{P}}$ is given by

$$\hat{f}_M(x, y) = -\hat{g}_m(x, y)\mathbf{1}_{\{(x,y)|y\geq 0, y\geq x\}}.$$

9. Carry out the following steps to find the price C_0^{ui} of an up-and-in call option for the case $S_0 < K < c$:

   (a) Show that $C_T^{ui} = (S_T - K)\mathbf{1}_B$, where

$$B = \{\hat{W}_T \geq a, M^{\hat{W}} \geq b\}, \quad a := \sigma^{-1}\ln(K/S_0), \quad b := \sigma^{-1}\ln(c/S_0).$$

(b) Show that

$$C_0^{ui} = e^{-(r+\beta^2/2)T}\left[S_0\hat{\mathbb{E}}\left(e^{(\beta+\sigma)\hat{W}_T}\mathbf{1}_B\right) - K\hat{\mathbb{E}}\left(e^{\beta\hat{W}_T}\mathbf{1}_B\right)\right],$$

where $\beta := r/\sigma - \sigma/2$.

(c) Use Exercise 8 and Lemma 15.1.2 to show that

$$\begin{aligned}\hat{\mathbb{E}}\left(e^{\lambda\hat{W}_T}\mathbf{1}_B\right) &= -\int_a^b\int_b^\infty e^{\lambda x}\hat{g}_m(x,y)\,dy\,dx - \int_b^\infty\int_x^\infty e^{\lambda x}\hat{g}_m(x,y)\,dy\,dx\\ &= e^{2b\lambda+\lambda^2T/2}\left\{\Phi\left(\frac{2b-a+\lambda T}{\sqrt{T}}\right) - \Phi\left(\frac{b+\lambda T}{\sqrt{T}}\right)\right\}\\ &\qquad + e^{\lambda^2T/2}\Phi\left(\frac{-b+\lambda T}{\sqrt{T}}\right).\end{aligned}$$

(d) Conclude from (b) and (c) that

$$\begin{aligned}C_0^{ui} = S_0\left(\frac{c}{S_0}\right)^{\frac{2r}{\sigma^2}+1}&[\Phi(e_1)-\Phi(e_3)] + S_0\Phi(e_5) - K^{-rT}\Phi(e_6)\\ &- K^{-rT}\left(\frac{c}{S_0}\right)^{\frac{2r}{\sigma^2}-1}[\Phi(e_2)-\Phi(e_4)],\end{aligned}$$

where

$$\begin{aligned}e_{1,2} &= \frac{\ln\left(c^2/(KS_0)\right) + (r\pm\sigma^2/2)T}{\sigma\sqrt{T}}\\ e_{3,4} &= \frac{\ln(c/S_0) + (r\pm\sigma^2/2)T}{\sigma\sqrt{T}}\\ e_{5,6} &= \frac{\ln(S_0/c) + (r\pm\sigma^2/2)T}{\sigma\sqrt{T}}.\end{aligned}$$

10. Find the price of an up-and-in call option for the case $S_0 < c < K$.

11. In the notation of Section 14.5, a *down-and-out quanto call option* has payoff $V_T = (S_T^e - K)\mathbf{1}_B$, where S_T^e and K are denominated in dollars. Carry out the following steps to find the cost V_0 of the option for the case $K > c$:

(a) Replace S_t in (15.4) by

$$S_t^e = S_0^e\exp\left[\sigma W^*(t) + \beta t\right], \quad \beta := r_e + \sigma_2^2 - \tfrac{1}{2}\sigma^2$$

(see (14.10)).

(b) Find a formula analogous to (15.12).

(c) Conclude from (b) that

$$V_0 = S_0^e e^{(s-r_d)T}\left[\Phi(d_1) - \left(\frac{c}{S_0}\right)^{\varrho+1}\Phi(\delta_1)\right] - Ke^{-r_dT}\left[\Phi(d_2) - \left(\frac{c}{S_0}\right)^{\varrho-1}\Phi(\delta_2)\right]$$

where $s = r_e + \sigma_2^2$, $\varrho = 2s/\sigma^2$,

$$d_{1,2} = \frac{\ln\left(S_0/K\right) + (s\pm\frac{1}{2}\sigma^2)T}{\sigma\sqrt{T}}, \quad \delta_{1,2} = \frac{\ln\left(c^2/(S_0K)\right) + (s\pm\frac{1}{2}\sigma^2)T}{\sigma\sqrt{T}}.$$

12. Let C_0^{do} be the price of a down-and-out barrier call option, as given by (15.1). Show that

$$\lim_{c \to S_0^-} C_0^{do} = 0 \quad \text{and} \quad \lim_{c \to 0+} C_0^{do} = C_0,$$

where C_0 is the cost of a standard call option.

13. In the notation of Theorems 15.3.1 and 15.3.4, show that

$$V_0 = \lim_{n \to \infty} V_0^{(n)}.$$

14. (*Barrier option on a stock with dividends*). Find a formula for the price of a down-and-out call option based on a stock that pays a continuous stream of dividends.

15. Referring to §15.1, find the probability under $\mathbb{P}$ that the barrier c is breached.

16. Use the methods of §15.1 to find the price V_0 of a derivative with payoff $V_T = (S_T - m^S)\mathbf{1}_A$, where $A = \{m^S \geq c\}$, $c < S_0$.

Appendix A

Basic Combinatorics

In this appendix we set out the principles of counting that are needed in the chapters on probability. As the objects to be counted are members of sets, we shall use the language and notation of sets in describing these principles. Thus we begin with a brief description of elementary set theory.

Set Theory

A *set* is a collection of objects called the *members* or *elements* of the set. Abstract sets are usually denoted by capital letters A, B, and so forth. If x is a member of the set A, we write $x \in A$; otherwise, we write $x \notin A$. The *empty set*, denoted by $\emptyset$, is the set with no members.

A set can be described either by words, by listing the elements, or by set-builder notation. Set builder notation is of the form $\{x \mid P(x)\}$, which is read "the set of all x such that $P(x)$," where $P(x)$ is a well-defined property that x must satisfy to belong to the set. For example, the set of all even positive integers can be described as $\{2, 4, 6, \ldots\}$ or as $\{n \mid n = 2m \text{ for some positive integer } m\}$.

A set A is *finite* if either A is the empty set or there is a one-to-one correspondence between A and the set $\{1, 2, \ldots, n\}$ for some positive integer n. In effect, this means that the members of A may be labeled with the numbers $1, 2, \ldots, n$, so that A may be described as, say, $\{a_1, a_2, \ldots, a_n\}$. A set is *countably infinite* if its members may be labeled with the positive integers $1, 2, 3, \ldots$, *countable* if it is either finite or countably infinite, and *uncountable* otherwise. The set $\mathbf{N}$ of all positive integers is obviously countably infinite, as is the set $\mathbf{Z}$ of *all* integers. Less obvious is the fact that set $\mathbf{Q}$ of all rational numbers is countably infinite. The set $\mathbf{R}$ of all real numbers is uncountable, as is any (nontrivial) interval of real numbers.

Sets A and B are said to be *equal*, written $A = B$, if every member of A is a member of B and vice versa. A is a *subset* of B, written $A \subseteq B$, if every member of A is a member of B. It follows that $A = B$ iff[1] $A \subseteq B$ and $B \subseteq A$. Note that the empty set is a subset of every set. Hereafter, we shall assume that all sets under consideration in any discussion are subsets of a larger set S, sometimes called the *universe (of discourse)* for the discussion.

[1]Read "if and only if."

The basic set operations are

$$\begin{array}{rcll} A \cup B & = & \{x \mid x \in A \text{ or } x \in B\}, & \text{the } \textit{union} \text{ of } A \text{ and } B; \\ A \cap B & = & \{x \mid x \in A \text{ and } x \in B\}, & \text{the } \textit{intersection} \text{ of } A \text{ and } B; \\ A' & = & \{x \mid x \in S \text{ and } x \notin A\}, & \text{the } \textit{complement} \text{ of } A; \\ A - B & = & \{x \mid x \in A \text{ and } x \notin B\}, & \text{the } \textit{difference} \text{ of } A \text{ and } B; \\ A \times B & = & \{(x, y) \mid x \in A \text{ and } y \in B\}, & \text{the } \textit{product} \text{ of } A \text{ and } B. \end{array}$$

Similar definitions may be given for the union, intersection, and product of three or more sets, or even for infinitely many sets. For example,

$$\bigcup_{n=1}^{\infty} A_n = A_1 \cup A_2 \cup \cdots = \{x \mid x \in A_n \text{ for some } n\},$$

$$\bigcap_{n=1}^{\infty} A_n = A_1 \cap A_2 \cap \cdots = \{x \mid x \in A_n \text{ for every } n\},$$

$$\prod_{n=1}^{\infty} A_n = A_1 \times A_2 \times \cdots = \{(a_1, a_2, \ldots) \mid a_n \in A_n,\ n = 1, 2, \ldots\}.$$

We usually omit the cap symbol in the notation for intersection, writing, for example, ABC instead of $A \cap B \cap C$. Similarly, we write AB' for $A - B$, etc.

A collection of sets is said to be *pairwise disjoint* if $AB = \emptyset$ for each pair of distinct members A and B of the collection. A *partition* of a set S is a collection of pairwise disjoint nonempty sets whose union is S.

The Addition Principle

The number of elements in a finite set A is denoted by $|A|$. In particular, $|\emptyset| = 0$. The following result is easily established by mathematical induction.

A.1 Theorem. *If $A_1, A_2, \ldots, A_r$ are pairwise disjoint finite sets, then*

$$|A_1 \cup A_2 \cup \cdots \cup A_r| = |A_1| + |A_2| + \ldots + |A_r|.$$

Theorem A.1 underlies the so-called *Addition Principle of Combinatorics* described as follows:

> *If there are n_1 ways of accomplishing one task, n_2 ways of accomplishing a second task, ..., and n_r ways to accomplish an rth task, and if no two tasks may be accomplished simultaneously, then there are $n_1 + n_2 + \cdots + n_r$ ways to choose one of the tasks.*

A.2 Corollary. *If A and B are finite sets, then*

$$|A \cup B| = |A| + |B| - |AB|.$$

Proof. Note that

$$A \cup B = AB' \cup A'B \cup AB, \quad A = AB' \cup AB, \quad \text{and} \quad B = A'B \cup AB,$$

where the sets comprising each of these unions are pairwise disjoint. By Theorem A.1,

$$|A \cup B| = |AB'| + |A'B| + |AB|, \quad |A| = |AB'| + |AB|, \quad \text{and} \quad |B| = |A'B| + |AB|.$$

Subtracting the second and third equations from the first and rearranging yields the desired formula. □

The Multiplication Principle

Here is the multiplicative analog of 2.2.1:

A.3 Theorem. *If $A_1, A_2, \ldots, A_n$ are finite sets then*

$$|A_1 \times A_2 \times \cdots \times A_n| = |A_1||A_2| \cdots |A_n|.$$

Proof. (By induction on n.) Consider the case $n = 2$. For each fixed $x_1 \in A_1$ there are $|A_2|$ elements of the form (x_1, x_2), where x_2 runs through the set A_2. Since all the members of $A_1 \times A_2$ may be listed in this manner it follows that $|A_1 \times A_2| = |A_1||A_2|$. Now suppose that the assertion of the theorem holds for $n = k - 1$ $(k > 1)$. Since $A_1 \times A_2 \times \cdots \times A_k = B \times A_k$, where $B = A_1 \times A_2 \times \cdots \times A_{k-1}$, the case $n = 2$ implies that $|B \times A_k| = |B||A_k|$. But by the induction hypothesis $|B| = |A_1||A_2| \cdots |A_{k-1}|$. Therefore, the assertion holds for $n = k$. □

Theorem A.3 is the basis of the so-called *Multiplication Principle of Combinatorics*:

> *When performing a task requiring r steps, if there are n_1 ways to complete step 1, and if for each of these there are n_2 ways to complete step 2, ..., and if for each of these there are n_r ways to complete step r, then there are $n_1 n_2 \cdots n_r$ ways to perform the task.*

A.4 Example. How many three-letter sequences can be formed from the letters of the word "formula" if no letter is used more than once?

Solution: We can describe the task as a three-step process. The first step is the selection of the first letter: 7 choices. The second step is the selection of the second letter: 6 choices, since the letter chosen in the preceding step is no longer available. The third step is the selection of the last letter: 5 choices. By the multiplication principle, there are a total of $7 \cdot 6 \cdot 5 = 210$ possible sequences. ◊

Permutations

The sequences in Example A.4 are called *permutations*, defined formally as follows: Let n and r be positive integers with $1 \le r \le n$. A *permutation of n items taken r at a time* is an ordered list of r items chosen from the n items.

The number of such lists is denoted by $(n)_r$. An argument similar to that in Example A.4 shows that

$$(n)_r = n(n-1)(n-2)\cdots(n-r+1).$$

We may also write this as

$$(n)_r = \frac{n!}{(n-r)!},$$

where the symbol $m!$, read "m factorial," is defined as

$$m! = m(m-1)(m-2)\cdots 2\cdot 1.$$

By convention, $0! = 1$, a choice that ensures consistency in combinatorial formulas. For example, the number of permutations of n items taken n at a time is, according to the above formula and convention, $\frac{n!}{0!} = n!$, which is what one obtains by directly applying the multiplication principle.

A.5 Example. In how many ways can a group of 4 women and 4 men line up for a photograph if no two adjacent people are of the same gender?

Solution: There are two alternatives, corresponding to the gender of, say, the first person on the left. For each alternative there are 4! arrangements of the women and 4! arrangements of the men. Thus there are a total of $2(4!)^2 = 1152$ arrangements. ◊

Combinations

In contrast to a permutation, which is an *ordered* list, a combination is an *unordered* list. Here is the formal definition: Let n and r be positive integers with $1 \le r \le n$. A *combination of n items taken r at a time* is a set of r items chosen from the n items. Note that each combination of r items gives rise to $r!$ permutations. Thus the number x of combinations of n things taken r at a time satisfies $xr! = (n)_r$. Solving for x, we arrive at the following formula for the number of combinations of n things taken r at a time:

$$\frac{(n)_r}{r!} = \frac{n!}{(n-r)!r!}.$$

The quotient on the right is called a *binomial coefficient* and is denoted by $\binom{n}{r}$, read "n choose r." Its name derives from the following result:

A.6 Theorem (Binomial Theorem). *Let a and b be real numbers and n a positive integer. Then*

$$(a+b)^n = \sum_{j=0}^{n} \binom{n}{j} a^j b^{n-j}.$$

Proof. The expression $(a+b)^n = (a+b)(a+b)\cdots(a+b)$ (n factors) is the sum terms of the form $x_1 x_2 \cdots x_n$, where $x_i = a$ or b. Collecting the $a's$ and $b's$ we can write each product as $a^j b^{n-j}$ for some $0 \le j \le n$. For each fixed j there are exactly $\binom{n}{j}$ such terms, corresponding to the number of ways one may choose exactly j of the n factors in $x_1 x_2 \cdots x_n$ to be a. □

A.7 Example. A restauranteur needs to hire a sauté chef, a fish chef, a vegetable chef, and three grill chefs. If there are 10 applicants equally qualified for the positions, in how many ways can the positions be filled?

Solution: We apply the multiplication principle: First, select a sauté chef: 10 choices; second, select a fish chef: 9 choices; third, select a vegetable chef: 8 choices; finally, select three grill chefs from the remaining 7 applicants: $\binom{7}{3} = 35$ choices. Thus there are a grand total of $10 \cdot 9 \cdot 8 \cdot 35 = 25,200$ choices. ◊

A.8 Example. A bag contains 5 red, 4 yellow, and 3 green marbles. In how many ways is it possible to draw 5 marbles at random from the bag with exactly 2 reds and no more than 1 green?

Solution: We have the following decision scheme:

Case 1: No green marbles.

Step 1: Choose the 2 reds: $\binom{5}{2} = 10$ possibilities.

Step 2: Choose 3 yellows: $\binom{4}{3} = 4$ possibilities.

Case 2: Exactly one green marble.

Step 1: Choose the green: 3 possibilities.

Step 2: Choose the 2 reds: $\binom{5}{2} = 10$ possibilities.

Step 3: Choose 2 yellows: $\binom{4}{2} = 6$ possibilities.

Thus there are a total of $40 + 180 = 220$ possibilities. ◊

A.9 Example. How many different 12-letter arrangements of the letters of the word “arrangements” are there?

Solution: Notice that there are duplicate letters, so the answer 12! is incorrect. We proceed as follows:

Step 1: Select positions for the a’s: $\binom{12}{2} = 66$ choices.

Step 2: Select positions for the r’s: $\binom{10}{2} = 45$ choices.

Step 3: Select positions for the e’s: $\binom{8}{2} = 28$ choices.

Step 4: Select positions for the n’s: $\binom{6}{2} = 15$ choices.

Step 5: Fill the remaining spots with the letters g, m, t, s: 4! choices.

Thus there are $66 \cdot 45 \cdot 28 \cdot 15 \cdot 24 = 29,937,600$ different arrangements. ◊

We conclude this section with the following application of the binomial theorem.

A.10 Theorem. *A set S with n members has 2^n subsets (including S and $\emptyset$).*

Proof. Since $\binom{n}{r}$ gives the number of subsets of size r, the total number of subsets of S is

$$\binom{n}{0} + \binom{n}{1} + \cdots + \binom{n}{n}.$$

By the binomial theorem, this quantity is $(1+1)^n = 2^n$. □

Remark. One can also prove Theorem A.10 by induction on n: The conclusion is obvious if $n = 0$ or 1. Assume the theorem holds for all sets with $n \geq 1$ members, and let S be a set with $n+1$ members. Choose a fixed element s from S. This produces two collections of subsets of S: those that contain s and those that don't. The latter are precisely the subsets of $S - \{s\}$, and, by the induction hypothesis, there are 2^n of these. But the two collections have the same number of subsets, since one collection may be obtained from the other by either removing or adjoining s. Thus, there are a total of $2^n + 2^n = 2^{n+1}$ subsets of S, completing the induction step. ◊

Appendix B

Solution of the BSM PDE

In this appendix we solve the Black-Scholes-Merton PDE

$$v_t + rsv_s + \tfrac{1}{2}\sigma^2 s^2 v_{ss} - rv = 0 \quad (s > 0,\ 0 \le t \le T), \tag{B.1}$$

with boundary conditions

$$v(T, s) = f(s)\ (s \ge 0) \quad \text{and} \quad v(t, 0) = 0 \quad (0 \le t \le T), \tag{B.2}$$

where f is continuous and satisfies suitable growth conditions (see footnote on page 257).

Reduction to a Diffusion Equation

As a first step, we simplify Equation (B.1) by making the substitutions

$$s = e^x, \quad t = T - \frac{2\tau}{\sigma^2}, \quad \text{and} \quad v(t, s) = u(\tau, x). \tag{B.3}$$

By the chain rule,

$$\begin{aligned} v_t(t, s) &= u_\tau(\tau, x)\tau_t = -\tfrac{1}{2}\sigma^2 u_\tau(\tau, x), \\ v_s(t, s) &= u_x(\tau, x)x_s = s^{-1}u_x(\tau, x), \\ v_{ss}(t, s) &= s^{-2}\left[u_{xx}(\tau, x) - u_x(\tau, x)\right]. \end{aligned}$$

Substituting these expressions into (B.1) produces the equation

$$-\tfrac{1}{2}\sigma^2 u_\tau(\tau, x) + ru_x(\tau, x) + \tfrac{1}{2}\sigma^2\left[u_{xx}(\tau, x) - u_x(\tau, x)\right] - ru(\tau, x) = 0.$$

Dividing by $\sigma^2/2$ and setting $k = 2r/\sigma^2$, we obtain

$$u_\tau(\tau, x) = (k-1)u_x(\tau, x) + u_{xx}(x, \tau) - ku(x, \tau). \tag{B.4}$$

In terms of u, the conditions in (B.2) become

$$u(0, x) = f(e^x), \quad \text{and} \quad \lim_{x \to -\infty} u(\tau, x) = 0, \quad 0 \le \tau \le T\sigma^2/2.$$

Equation (B.4) is an example of a *diffusion equation*.

Reduction to the Heat Equation

The diffusion equation (B.4) may be reduced to a simpler form by the substitution

$$u(\tau,x) = e^{ax+b\tau}w(\tau,x) \tag{B.5}$$

for suitable constants a and b. To determine a and b, we compute the partial derivatives

$$\begin{aligned}
u_\tau(\tau,x) &= e^{ax+b\tau}[bw(\tau,x) + w_\tau(\tau,x)]\\
u_x(\tau,x) &= e^{ax+b\tau}[aw(\tau,x) + w_x(\tau,x)]\\
u_{xx}(\tau,x) &= e^{ax+b\tau}[aw_x(\tau,x) + w_{xx}(\tau,x)] + ae^{ax+b\tau}[aw(\tau,x) + w_x(\tau,x)]\\
&= e^{ax+b\tau}[a^2w(\tau,x) + 2aw_x(\tau,x) + w_{xx}(\tau,x)].
\end{aligned}$$

Substituting these expressions into (B.4) and dividing by $e^{ax+b\tau}$ yields

$$\begin{aligned}
bw(\tau,x) + w_\tau(\tau,x) = (k-1)\big[aw(\tau,x) + w_x(\tau,x)\big] + a^2w(\tau,x)\\
+ 2aw_x(\tau,x) + w_{xx}(\tau,x) - kw(\tau,x),
\end{aligned}$$

which simplifies to

$$w_\tau(\tau,x) = \big[a(k-1) + a^2 - k - b\big]w(\tau,x) + [2a + k - 1]w_x(\tau,x) + w_{xx}(\tau,x).$$

The terms involving $w(\tau,x)$ and $w_x(\tau,x)$ may be eliminated by choosing

$$a = \tfrac{1}{2}(1-k) \quad\text{and}\quad b = a(k-1) + a^2 - k = -\tfrac{1}{4}(k+1)^2.$$

With these values of a and b, we obtain the PDE

$$w_\tau(\tau,x) = w_{xx}(\tau,x), \quad \tau > 0. \tag{B.6}$$

Since $w(0,x) = e^{-ax}u(0,x)$, the boundary condition $u(0,x) = f(e^x)$ becomes

$$w_0(x) := w(0,x) = e^{-ax}f(e^x). \tag{B.7}$$

Equation (B.6) is the well-known *heat equation* of mathematical physics.

Solution of the Heat Equation

To solve the heat equation, we begin with the observation that the function

$$\kappa(\tau,x) = \frac{1}{2\sqrt{\pi\tau}}e^{-x^2/4\tau} = \frac{1}{\sqrt{2\tau}}\varphi\left(\frac{x}{\sqrt{2\tau}}\right)$$

is a solution of (B.6), as may be readily verified. The function κ is called the *kernel* of the heat equation. To construct a solution of (B.6) that satisfies (B.7), we form the *convolution* of κ with w_0:

$$w(\tau,x) = \int_{-\infty}^{\infty} w_0(y)\kappa(\tau,x-y)\,dy. \tag{B.8}$$

Differentiating inside the integral,[1] we obtain

$$w_\tau(\tau,x) = \int_{-\infty}^{\infty} w_0(y)\kappa_\tau(\tau, x-y)\,dy,$$

$$w_x(\tau,x) = \int_{-\infty}^{\infty} w_0(y)\kappa_x(\tau, x-y)\,dy, \quad \text{and}$$

$$w_{xx}(\tau,x) = \int_{-\infty}^{\infty} w_0(y)\kappa_{xx}(\tau, x-y)\,dy.$$

Since $\kappa_t = \kappa_{xx}$ we see that w satisfies (B.6). Also,

$$w(\tau,x) = \frac{1}{\sqrt{2\tau}}\int_{-\infty}^{\infty} w_0(y)\varphi\left(\frac{y-x}{\sqrt{2\tau}}\right)dy = \int_{-\infty}^{\infty} w_0\left(x+z\sqrt{2\tau}\right)\varphi(z)\,dz,$$

hence,

$$\begin{aligned}\lim_{\tau\to 0+} w(\tau,x) &= \int_{-\infty}^{\infty} \lim_{\tau\to 0+} w_0\left(x+z\sqrt{2\tau}\right)\varphi(z)\,dz\\ &= \int_{-\infty}^{\infty} w_0(x)\varphi(z)\,dz\\ &= w_0(x).\end{aligned}$$

Therefore, w has a continuous extension to $\mathbf{R}^+ \times \mathbf{R}$ that satisfies the initial condition (B.7).

From (B.7) we see that the solution w in (B.8) may now be written

$$w(\tau,x) = \frac{1}{2\sqrt{\tau\pi}}\int_{-\infty}^{\infty} f(e^y)\exp\left\{-ay - \frac{1}{2}\left(\frac{y-x}{\sqrt{2\tau}}\right)^2\right\}dy.$$

Rewriting the exponent in the integrand as $-ax + a^2\tau - \frac{1}{4\tau}(y-x+2a\tau)^2$ and making the substitution $z = e^y$ we obtain

$$w(\tau,x) = \frac{e^{-ax+a^2\tau}}{2\sqrt{\tau\pi}}\int_0^{\infty} f(z)\exp\left\{-\frac{1}{4\tau}\{\ln z - x + 2a\tau\}^2\right\}\frac{dz}{z}. \tag{B.9}$$

Back to the BSM PDE

The final step is to unravel the substitutions that led to the heat equation. From (B.3), (B.5), and (B.9) we have

$$\begin{aligned}v(t,s) = u(\tau,x) &= e^{ax+b\tau}w(\tau,x)\\ &= \frac{e^{(a^2+b)\tau}}{2\sqrt{\tau\pi}}\int_0^{\infty} f(z)\exp\left\{-\frac{1}{4\tau}\{\ln z - x + 2a\tau\}^2\right\}\frac{dz}{z}.\end{aligned}$$

[1] That limit operations such as differentiation may be moved inside the integral is justified by a theorem in real analysis, which is applicable in the current setting provided that w_0 does not grow too rapidly. In the case of a call option, for example, one can show that $|w_0(x)| \le Me^{N|x|}$ for suitable positive constants M and N and for all x. This inequality is sufficient to ensure that the appropriate integrals converge, allowing the interchange of limit operation and integral.

Recalling that

$$k = \frac{2r}{\sigma^2}, \quad a = \frac{1-k}{2} = \frac{\sigma^2 - 2r}{2\sigma^2}, \quad b = -\frac{(k+1)^2}{4}, \quad \text{and} \quad \tau = \frac{\sigma^2(T-t)}{2},$$

we have

$$\begin{aligned}(a^2 + b)\tau &= \frac{(k-1)^2 - (k+1)^2}{4}\tau = -k\tau = -r(T-t) \quad \text{and} \\ 2a\tau &= \frac{\sigma^2 - 2r}{\sigma^2} \cdot \frac{\sigma^2(T-t)}{2} = \frac{(\sigma^2 - 2r)(T-t)}{2} = -(r - \tfrac{1}{2}\sigma^2)(T-t).\end{aligned}$$

Since $x = \ln s$, we obtain the following solution for the general Black-Scholes-Merton PDE

$$v(t,s) = \frac{e^{-r\tau}}{\sigma\sqrt{2\pi\tau)}} \int_0^\infty f(z) \exp\left(-\frac{1}{2}\left\{\frac{\ln(z/s) - (r - \sigma^2/2)\tau}{\sigma\sqrt{\tau}}\right\}^2\right)\frac{dz}{z},$$

where $\tau := T - t$. Making the substitution

$$y = \frac{\ln(z/s) - (r - \sigma^2/2)\tau}{\sigma\sqrt{\tau}}$$

and noting that

$$z = s\exp\left\{y\sigma\sqrt{\tau} + (r - \sigma^2/2)\tau\right\} \quad \text{and} \quad \frac{dz}{z} = \sigma\sqrt{\tau}\,dy,$$

we arrive at the desired solution

$$v(t,s) = e^{-r\tau}\int_{-\infty}^{\infty} f\left(s\exp\left\{\sigma\sqrt{\tau}\,y + (r - \sigma^2/2)\tau\right\}\right)\varphi(y)\,dy.$$

It may be shown that the solution to the BSM PDE is unique within a class of functions that do not grow too rapidly. (See, for example, [20].)

Appendix C

Properties of the BSM Call Function

Recall that the Black-Scholes-Merton call function is defined as

$$C = C(\tau, s, k, \sigma, r) = s\Phi(d_1) - ke^{-r\tau}\Phi(d_2), \quad \tau, s, k, \sigma, r > 0,$$

where

$$d_{1,2} = d_{1,2}(\tau, s, k, \sigma, r) = \frac{\ln(s/k) + (r \pm \sigma^2/2)\tau}{\sigma\sqrt{\tau}}.$$

$C(\tau, s, k, \sigma, r)$ is the price of a call option with strike price k, maturity τ, and underlying stock price s. In this appendix, we prove Theorems 12.4.1 and 12.4.2, which summarize the main analytical properties of C.

Preliminary Lemmas

C.1 Lemma. $\displaystyle\int_{-d_2}^{\infty} e^{\sigma\sqrt{\tau}z}\varphi(z)\,dz = e^{\sigma^2\tau/2}\Phi(d_1).$

Proof. The integral may be written

$$\begin{aligned}
\frac{1}{\sqrt{2\pi}}\int_{-d_2}^{\infty} e^{\sigma\sqrt{\tau}z - z^2/2}\,dz &= \frac{e^{\sigma^2\tau/2}}{\sqrt{2\pi}}\int_{-d_2}^{\infty} e^{-(z-\sigma\sqrt{\tau})^2/2}\,dz \\
&= \frac{e^{\sigma^2\tau/2}}{\sqrt{2\pi}}\int_{-d_2-\sigma\sqrt{\tau}}^{\infty} e^{-x^2/2}\,dx \\
&= e^{\sigma^2\tau/2}\Phi\left(-d_2 - \sigma\sqrt{\tau}\right),
\end{aligned}$$

where we have made the substitution $x = z - \sigma\sqrt{\tau}$. Since $d_2 + \sigma\sqrt{\tau} = d_1$, the conclusion of the lemma follows. (Or use Exercise 12.12.) □

C.2 Lemma. $\displaystyle\int_{-d_2}^{\infty} ze^{\sigma\sqrt{\tau}z}\varphi(z)\,dz = e^{\sigma^2\tau/2}\left[\sigma\sqrt{\tau}\,\Phi(d_1) + \varphi(d_1)\right].$

Proof. Arguing as in the proof Lemma C.1, we have

$$\begin{aligned}
\int_{-d_2}^{\infty} ze^{\sigma\sqrt{\tau}z}\varphi(z)\,dz &= \frac{e^{\sigma^2\tau/2}}{\sqrt{2\pi}}\int_{-d_2}^{\infty} ze^{-(z-\sigma\sqrt{\tau})^2/2}\,dz \\
&= \frac{e^{\sigma^2\tau/2}}{\sqrt{2\pi}}\int_{-d_1}^{\infty} \left(x+\sigma\sqrt{\tau}\right)e^{-x^2/2}\,dx \\
&= \frac{e^{\sigma^2\tau/2}}{\sqrt{2\pi}}\int_{-d_1}^{\infty} xe^{-x^2/2}\,dx + \frac{\sigma\sqrt{\tau}e^{\sigma^2\tau/2}}{\sqrt{2\pi}}\int_{-d_1}^{\infty} e^{-x^2/2}\,dx \\
&= \frac{e^{\sigma^2\tau/2}}{\sqrt{2\pi}}\int_{d_1^2/2}^{\infty} e^{-y}\,dy + \sigma\sqrt{\tau}e^{\sigma^2\tau/2}\left[1-\Phi(-d_1)\right] \\
&= e^{\sigma^2\tau/2}\varphi(d_1) + \sigma\sqrt{\tau}e^{\sigma^2\tau/2}\Phi(d_1).
\end{aligned}$$

□

C.3 Lemma. *For positive τ, s, k, σ, r, define*

$$g(z) = g(\tau, s, k, \sigma, r, z) := se^{\sigma\sqrt{\tau}z+(r-\sigma^2/2)\tau} - k.$$

Then

$$C = e^{-r\tau}\int_{-d_2}^{\infty} g(z)\varphi(z)\,dz = e^{-r\tau}\int_{-\infty}^{\infty} g^+(z)\varphi(z)\,dz.$$

Proof. By Lemma C.1,

$$\begin{aligned}
\int_{-d_2}^{\infty} g(z)\varphi(z)\,dz &= se^{(r-\sigma^2/2)\tau}\int_{-d_2}^{\infty} e^{\sigma\sqrt{\tau}z}\varphi(z)\,dz - k\int_{-d_2}^{\infty}\varphi(z)\,dz \\
&= se^{r\tau}\Phi(d_1) - k\Phi(d_2) \\
&= e^{r\tau}C.
\end{aligned}$$

Since $g^+(z)$ is increasing, equals 0 if $z \le -d_2$, and equals g otherwise, the assertion follows. □

C.4 Lemma. *With g as in Lemma C.3,*

$$\frac{\partial}{\partial x}\int_{-d_2(\tau,s,k,\sigma,r)}^{\infty} g(\tau, s, k, \sigma, r, z)\varphi(z)\,dz = \int_{-d_2}^{\infty} g_x(z)\varphi(z)\,dz,$$

where x denotes any of the variables τ, s, k, σ, r.

Proof. Suppose that $x = s$. Fix the variables τ, k, σ, and r, and define

$$h(s) = d_2(\tau, s, k, \sigma, r) \quad \text{and} \quad F(s_1, s_2) = \int_{-h(s_1)}^{\infty} g(\tau, s_2, k, \sigma, r, z)\varphi(z)\,dz.$$

By the chain rule for functions of several variables, the left side of the equation in the assertion of the lemma for $x = s$ is

$$\frac{d}{ds}F(s, s) = F_1(s, s) + F_2(s, s).$$

Since

$$F_1(s_1, s_2) = g\big(\tau, s_2, k, \sigma, r, -h(s_1)\big)\varphi\big(-h(s_1)\big)h'(s_1)$$

and $g\big(\tau, s, k, \sigma, r, -h(s)\big) = 0$, we see that $F_1(s, s) = 0$. Noting that

$$F_2(s_1, s_2) = \int_{-h(s_1)}^{\infty} g_s(\tau, s_2, k, \sigma, r, z)\varphi(z)\,dz$$

we now have

$$\frac{d}{ds}F(s, s) = F_2(s, s) = \int_{-d_2(\tau,s,k,\sigma,r)}^{\infty} g_s(\tau, s, k, \sigma, r, z)\varphi(z)\,dz,$$

which is the assertion of the lemma for the case $x = s$. A similar argument works for the variables τ, k, σ, and r. □

Proof of Theorem 12.4.1

(i) $\dfrac{\partial C}{\partial s} = \Phi(d_1)$: By Lemmas C.1, C.3, and C.4,

$$\begin{aligned}\frac{\partial C}{\partial s} &= e^{-r\tau}\int_{-d_2}^{\infty} g_s(z)\varphi(z)\,dz = e^{-\tau\sigma^2/2}\int_{-d_2}^{\infty} e^{\sigma\sqrt{\tau}z}\varphi(z)\,dz \\ &= \Phi(d_1).\end{aligned}$$

(ii) $\dfrac{\partial^2 C}{\partial s^2} = \dfrac{\varphi(d_1)}{s\sigma\sqrt{\tau}}$: This follows from the chain rule and **(i)**.

(iii) $\dfrac{\partial C}{\partial \tau} = \dfrac{\sigma s}{2\sqrt{\tau}}\varphi(d_1) + kre^{-r\tau}\Phi(d_2)$: By Lemmas C.3 and C.4,

$$\begin{aligned}\frac{\partial C}{\partial \tau} &= e^{-r\tau}\int_{-d_2}^{\infty} g_\tau(z)\varphi(z)\,dz - re^{-r\tau}\int_{-d_2}^{\infty} g(z)\varphi(z)\,dz \\ &= A - B, \quad \text{say.}\end{aligned}$$

By Lemma C.3

$$B = rC = rs\Phi(d_1) - rke^{-r\tau}\Phi(d_2),$$

and by Lemmas C.1 and C.2

$$\begin{aligned}A &= \frac{s}{e^{\sigma^2\tau/2}}\left\{\frac{\sigma}{2\sqrt{\tau}}\int_{-d_2}^{\infty} ze^{\sigma\sqrt{\tau}z}\varphi(z)\,dz + (r - \sigma^2/2)\int_{-d_2}^{\infty} e^{\sigma\sqrt{\tau}z}\varphi(z)\,dz\right\} \\ &= \frac{s}{e^{\sigma^2\tau/2}}\left\{\frac{\sigma e^{\sigma^2\tau/2}}{2\sqrt{\tau}}\big(\sigma\sqrt{\tau}\Phi(d_1) + \varphi(d_1)\big) + (r - \sigma^2/2)e^{\sigma^2\tau/2}\Phi(d_1)\right\} \\ &= \frac{s\sigma}{2\sqrt{\tau}}\varphi(d_1) + rs\Phi(d_1).\end{aligned}$$

(iv) $\dfrac{\partial C}{\partial \sigma} = s\sqrt{\tau}\varphi(d_1)$: By Lemmas C.1–C.4,

$$\begin{aligned}
\frac{\partial C}{\partial \sigma} &= e^{-r\tau}\int_{-d_2}^{\infty} g_\sigma(z)\varphi(z)\,dz \\
&= se^{-\sigma^2\tau/2}\int_{-d_2}^{\infty} \left(z\sqrt{\tau}-\sigma\tau\right)e^{\sigma\sqrt{\tau}z}\varphi(z)\,dz \\
&= se^{-\sigma^2\tau/2}\left\{\sqrt{\tau}e^{\sigma^2\tau/2}\left(\sigma\sqrt{\tau}\Phi(d_1)+\varphi(d_1)\right)-\sigma\tau e^{\sigma^2\tau/2}\Phi(d_1)\right\} \\
&= s\sqrt{\tau}\,\varphi(d_1).
\end{aligned}$$

(v) $\dfrac{\partial C}{\partial r} = k\tau e^{-r\tau}\Phi(d_2)$: By Lemmas C.1, C.3, and C.4,

$$\begin{aligned}
\frac{\partial C}{\partial r} &= e^{-r\tau}\int_{-d_2}^{\infty} g_r(z)\varphi(z)\,dz - \tau e^{-r\tau}\int_{-d_2}^{\infty} g(z)\varphi(z)\,dz \\
&= e^{-r\tau}\int_{-d_2}^{\infty} g_r(z)\varphi(z)\,dz - \tau C \\
&= \tau se^{-\sigma^2/2t}\int_{-d_2}^{\infty} e^{\sigma\sqrt{\tau}z}\varphi(z)\,dz - \tau s\Phi(d_1) + k\tau e^{-r\tau}\Phi(d_2) \\
&= k\tau e^{-r\tau}\Phi(d_2).
\end{aligned}$$

(vi) $\dfrac{\partial C}{\partial k} = -e^{-r\tau}\Phi(d_2)$: By Lemmas C.3 and C.4,

$$\frac{\partial C}{\partial k} = e^{-r\tau}\int_{-d_2}^{\infty} g_k(z)\varphi(z)\,dz = -e^{-r\tau}\int_{-d_2}^{\infty}\varphi(z)\,dz = -e^{-r\tau}\Phi(d_2).$$

Proof of Theorem 12.4.2

The proofs of the limit formulas make use of

$$\lim_{z\to\infty}\Phi(z) = 1, \qquad \lim_{z\to-\infty}\Phi(z) = 0,$$

and the limit properties of $d_{1,2}$.

(i) $\lim_{s\to+\infty}\left[C(\tau,s,k,\sigma,r) - (s - ke^{-r\tau})\right] = 0$: Note first that

$$C(\tau,s,k,\sigma,r) - s + ke^{-r\tau} = s\left(\Phi(d_1)-1\right) - ke^{-r\tau}\left(\Phi(d_2)-1\right).$$

Since $\lim_{s\to\infty} d_1 = \lim_{s\to\infty} d_2 = \infty$, we have $\lim_{s\to\infty}\Phi(d_{1,2}) = 1$. It remains to show that $\lim_{s\to\infty} s\left(\Phi(d_1)-1\right) = 0$ or, by L'Hôpital's rule,

$$\lim_{s\to+\infty} s^2\frac{\partial}{\partial s}\Phi(d_1) = 0. \tag{†}$$

Since $\dfrac{\partial d_1}{\partial s} = (s\sigma\sqrt{\tau})^{-1}$,

$$s^2 \frac{\partial}{\partial s}\Phi(d_1) = s^2\varphi(d_1)\frac{\partial d_1}{\partial s} = \frac{se^{-d_1^2/2}}{\sigma\sqrt{2\pi\tau}}.$$

Now, d_1 is of the form $(\ln(s/k) + b)/a$ for suitable constants a and b hence $s = ke^{\ln(s/k)} = ke^{ad_1 - b}$. It follows that

$$\lim_{s\to+\infty} se^{-d_1^2/2} = ke^{-b} \lim_{s\to+\infty} e^{ad_1 - d_1^2/2} = 0,$$

verifying (†) and completing the proof of **(i)**.

(ii) $\lim_{s\to 0^+} C(\tau, s, k, \sigma, r) = 0$: Immediate from $\lim_{s\to 0^+} d_{1,2} = -\infty$.

(iii) $\lim_{\tau\to\infty} C(\tau, s, k, \sigma, r) = s$: Follows from $\lim_{\tau\to\infty} d_{1,2} = \pm\infty$.

(iv) $\lim_{\tau\to 0^+} C(\tau, s, k, \sigma, r) = (s-k)^+$: Since $\lim_{\tau\to 0^+} g^+ = (s-k)^+$, Lemma C.3 implies that

$$\lim_{\tau\to 0^+} C(\tau, s, k, \sigma, r) = \int_{-\infty}^{\infty} (s-k)^+\varphi(z)\,dz = (s-k)^+.$$

(v) $\lim_{k\to\infty} C(\tau, s, k, \sigma, r) = 0$: This follows from Lemma C.3 and from $\lim_{k\to\infty} g^+ = 0$.

(vi) $\lim_{k\to 0^+} C(\tau, s, k, \sigma, r) = s$: Follows from $\lim_{k\to 0^+} d_1 = +\infty$.

(vii) $\lim_{\sigma\to\infty} C(\tau, s, k, \sigma, r) = s$: Immediate from $\lim_{\sigma\to\infty} d_{1,2} = \pm\infty$.

(viii) $\lim_{\sigma\to 0^+} C(\tau, s, k, \sigma, r) = (s - e^{-r\tau}k)^+$: Since $\lim_{\sigma\to 0^+} g^+ = (se^{r\tau} - k)^+$, by Lemma C.3 we have

$$\lim_{\sigma\to 0^+} C(\tau, s, k, \sigma, r) = e^{-r\tau}\int_{-\infty}^{\infty} (se^{r\tau} - k)^+\varphi(z)\,dz = (s - e^{-r\tau}k)^+.$$

(ix) $\lim_{r\to\infty} C(\tau, s, k, \sigma, r) = s$: Follows immediately from $\lim_{r\to\infty} d_1 = +\infty$.

Appendix D

Solutions to Odd-Numbered Problems

Chapter 1

1. If interest is compounded quarterly, then $2A(0) = A(0)(1+r/4)^{6\cdot 4}$, hence $r = 4(2^{1/24} - 1) \approx .1172$. If interest is compounded continuously, then $2A(0) = A(0)e^{6r}$, hence $r = (\ln 2)/6 \approx 0.1155$.

3. By (1.4), $A_5 = \$29{,}391$ and $A_{10} = \$73{,}178$.

5. By (1.4), the number of payments n must satisfy

$$400\,\frac{(1.005)^n - 1}{.005} \geq 30,000.$$

   Using a spreadsheet, one determines that the smallest value of n satisfying this inequality is 63 months.

7. Let

$$x = \frac{\ln(P) - \ln(P - iA_0)}{\ln(1+i)}.$$

   By (1.6), $N = \lfloor x \rfloor + 1$ unless x is an integer, in which case $x = N$. The number for the second part is $\lfloor 138.976 \rfloor + 1 = 139$. (The 138th withdrawal leaves \$1,941.85 in the account.)

9. The number of withdrawals n must satisfy

$$A_n = \frac{(1.005)^n[300,000(.005) - 5000] + 5000}{.005} \leq \$150,000.$$

   Using a spreadsheet one determines that the smallest value of n satisfying this inequality is 39.

11. The time-n value of the withdrawal made at time $n + j$ is $Pe^{-rj/12}$. Since the account supports exactly N withdrawals, j takes on the values $j = 1, 2, \ldots, N - n$. Adding these amounts produces the desired result.

13. By (1.7), $i = r/12$ must satisfy

$$1800 = 300,000\,\frac{i}{1 - (1+i)^{-360}}.$$

By entering experimental values of i into a spreadsheet, one determines that $r \approx .06$.

15. By (1.6),

$$A_0 = 1000\,\frac{1 - 1.0125^{-60}}{0.0125} = \$42,035.$$

17. If you pay \$6000 now and invest \$2000 for 10 years you will have $\$2000e^{10r}$ with which to pay off the remaining \$6000. The rate r_0 that would allow you to cover that amount exactly satisfies the equation $2000e^{10r_0} = 6000$, so $r_0 = \frac{\ln 3}{10} \approx 0.11$. You will have money left over if $r > r_0$ but will have to come up with the difference $6000 - 2000e^{10r}$ if $r < r_0$.

19. From (1.12)

$$P_1 = iA_0\,\frac{(1+i)^{-N}}{1-(1+i)^{-N}}.$$

Dividing P_n by P_1 gives

$$\frac{P_n}{P_1} = (1+i)^{n-1}.$$

Similarly for I_n.

21. The time-t value of the coupon bond is the sum of the discounted payments after time m:

$$B_t = \sum_{n=m+1}^{N} e^{-r(t_n-t)}C_n + Fe^{-r(T-t)}.$$

23. The rate of return for plan A is 22.68%, while that for plan B is 22.82%. Therefore plan B is slightly better.

25. The bank is the better bet: The total present value of the payments, as calculated by your bank's interest, is

$$3500e^{-.06} + 2500e^{-.12} + 2500e^{-.18} + 2500e^{-.24} = \$9568.22.$$

Thus for about \$430.00 less you would receive the same return as with the investment scheme.

27. Let $\widetilde{A}_n$ denote the value of the account *after* the nth withdrawal. Then $A_n = \widetilde{A}_n - P$, so by (1.5),

$$\begin{aligned}\widetilde{A}_n &= P + (1+i)^n(A_0 - P) + P\,\frac{1-(1+i)^n}{i}\\ &= (1+i)^n A_0 + P\,\frac{(1+i)-(1+i)^{n+1}}{i}.\end{aligned}$$

29. In (1.14), take $P = B_0$, $A_n = C$ $(n = 1, 2, \ldots, N-1)$, $A_N = F$, and sum a geometric series.

Chapter 2

1. The first inequality follows directly from the inclusion-exclusion rule.

3. Let A_j be the event that Jill wins in j races, $j = 3, 4, 5$. The event A_3 occurs in only one way by Jill winning all three races. The event A_4 occurs in 3 ways, corresponding to Jill's single loss occurring in race 1, 2, or 3. The event A_5 occurs in 6 ways, corresponding to Jill's three wins and two losses appearing as

$$\text{LLWWW, LWLWW, LWWLW, WLLWW, WLWLW, WWLLW.}$$

Therefore, $\mathbb{P}(A_3) = q^3$, $\mathbb{P}(A_4) = 3pq^3$, and $\mathbb{P}(A_5) = 6p^2q^3$, hence Jill wins with probability $q^3(1 + 3p + 6p^2)$ $(= 1/2$ if $p = q)$.

5. In part A, if you answer the true-false questions there are $2^{10} = 1024$ different answers; if you answer the multiple choice questions there are $5^8 = 390,625$ different answers. Together, there are a possible $1024 + 390,625 = 391,649$ different choices in part A.

In part B, if you answer the true-false questions there are $2^{12} = 4096$ different answers; if you answer the multiple choice questions there are $3^6 = 729$ different answers. Together, there are a possible $4096 + 729 = 4825$ different choices in part B.

Thus the total number of possible sets of answers for A and B is

$$391,649 \times 4825 = 1,889,706,425,$$

hence the probability of a perfect score is

$$\frac{1}{1,889,706,425} = 5.3 \times 10^{-10} \quad \text{(rounded)}.$$

7. $\mathbb{P}(1) = \mathbb{P}(2) = \cdots = \mathbb{P}(5) = x$, $\mathbb{P}(6) = 3x$, hence $8x = 1$. Therefore $\mathbb{P}(\text{even}) = \frac{2}{8} + \frac{3}{8} = \frac{5}{8}$.

9. Let n be the number of balls thrown. The sample space consists of all n-tuples $(m_1, m_2, \ldots, m_n)$, where m_j is the number of the jar containing the jth ball, $1 \le m_j \le 30$. The event that no two balls land in the same jar is the set consisting of all n-tuples with distinct coordinates m_j. Therefore, the probability p_n that at least two balls land in the same jar is

$$1 - \frac{30 \cdot 29 \cdot \cdots \cdot (30 - n + 1)}{(30)^n}.$$

From a spreadsheet, $p_7 \approx .53$ and $p_8 \approx .64$ so that at least 8 throws are needed.

11. Let C be the event that both tosses are heads, A the event that at least one toss comes up heads, and B the event that the first toss comes up heads. For $q = 1 - p$ we have

$$\mathbb{P}(AC) = \mathbb{P}(BC) = \mathbb{P}(C) = p^2, \quad \mathbb{P}(A) = 1 - q^2, \quad \text{and} \quad \mathbb{P}(B) = p,$$

hence

$$\mathbb{P}(C|A) = \frac{p^2}{1-q^2} = \frac{p}{1+q} \quad \text{and} \quad \mathbb{P}(C|B) = \frac{p^2}{p} = p.$$

The probabilities are not the same since $\dfrac{p}{1+q} = p$ implies $p = 0$ or 1.

13. Let A be the event that the slip numbered 1 was drawn twice and B the event that the sum of the numbers on the three slips drawn is 8. Since AB consists of the outcomes $(1,1,6)$, $(1,6,1)$, and $(6,1,1)$, $\mathbb{P}(AB) = 3/6^3$. The distinct sets of three numbers that add up to 8 are $\{1,1,6\}$, $\{1,2,5\}$, $\{1,3,4\}$, $\{2,2,4\}$, and $\{2,3,3\}$. There are 3 ways that each of the first, fourth, and fifth sets can be drawn, and 6 ways for the second and third. Therefore, $\mathbb{P}(B) = 21/6^3$, so $\mathbb{P}(A|B) = 3/21$.

15. Since $x \in AB$ iff $0 \le d_1 \le 4$ and $d_2 = d_3 = 0$, we have $\mathbb{P}(AB) = (.5)(.1)^2$. Since $\mathbb{P}(A) = .5$ and $\mathbb{P}(B) = (.1)^2$, the events are independent. Changing the inequality to $x < .49$ makes the events dependent.

17. For (c),

$$\begin{aligned}\mathbb{P}(A'B') &= \mathbb{P}[(A \cup B)'] = 1 - \mathbb{P}(A \cup B) = 1 - [\mathbb{P}(A) + \mathbb{P}(B) - \mathbb{P}(AB)] \\ &= [1 - \mathbb{P}(A)][1 - \mathbb{P}(B)] = \mathbb{P}(A')\mathbb{P}(B').\end{aligned}$$

19. (a) If H_1 comes in first, then the bettor wins $o_1 b_1$ but loses her wagers b_2 and b_3, giving her a net win of

$$o_1 b_1 - b_2 - b_3 = (o_1 + 1)b_1 - (b_1 + b_2 + b_3).$$

(b) The system of inequalities $W_b(i) > 0$ may be written

$$\begin{array}{cccc} -o_1 b_1 & +b_2 & +b_3 & < 0 \\ b_1 & -o_2 b_2 & +b_3 & < 0 \\ b_1 & +b_2 & -o_3 b_3 & < 0, \end{array}$$

which is equivalent to the existence of negative numbers s_i satisfying the displayed system of equations.

(e) Take $s = (-|D|, -|D|, -|D|)$ in (d).

(f) A little algebra shows that $(1 + o_1)^{-1} + (1 + o_2)^{-1} + (1 + o_3)^{-1} = 1$ iff $D = 0$, that is, iff the system in (b) has no solution.

21. Let $N = r + g$. By the total probability law,

$$\begin{aligned}\mathbb{P}(R_2) &= \mathbb{P}(R_1)\mathbb{P}(R_2|R_1) + \mathbb{P}(G_1)\mathbb{P}(R_2|G_1)\\ &= \frac{r}{N}\frac{r-1}{N-1} + \frac{g}{N}\frac{r}{N-1}\\ &= \frac{r}{N}.\end{aligned}$$

Therefore, $\mathbb{P}(R_2) = \mathbb{P}(R_1)$. Since the third drawing begins with a similar urn, $\mathbb{P}(R_3) = \mathbb{P}(R_2)$. Continuing in this way, we arrive at the assertion.

23. Let A be the first event and B the second. Then $\mathbb{P}(A) = 1/3 = \mathbb{P}(B)$ and $\mathbb{P}(AB) = \int_{\sqrt{2}/2}^{1} 2x^2 - 1\,dx \approx .1381$, hence $\mathbb{P}(A|B) \approx 0.4142$. The events are not independent.

25. Let J_p (M_p) be the event that John (Mary) gets a plain, etc., and let A be the event that neither gets a plain and at least one gets an anchovy. Then

$$A = J_p'M_p'(J_a \cup M_a) = (J_p'M_p'J_a) \cup (J_p'M_p'M_a) = (J_aM_p') \cup (J_p'M_a)$$

hence, by independence,

$$\begin{aligned}\mathbb{P}(A) &= \mathbb{P}(J_aM_p') + \mathbb{P}(J_p'M_a) - \mathbb{P}(J_aM_a)\\ &= \mathbb{P}(J_a)\mathbb{P}(M_p') + \mathbb{P}(J_p')\mathbb{P}(M_a) - \mathbb{P}(J_a)\mathbb{P}(M_a)\\ &= \mathbb{P}(J_a)\left[\mathbb{P}(M_p') - \mathbb{P}(M_a)\right] + \mathbb{P}(J_p')\mathbb{P}(M_a)\\ &= \mathbb{P}(J_a)\mathbb{P}(M_s) + \mathbb{P}(J_p')\mathbb{P}(M_a)\\ &= (.2)(.3) + (.9)(.4)\\ &= .42.\end{aligned}$$

Chapter 3

1. For (b), use the inclusion-exclusion principle. For (c), suppose first that $\mathbf{1}_A \le \mathbf{1}_B$. Then $\omega \in A \Rightarrow \mathbf{1}_A(\omega) = 1 \Rightarrow \mathbf{1}_B(\omega) = 1 \Rightarrow \omega \in B$, hence $A \subseteq B$. Conversely, if $A \subseteq B$ then $\mathbf{1}_A(\omega) = 1 \Rightarrow \omega \in A \Rightarrow x \in B \Rightarrow \mathbf{1}_B(\omega) = 1$ hence $\mathbf{1}_A \le \mathbf{1}_B$.

3. Let Y_n be the number of heads in n tosses. We seek the smallest n for which $\mathbb{P}(Y_n \ge 2) \ge .99$ or, equivalently, $\mathbb{P}(Y_n < 2) < .01$. Since $Y_n \sim B(n, .5)$, the last inequality may be written

$$\mathbb{P}(Y_n = 0) + \mathbb{P}(Y_n = 1) = 2^{-n}(1 + n) < .01.$$

Using a spreadsheet, one calculates that $n = 11$.

5. The pmf of X may be written $p_X(k) = \binom{n}{k} A_N B_N$, where

$$A_N = \left(\frac{Np}{N}\right)\left(\frac{Np-1}{N-1}\right)\cdots\left(\frac{Np-k+1}{N-k+1}\right) \quad \text{and}$$
$$B_N = \left(\frac{Nq}{N-k}\right)\left(\frac{Nq-1}{N-k-1}\right)\cdots\left(\frac{Nq-n+k+1}{N-n+1}\right).$$

Since

$$\lim_{N\to\infty} A_N = p^k \quad \text{and} \quad \lim_{N\to\infty} B_N = q^{n-k},$$

the conclusion follows.

7. For $a > 0$,

$$F_Y(y) = \mathbb{P}\left(X \le \frac{y-b}{a}\right) = F_X\left(\frac{y-b}{a}\right).$$

Differentiating yields

$$f_Y(y) = a^{-1} f_X\left(\frac{y-b}{a}\right).$$

If $a < 0$, similar calculations show that

$$f_Y(y) = -a^{-1} f_X\left(\frac{y-b}{a}\right).$$

9. Since $[\Phi(x) + \Phi(-x)]' = \varphi(x) - \varphi(-x) = 0$, the function $\Phi(x) + \Phi(-x)$ is constant hence $\Phi(x) + \Phi(-x) = 2\Phi(0) = 1$. For the second part, if $X \sim N(0,1)$ then

$$\mathbb{P}(-X \le x) = \mathbb{P}(X \ge -x) = 1 - \mathbb{P}(X \le -x) = 1 - \Phi(-x) = \Phi(x),$$

which shows that $-X \sim N(0,1)$.

11. By Exercises 9 and 10, $X \sim N(\mu, \sigma^2)$ iff $\dfrac{X-\mu}{\sigma} \sim N(0,1)$ iff $\dfrac{\mu - X}{\sigma} \sim N(0,1)$ iff $2\mu - X \sim N(\mu, \sigma^2)$.

13. For $r \le 0$, $F_Z(r) = 0$. For $r > 0$, $F_Z(r)$ is the area of the intersection of the dartboard with the circle of radius r and center $(0,0)$. Therefore,

$$F_Z(r) = \begin{cases} 1, & r \ge \sqrt{2}, \\ \frac{\pi r^2}{4}, & 0 \le r \le 1, \\ r^2 \arcsin\left(\frac{1}{r}\right) - \frac{\pi r^2}{4} + \sqrt{r^2 - 1}, & 1 \le r \le \sqrt{2}. \end{cases}$$

It follows that $f_Z = g_Z \mathbf{1}_{[0,\sqrt{2}]}$, where

$$g_Z(r) = \begin{cases} \frac{\pi r}{2}, & 0 \le r \le 1 \\ 2r \arcsin\left(\frac{1}{r}\right) - \frac{\pi r}{2}, & 1 \le r \le \sqrt{2}. \end{cases}$$

15. By independence,

$$\mathbb{P}(\max(X,Y) \le z) = \mathbb{P}(X \le z, Y \le z) = \mathbb{P}(X \le z)\mathbb{P}(Y \le z)$$

and

$$\mathbb{P}(\min(X,Y) > z) = \mathbb{P}(X > z, Y > z) = \mathbb{P}(X > z)\mathbb{P}(Y > z).$$

Thus, $F_M = F_X F_Y$ and $1 - F_m = (1 - F_X)(1 - F_Y)$, so

$$F_m = F_X + F_Y - F_X F_Y = F_X + F_Y - F_M.$$

17. Let $\alpha = \Phi\left(\mu\sigma^{-1}\right)$.

(a) There are $\binom{n}{k}$ choices of times for the k increases, each of which has probability

$$\mathbb{P}(Z_1 > 1, \ldots, Z_k > 1, Z_{k+1} < 1, \ldots, Z_n < 1) = \alpha^k(1-\alpha)^{n-k}.$$

Therefore, the desired probability is $\dbinom{n}{k}\alpha^k(1-\alpha)^{n-k}$.

(b) The k consecutive increases can start at the times $1, 2, \ldots, n-k+1$ hence the required probability is $(n-k+1)\alpha^k(1-\alpha)^{n-k}$.

(c) Since $k \ge n/2$, there can be only one such run. Assuming that $k \le n-1$ the event in question is the union of the mutually exclusive events

$$\begin{aligned}
&\{Z_1 > 1, Z_2 > 1, \ldots, Z_k > 1, Z_{k+1} < 1\},\\
&\{Z_{n-k} < 1, Z_{n-k+1} > 1, \ldots, Z_n > 1\}, \text{ and}\\
&\{Z_j < 1, Z_{j+1} > 1, \ldots, Z_{j+k} > 1, Z_{j+k+1} < 1\},
\end{aligned}$$

where $j = 1, 2, \ldots, n-k-1$ (the last sets being empty if $k = n-1$). Therefore, the required probability is

$$2\alpha^k(1-\alpha) + (n-k-1)\alpha^k(1-\alpha)^2.$$

If $k = n$ the probability is α^n.

19. For integers y, by independence,

$$\begin{aligned}
\mathbb{P}(Y_{\min} \le y) &= \mathbb{P}(X_1 \le y, \ldots X_n \le y) = \mathbb{P}(X_1 \le y) \cdots \mathbb{P}(X_n \le y)\\
&= F^n(y).
\end{aligned}$$

Since $Y_{\min}$ is integer-valued,

$$p_{Y_{\min}}(y) = \mathbb{P}(Y_{\min} \le y) - \mathbb{P}(Y_{\min} \le y-1) = F^n(y) - F^n(y-1).$$

Similarly,

$$\begin{aligned}\mathbb{P}(Y_{\max} \le y) &= 1 - \mathbb{P}(Y_{\max} \ge y+1) \\ &= 1 - \mathbb{P}(X_1 \ge y+1) \cdots \mathbb{P}(X_n \ge y+1) \\ &= 1 - \big(1 - F(y)\big)^n,\end{aligned}$$

hence

$$p_{Y_{\max}}(y) = \mathbb{P}(Y_{\max} \le y) - \mathbb{P}(Y_{\max} \le y-1) = \big(1-F(y-1)\big)^n - \big(1-F(y)\big)^n.$$

21. The event is $\{2 < X < 3\} \cup \{-3 < X < -2\}$ or

$$\begin{aligned}&\left\{\frac{2-\mu}{\sigma} < \frac{X-\mu}{\sigma} < \frac{3-\mu}{\sigma}\right\} \cup \left\{\frac{-3-\mu}{\sigma} < \frac{X-\mu}{\sigma} < \frac{-2-\mu}{\sigma}\right\} \\ &\quad = \left\{\frac{1}{2} < \frac{X-\mu}{\sigma} < 1\right\} \cup \left\{-2 < \frac{X-\mu}{\sigma} < -\frac{3}{2}\right\}.\end{aligned}$$

The probability is therefore

$$\Phi(1) - \Phi(.5) + \Phi(-1.5) - \Phi(-2) \approx 0.194.$$

23. By independence, the event has probability one fourth of the integral

$$\iint_{1/2<y/x<2} \mathbf{1}_{[1,3]}(x)\mathbf{1}_{[1,3]}(y)\,dx\,dy,$$

which is the area of the part of the rectangle $[1,3] \times [1,3]$ between the lines $y = x/2$ and $y = 2x$. By a straightforward calculation, this area is $7/2$, so the probability is $7/8$.

25. $\mathbb{P}(X+Y=m) = \sum_{k=0}^{m} \mathbb{P}(Y=k, X=m-k) = \sum_{k=0}^{m} q^k p q^{m-k} p = (m+1)p^2 q^m$, hence

$$\mathbb{P}(X=n \mid X+Y=m) = \frac{\mathbb{P}(X=n)\,\mathbb{P}(Y=m-n)}{\mathbb{P}(X+Y=m)} = \frac{q^n p q^{m-n} p}{(m+1)p^2 q^m},$$

which reduces to $1/(m+1)$.

27. If $x, y > 0$, then $\mathbb{P}(1/X < x, 1/Y < y) = \mathbb{P}(X > 1/x, Y > 1/y) = \mathbb{P}(X > 1/x)\mathbb{P}(Y > 1/y) = \mathbb{P}(1/X < x)\mathbb{P}(1/Y < y)$.

29. (a) By independence,

$$\begin{aligned}&\mathbb{P}(Y_1 = k_1, Y_2 = k_2, \ldots, Y_n = k_n) \\ &\qquad = \mathbb{P}(X_1 = k_1, X_2 = k_2 - k_1 \cdots, X_n = k_n - k_{n-1}) \\ &\qquad = \mathbb{P}(X_1 = k_1)\mathbb{P}(X_2 = k_2 - k_1) \cdots \mathbb{P}(X_n = k_n - k_{n-1}).\end{aligned}$$

Note that the probability is zero unless $0 \le k_1 \le k_2 \le \cdots \le k_n \le n$, and $0 \le k_j - k_{j-1} \le 1$. In this case, the number of factors $p_X(k_j - k_{j-1})$ that equal p is $\sum_{j=1}^{n}(k_j - k_{j-1}) = k_n$, hence

$$\mathbb{P}(Y_1 = k_1,\, Y_2 = k_2, \ldots, Y_n = k_n) = p^{k_n} q^{n-k_n}.$$

(b) By (a),

$$p_{Z_n}(k) = \mathbb{P}(Y_1 + \cdots + Y_n = k) = \sum \prod_{j=1}^{n} p_X(k_j - k_{j-1}),$$

where the sum is taken over all nonnegative integers $k_1, k_2, \ldots, k_n$ such that $k_1 + \cdots + k_n = k$ and $0 \le k_j - k_{j-1} \le 1$. Since

$$p_X(k_j - k_{j-1}) = \begin{cases} p & \text{if } k_j - k_{j-1} = 1, \\ q & \text{if } k_j - k_{j-1} = 0, \end{cases}$$

and since $k_n = \sum_{j=1}^{n}(k_j - k_{j-1})$ is the number of factors $p_X(k_j - k_{j-1})$ that are equal to p,

$$\prod_{j=1}^{n} p_X(k_j - k_{j-1}) = p^{k_n} q^{n-k_n}.$$

(c) $\mathbb{P}(Z_n = k,\, Y_n = j) = \mathbb{P}(Y_1 + \cdots + Y_{n-1} = k - j,\, Y_n = j)$
and for $0 \le j \le k$, $k_0 = 0 \le k_1 \le k_2 \le \cdots \le k_{n-1} \le j$, $k_j - k_{j-1} \le 1$, and $k_1 + k_2 + \cdots + k_{n-1} = k - j$, we have

$$\begin{aligned}
&\mathbb{P}(Y_1 = k_1,\, Y_2 = k_2, \ldots, Y_{n-1} = k_{n-1}, Y_n = j) \\
&\quad = \mathbb{P}(X_1 = k_1, X_2 = k_2 - k_1, \cdots, X_{n-1} = k_{n-1} - k_{n-2}, X_n = j - k_{n-1}) \\
&\quad = p_X(j - k_{n-1}) \prod_{i=1}^{n-1} p_X(k_i - k_{i-1}) \\
&\quad = p_X(j - k_{n-1}) p^{k_{n-1}} q^{n-1-k_{n-1}}.
\end{aligned}$$

The assertion follows.

(d) Let $P_k = \mathbb{P}(Y_1 + Y_2 + Y_3 = k)$.

TABLE D.1: Probability distribution of $Y_1 + Y_3 + Y_3$.

k	0	1	2	3	4	5	6
P_k	q^3	q^2p	q^2p	$qp^2 + q^2p$	qp^2	qp^2	p^3

31. Follows from

$$\mathbb{P}(Y = k) = \mathbb{P}(n - Y = n - k) \quad \text{and}$$

$$\binom{n}{k} p^k (1-p)^{n-k} = \binom{n}{n-k} q^{n-k} (1-q)^k.$$

Chapter 4

1. We need only show that $C_0 \le S_0$; the remaining inequalities follow from the put-call parity formula. Suppose that $C_0 > S_0$. We then buy the security for S_0, write a call option, and place the proceeds $C_0 - S_0$ into a risk-free account yielding $e^{rT}(C_0 - S_0)$ at maturity T. If $S_T > K$ the call option we sold will be exercised, requiring us to sell the security for K. If $S_T \le K$ we sell the security for S_T. Our total proceeds from these transactions are $e^{rT}(C_0 - S_0) + \min(S_T, K)$. Since this amount is positive, the strategy constitutes an arbitrage. Therefore $C_0 \le S_0$.

3. For (a), assume that $X_t < Y_t$ for some $t < T$. At this time, take a short position on $\mathcal{Y}$ and a long position on $\mathcal{X}$. This will yield cash $Y_t - X_t > 0$, which you deposit into an account. At time T, after covering your obligation, you have

$$e^{r(T-t)}(Y_t - X_t) + X_T - Y_T > 0,$$

an arbitrage. Part (b) is similar.

5. $\mathcal{P}$ can be exercised at any time in the interval $[0, T]$, while $\mathcal{P}'$ can be exercised only at times in the subinterval $[0, T']$. This gives $\mathcal{P}$ greater flexibility and hence greater value.

7. The payoff for (a) is $(S_T - K)^+ + 2(K - S_T)^+$, which has values $2(K - S_T)$ if $S_T < K$ and $S_T - K$ if $S_T \ge K$. The payoff for (b) is $2(S_T - K)^+ + (K - S_T)^+$, which has values $K - S_T$ if $S_T < K$ and $2(S_T - K)$ if $S_T \ge K$. (Figure D.1.)

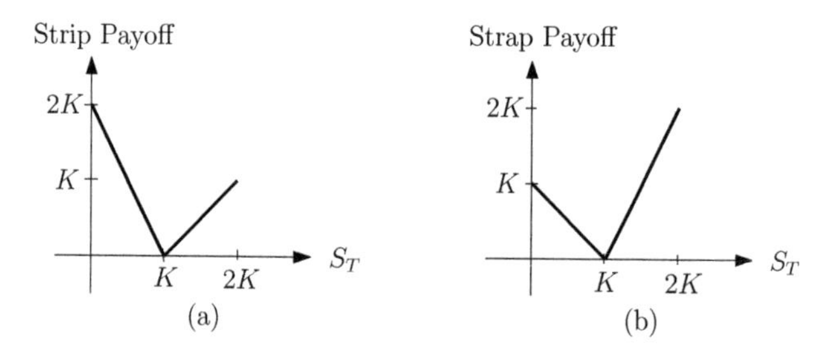

FIGURE D.1: Illustrations for Exercise 7.

9. The payoff is $(K_1 - S_T)^+ + (S_T - K_2)^+$, which has values $K_1 - S_T$ if $S_T < K_1$, 0 if $K_1 \le S_T \le K_2$, and $S_T - K_2$ if $S_T > K_2$. (Figure D.2.)

11. If $C_0 < C_0'$, we buy the lower priced option and sell the higher priced one for a cash profit of $C_0' - C_0$. At maturity there are three possibilities:

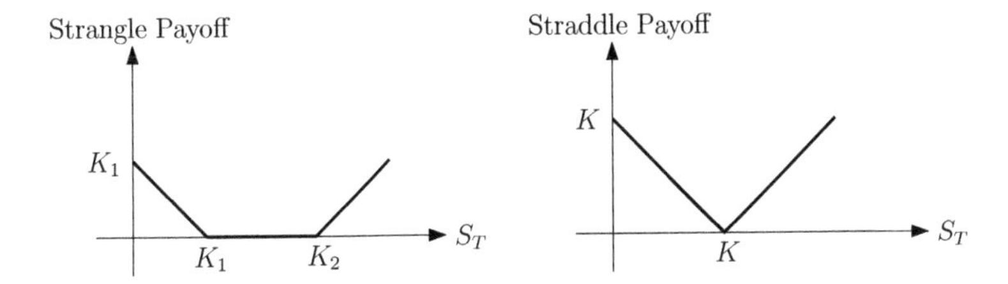

FIGURE D.2: Illustrations for Exercises 9 and 10.

(a) $S_T \leq K$: neither option is in the money.

(b) $K < S_T \leq K'$: we exercise the option we hold, giving us a payoff of $S_T - K > 0$.

(c) $K' < S_T$: both options are in the money. We exercise the option we hold and fulfill our obligation as writer of the other option. This gives us a net payoff of $(S_T - K) - (S_T - K') = K' - K > 0$.

Each case results in a nonnegative payoff so our profit is at least $C_0' - C_0 > 0$, contradicting the no-arbitrage assumption.

That $C_0 \geq C_0'$ is to be expected, since a smaller strike price gives a larger payoff.

13. If $S_T < K'$, all options are worthless. If $K' \leq S_T < K$, exercise the only option in the money, giving you a payoff of $S_T - K'$. If $K \leq S_T \leq K''$, exercise the option with strike price K' and fulfill your obligation, giving you a payoff of $S_T - K' - 2(S_T - K) = 2K - K' - S_T = K'' - S_T \geq 0$. If $K'' < S_T$, exercise the options you hold and fulfill your obligation giving you a payoff of $S_T - K' + S_T - K'' - 2(S_T - K) = 0$. Each case results in positive profit of at least $2C_0 - C_0' - C_0''$, contradicting the no-arbitrage assumption. The graph of the portfolio is shown in Figure D.3.

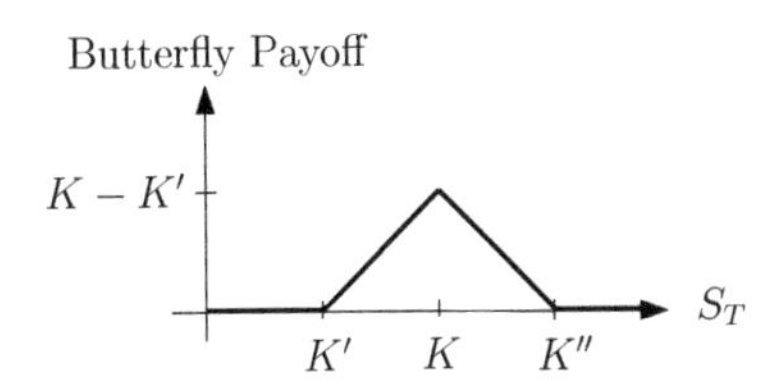

FIGURE D.3: Illustration for Exercise 13.

15. The payoff of a capped call option is

$$\min\left((S_T - K)^+, A\right) = \begin{cases} 0 & \text{if } S_T \le K, \\ S_T - K & \text{if } K \le S_T \le K + A, \\ A & \text{if } S_T > K + A \end{cases}$$
$$= (S_T - K)^+ - (S_T - (K + A))^+,$$

which is the payoff of a bull spread. Therefore, by the law of one price, the time-t value of a capped call option is that of a bull spread.

17. The portfolio may be inferred from payoff

$$1.67(S_T - 3)^+ + .50(S_T - 12)^+ + (S_T - 20)^+ - 1.17(S_T - 6)^+ - 2.00(S_T - 16)^+ - 3.$$

19. Since $(S_T - K)^+ = S_T - K$ and $(K - S_T)^+ = 0$, the profit is

$$\begin{aligned} e^{rT}(S_0 - 2C_0 + P_0) &+ 2(S_T - K)^+ - (K - S_T)^+ - S_T \\ &= e^{rT}(S_0 - 2C_0 + P_0) + S_T - 2K \\ &= e^{rT}(Ke^{-rT} - C_0) + S_T - 2K \quad \text{(put-call parity)} \\ &= S_T - K - e^{rT}C_0. \end{aligned}$$

21. Suppose that $\mathbb{P}(S_T \ge S_0 e^{rT}) < 1$, so $S_T < S_0 e^{rT}$ with probability $p > 0$. Sell $\mathcal{S}$ short and invest the cash S_0 in a risk-free account. The transaction costs nothing. At time T, after returning the stock you have $S_0 e^{rt} - S_T$, which is positive with positive probability.

23. Suppose $\mathbb{P}(S_T \ge K + C_0 e^{rT}) < 1$ so $S_T < K + C_0 e^{rT}$ with probability > 0. Write a call and invest the proceeds in a risk-free account. At maturity you have a minimum profit of $C_0 e^{rT} - (S_T - K) > 0$ with positive probability, an arbitrage.

Chapter 5

1. The sample space consists of the six permutations of the numbers 1, 2, and 3. $\mathcal{F}_1$ is generated by the sets $\{(1,2,3),(1,3,2)\}$, $\{(2,1,3),(2,3,1)\}$, and $\{(3,1,2),(3,2,1)\}$; $\mathcal{F}_2 = \mathcal{F}_3$ contains all subsets of Ω.

3. • $\mathcal{F}_0 = \{\emptyset, \Omega\}$
   • $\mathcal{F}_1$ is generated by the sets

$$A_{\mathrm{R}} = \{\mathrm{RWRWR}, \mathrm{RWRWW}, \mathrm{RWWRR}, \mathrm{RWWRW}\},$$
$$A_{\mathrm{W}} = \{\mathrm{WRWRW}, \mathrm{WRWRR}, \mathrm{WRRWW}, \mathrm{WRRWR}\}.$$

- $\mathcal{F}_2 = \mathcal{F}_1$
- $\mathcal{F}_3$ is generated by

$$A_{\text{RWR}} = \{\text{RWRWR}, \text{RWRWW}\}, \ A_{\text{RWW}} = \{\text{RWWRR}, \text{RWWRW}\},$$
$$A_{\text{WRW}} = \{\text{WRWRW}, \text{WRWRR}\}, \ A_{\text{WRR}} = \{\text{WRRWW}, \text{WRRWR}\}.$$

- $\mathcal{F}_4 = \mathcal{F}_3$
- $\mathcal{F}_5$ is the collection of all subsets of $A_R \cup A_W$.

5. If (e) holds, then for $n = 1, 2, \ldots, N-1$ we have

$$\phi_{n+1} = V_0 - \sum_{j=0}^{n} \widetilde{S}_j \Delta\theta_j \quad \text{and} \quad \phi_n = V_0 - \sum_{j=0}^{n-1} \widetilde{S}_j \Delta\theta_j.$$

Subtracting these equations yields $\Delta\phi_n = -\widetilde{S}_n \Delta\theta_n$, which is Equation (5.4).

7. We take the case $d = 1$. If the portfolio is self-financing, then, by Theorem 5.4.1(b)

$$\Delta V_{j-1} = \phi_j \Delta B_{j-1} + \theta_j \Delta S_{j-1}, \ j \geq 1,$$

hence

$$G_n = \sum_{j=1}^{n} \Delta V_{j-1} = V_n - V_0.$$

Conversely, if $V_n = V_0 + G_n$ for all n then

$$\begin{aligned}\Delta V_n = \Delta G_n &= \sum_{j=1}^{n+1} (\phi_j \Delta B_{j-1} + \theta_j \Delta S_{j-1}) - \sum_{j=1}^{n} (\phi_j \Delta B_{j-1} + \theta_j \Delta S_{j-1}) \\ &= \phi_{n+1} \Delta B_n + \theta_{n+1} \Delta S_n,\end{aligned}$$

hence the portfolio is self-financing.

Chapter 6

1. Let X_1 be the number of red marbles drawn before the first white one and X_2 the number of reds between the first two whites. Then $Y = X_1 + X_2$, and $X_1 + 1$ and $X_2 + 1$ are geometric with parameter $p = w/(r + w)$ (Example 3.7.1). Therefore,

$$\mathbb{E}\, Y = \mathbb{E}\, X_1 + \mathbb{E}\, X_2 = (2/p) - 2 = 2r/w$$

(Example 6.1.1(c)).

3. By Example 6.1.1(c), $\mathbb{E}\,X = q/p$. Also,

$$\mathbb{E}[X(X-1)] = \sum_{n=2}^{\infty} n(n-1)q^n p = pq^2 \sum_{n=2}^{\infty} \frac{d^2}{dq^2} q^n = pq^2 \frac{d^2}{dq^2}\frac{q^2}{p} = \frac{2q^2}{p^2}.$$

Therefore,

$$\mathbb{V}\,X = \mathbb{E}[X(X-1)] + \mathbb{E}\,X - \mathbb{E}^2 X = \frac{2q^2}{p^2} + \frac{q}{p} - \frac{q^2}{p^2} = \frac{q}{p^2}.$$

5. Let $N = r + w$. The number X of marbles drawn is either 2 or 3. If $X = 2$, the marbles are either both red or both white hence

$$\mathbb{P}(X=2) = \begin{cases} \dfrac{r^2+w^2}{N^2} & \text{for (a)} \\ \dfrac{r(r-1)+w(w-1)}{N(N-1)} & \text{for (b).} \end{cases}$$

The event $\{X = 3\}$ consists of the four outcomes RWR, RWW, WRR, and WRW, hence

$$\mathbb{P}(X=3) = \begin{cases} \dfrac{2r^2w+2rw^2}{N^3} & \text{for (a)} \\ \dfrac{2r(r-1)w+2w(w-1)r}{N(N-1)(N-2)} & \text{for (b).} \end{cases}$$

Therefore, for case (a),

$$\mathbb{E}\,X = 2 \cdot \frac{r^2+w^2}{N^2} + 6 \cdot \frac{r^2w+rw^2}{N^3},$$

and for case (b)

$$\mathbb{E}\,X = 2 \cdot \frac{r(r-1)+w(w-1)}{N(N-1)} + 6 \cdot \frac{r(r-1)w+w(w-1)r}{N(N-1)(N-2)}.$$

7. For any $A \in \mathcal{F}$,

$$\mathbb{V}\,\mathbf{1}_A = \mathbb{E}\,\mathbf{1}_A^2 - \mathbb{E}^2\mathbf{1}_A = \mathbb{P}(A) - \mathbb{P}^2(A) = \mathbb{P}(A)\mathbb{P}(A').$$

The desired result now follows from independence and Theorem 6.4.2.

9. By independence, $f_{X,Y}(x,y) = \mathbf{1}_{[0,1]}(x)\mathbf{1}_{[0,1]}(y)$, hence

$$\begin{aligned}
\mathbb{E}\left(\frac{XY}{X^2+Y^2+1}\right) &= \int_0^1\int_0^1 \frac{xy}{x^2+y^2+1}\,dy\,dx \\
&= \frac{1}{4}\int_0^1 2x\left[\ln(x^2+2) - \ln(x^2+1)\right]\,dx \\
&= \frac{1}{4}\left\{\int_2^3 \ln u\,du - \int_1^2 \ln u\,du\right\} \\
&= \frac{1}{4}\ln(27/16).
\end{aligned}$$

11. By linearity and independence,

$$\mathbb{E}(X+Y)^2 = \mathbb{E}X^2 + \mathbb{E}Y^2 + 2(\mathbb{E}X)(\mathbb{E}Y) = \mathbb{E}X^2 + \mathbb{E}Y^2$$

and

$$\mathbb{E}(X+Y)^3 = \mathbb{E}X^3 + \mathbb{E}Y^3 + 3(\mathbb{E}X^2)(\mathbb{E}Y) + 3(\mathbb{E}X)(\mathbb{E}Y^2) = \mathbb{E}X^3 + \mathbb{E}Y^3.$$

For higher powers there are additional terms. For example,

$$\mathbb{E}(X+Y)^4 = \mathbb{E}X^4 + 6(\mathbb{E}X^2)(\mathbb{E}Y^2) + \mathbb{E}Y^4.$$

13. For (a), complete the square to obtain

$$\int_a^b e^{\alpha x}\varphi(x)\,dx = e^{\alpha^2/2}\int_a^b \varphi(x-\alpha)\,dx = e^{\alpha^2/2}\left[\Phi(b-\alpha) - \Phi(a-\alpha)\right].$$

For (b), integrate by parts and use (a) to obtain

$$\begin{aligned}\int_a^b e^{\alpha x}\Phi(x)\,dx &= \frac{1}{\alpha}e^{\alpha x}\Phi(x)\Big|_a^b - \frac{1}{\alpha}\int_a^b e^{\alpha x}\varphi(x)\,dx \\ &= \frac{1}{\alpha}\left(e^{\alpha x}\Phi(x) - e^{\alpha^2/2}\Phi(x-\alpha)\right)\Big|_a^b.\end{aligned}$$

15. $\mathbb{V}X = \mathbb{E}X^2 - \mathbb{E}^2X$, where $\mathbb{E}X = \frac{1}{2}(a+b)$ (Example 6.2.1). Since

$$\mathbb{E}X^2 = (b-a)^{-1}\int_a^b x^2\,dx = \tfrac{1}{3}(a^2+ab+b^2),$$

we have

$$\mathbb{V}X = \tfrac{1}{3}(a^2+ab+b^2) - \tfrac{1}{4}(a+b)^2 = \tfrac{1}{12}(a-b)^2.$$

17. $\mathbb{E}^2X = \mathbb{E}X^2 - \mathbb{V}X \le \mathbb{E}X^2$, since $\mathbb{V}X \ge 0$.

19. Referring to (3.4) the expectation of a hypergeometric random variable X is seen to be

$$\mathbb{E}X = \binom{N}{z}^{-1}\sum_x \binom{m}{x}\binom{n}{z-x}x,$$

where $m = Mp$, $n = Mq$ and where the sum is taken over all integers x satisfying the inequalities $\max(z-n,1) \le x \le \min(z,m)$. The sum may be written as

$$m\sum_x \binom{m-1}{x-1}\binom{n}{z-x} = m\sum_y \binom{m-1}{y}\binom{n}{z-1-y}$$

where $\max(z-1-n,0) \le y \le \min(z-1,m-1)$. Referring to the hypergeometric distribution again we see that

$$\mathbb{E}X = m\binom{N}{z}^{-1}\binom{N-1}{z-1} = \frac{mz}{N} = pz.$$

21. (a) By Equation (6.9)

$$\mathbb{P}(Y = 50) \approx \Phi(.1) - \Phi(-.1) \approx 0.07966$$

The exact probability is $\binom{100}{50}2^{-100} = 0.07959$.

(b) By Equation (6.9) with $p = .5$,

$$\mathbb{P}(40 \le Y \le 60) = \mathbb{P}\left(-2 \le \frac{2Y - 100}{10} \le 2\right) \approx \Phi(2) - \Phi(-2) \approx .95.$$

23.

$$\begin{aligned}\mathbb{E}\, X(X-1)\cdots(X-m) &= \sum_{k=m+1}^{n} k(k-1)(k-2)\cdots(k-m)\binom{n}{k}p^k q^{n-k}\\ &= n(n-1)\cdots(n-m)p^{m+1}\sum_{k=m+1}^{n}\binom{n-(m+1)}{k-(m+1)}p^{k-(m+1)}q^{n-k}\\ &= n(n-1)\cdots(n-m)p^{m+1}.\end{aligned}$$

25. (a)

$$\begin{aligned}\mathbb{E}(X-Y)^2 &= 2\big[\mathbb{E}(X^2) - \mathbb{E}(XY)\big] = 2\big[\mathbb{E}(X^2) - \mathbb{E}^2(X)\big]\\ &= 2\int_0^1 x^2 - 2\left(\int_0^1 x\right)^2\\ &= 2/3 - 2(1/2)^2 = 1/6.\end{aligned}$$

(b)

$$\mathbb{E}\left(\frac{Y}{1+XY}\right) = \int_0^1\int_0^1 \frac{y}{1+xy}\,dx\,dy.$$

The inner integral is $\int_1^{1+y}\frac{dz}{z} = \ln(1+y)$ so the expectation is

$$\int_1^2 \ln y\,dy = (y\ln y - y)\Big|_1^2 = \ln 4 - 1.$$

27. Follows from

$$\mathbb{E}\, f(Y_1,\ldots,Y_n) = \sum_{(k_1,\ldots,k_n)} f(k_1,\ldots,k_n)\mathbb{P}(Y_1 = k_1,\ldots,Y_n = k_n)$$

29. (a) By the binomial theorem

$$\mathbb{E}\, e^{\lambda X} = \sum_{k=0}^{n} e^{\lambda k}\binom{n}{k}p^k q^{n-k} = \left(pe^{\lambda} + q\right)^n.$$

(b) Summing a geometric series gives

$$\mathbb{E}\, e^{\lambda X} = \sum_{n=1}^{\infty} e^{\lambda n} p q^{n-1} = p e^{\lambda} \sum_{n=0}^{\infty} e^{\lambda n} q^{n} = \frac{p e^{\lambda}}{1 - q e^{\lambda}}.$$

Chapter 7

1. If A denotes the Cartesian product, then

$$\begin{aligned}\mathbb{P}(A) &= \sum_{\omega \in A} \mathbb{P}_1(\omega_1)\mathbb{P}_2(\omega_2)\cdots\mathbb{P}_N(\omega_N) \\ &= \sum_{\omega_1 \in A_1} \cdots \sum_{\omega_N \in A_N} \mathbb{P}_1(\omega_1)\cdots\mathbb{P}_N(\omega_N) \\ &= \mathbb{P}_1(A_1)\mathbb{P}_2(A_2)\cdots\mathbb{P}_N(A_N).\end{aligned}$$

3. Part (a) follows immediately from the definitions of Z_n and X_n, and (b) is a restatement of (7.1). For (c) use the fact that $S_n(\omega) = \omega_n S_{n-1}(\omega)$, so that the right side of (c) reduces to $(\omega_n - d)/(u - d)$.

   Part (c) implies that X_n is $\mathcal{F}_n^S$-measurable hence also $\mathcal{F}_m^S$-measurable, $n = 1, 2, \ldots, m$. Therefore, by definition of $\mathcal{F}_m^X$, $\mathcal{F}_m^X \subseteq \mathcal{F}_m^S$. On the other hand, (7.3) implies that S_n is $\mathcal{F}_n^X$-measurable hence $\mathcal{F}_m^X$-measurable, $n = 1, 2, \ldots, m$, so $\mathcal{F}_m^S \subseteq \mathcal{F}_m^X$.

5. If $S_0 d < K < S_0 u$ then

$$C_0 = (1+i)^{-1}(S_0 u - K)p^* = \frac{(1+i-d)(S_0 u - K)}{(1+i)(u-d)},$$

   hence

$$\frac{\partial C_0}{\partial u} = \frac{(1+i-d)(K - S_0 d)}{(1+i)(u-d)^2} > 0$$

   and

$$\frac{\partial C_0}{\partial d} = \frac{(1+i-u)(S_0 u - K)}{(1+i)(u-d)^2} < 0.$$

   If $d \geq K/S_0$, then

$$C_0 = \frac{(S_0 u - K)p^* + (S_0 d - K)q^*}{1+i} = S_0 - \frac{K}{1+i}.$$

7.

u	1.1	1.2	1.3	1.5
$\mathbb{P}(S_5 > 20)$	0	.031	.186	.500

9. Since $k \geq N/2$, there can be only one k-run. Let A denote the event that there was a k-run of u's, and A_j the event that the run started at time $j = 1, 2, \ldots, N-k+1$. The events A_j are mutually exclusive with union A, $\mathbb{P}(A_1) = \mathbb{P}(A_{N-k+1}) = p^k q$, and, assuming that $k < N-1$, $\mathbb{P}(A_j) = p^k q^2$, $j = 2, \ldots, N-k$. Therefore, $\mathbb{P}(A) = p^k q[2 + (N-k-1)q]$. The formula still holds if $k = N-1$, in which case there are only two sets A_j.

11. By Theorem 7.3.1 with $f(x) = x\mathbf{1}_{(K,\infty)}(x)$,

$$V_0 = (1+i)^{-N} \sum_{j=m}^{N} \binom{N}{j} S_0 u^j d^{N-j} p^{*j} q^{*N-j} = S_0 \sum_{j=m}^{N} \binom{N}{j} \hat{p}^j \hat{q}^{N-j}.$$

13. Since

$$\begin{aligned}(S_N - S_M)^+ &= \left(S_0 u^{U_N} d^{N-U_N} - S_0 u^{U_M} d^{M-U_M}\right)^+ \\ &= S_0 u^{U_M} d^{M-U_M} \left(u^{U_N-U_M} d^{L-(U_N-U_M)} - 1\right)^+,\end{aligned}$$

Corollary 7.2.3 and independence imply that

$$\begin{aligned}(1+i)^{-N} V_0 &= \mathbb{E}^*(S_N - S_M)^+ \\ &= \mathbb{E}^*\left(S_0 u^{U_M} d^{M-U_M}\right) \mathbb{E}^*\left(u^{U_N-U_M} d^{L-(U_N-U_M)} - 1\right)^+ \\ &= \mathbb{E}^*(S_M) \mathbb{E}^*\left(u^{U_N-U_M} d^{L-(U_N-U_M)} - 1\right)^+. \qquad (\alpha)\end{aligned}$$

By Remark 7.2.4(b)

$$\mathbb{E}^*(S_M) = (1+i)^M S_0. \qquad (\beta)$$

Since $U_N - U_M = X_{M+1} + \ldots + X_N$ is binomial with parameters (L, p^*),

$$\mathbb{E}^*\left(u^{U_N-U_M} d^{L-(U_N-U_M)} - 1\right)^+ = \mathbb{E}^*\left(u^{U_L} d^{L-U_L} - 1\right)^+.$$

The last expression is $(1+i)^L$ times the cost of a call option with maturity L, strike price one unit, and initial stock value one unit. Therefore, by the CRR formula,

$$\mathbb{E}^*\left(u^{U_N-U_M} d^{L-(U_N-U_M)} - 1\right)^+ = (1+i)^L \Psi(k, L, \hat{p}) - \Psi(k, L, p^*), \quad (\gamma)$$

where k is the smallest nonnegative integer for which $u^k d^{L-k} > 1$. The desired expression for V_0 now follows from (α), (β) and (γ).

15. By Theorem 7.3.1 with $f(x) = x(x-K)^+$,

$$\begin{aligned} V_0 &= (1+i)^{-N} \sum_{j=m}^{N} \binom{N}{j} S_0 u^j d^{N-j} (S_0 u^j d^{N-j} - K) p^{*j} q^{*N-j} \\ &= \frac{S_0^2}{(1+i)^N} \sum_{j=m}^{N} \binom{N}{j} (u^2 p^*)^j (d^2 q^*)^{N-j} \\ &\qquad - \frac{K S_0}{(1+i)^N} \sum_{j=m}^{N} \binom{N}{j} (up^*)^j (dq^*)^{N-j} \\ &= S_0^2 \left(\frac{v}{1+i}\right)^N \sum_{j=m}^{N} \binom{N}{j} \tilde{p}^j \tilde{q}^{N-j} - K S_0 \sum_{j=m}^{N} \binom{N}{j} \hat{p}^j \hat{q}^{N-j}, \end{aligned}$$

where $\tilde{q} = q^* d^2/v$ and $\hat{q} = 1 - \hat{p}$. Since $u^2 p^* + d^2 q^* = v$, $(\tilde{p}, \tilde{q})$ is a probability vector and the desired formula follows.

17. By Corollary 7.2.3 and the law of the unconscious statistician,

$$\begin{aligned} a^N V_0 &= \mathbb{E}^* f(S_m, S_n) \\ &= \sum_{j=0}^{m} \sum_{k=0}^{n} f(S_0 u^j d^{m-j}, S_0 u^k d^{n-k}) \mathbb{P}^*(U_m = j, U_n = k) \\ &= \sum_{j=0}^{m} \sum_{k=j}^{n-m+j} f(S_0 u^j d^{m-j}, S_0 u^k d^{n-k}) \binom{m}{j} \binom{n-m}{k-j} p^{*k} q^{*n-k}, \end{aligned}$$

the last equality by Exercise 3.18.

19. By Exercise 17 with $f(x,y) = \left(\frac{1}{2}(x+y) - K\right)^+$, $m = 1$ and $n = N$, we have $(1+i)^N V_0 = \frac{1}{2}(A_0 + A_1)$, where

$$\begin{aligned} A_0 &:= \sum_{k=0}^{N-1} \binom{N-1}{k} p^{*k} q^{*N-k} (S_0 d + S_0 u^k d^{N-k} - 2K)^+ \quad \text{and} \\ A_1 &:= \sum_{k=1}^{N} \binom{N-1}{k-1} p^{*k} q^{*N-k} (S_0 u + S_0 u^k d^{N-k} - 2K)^+ \\ &= p^* \sum_{k=0}^{N-1} \binom{N-1}{k} p^{*k} q^{*N-1-k} (S_0 u + S_0 u^{k+1} d^{N-1-k} - 2K)^+. \end{aligned}$$

The hypothesis implies that $S_0 d + S_0 u^k d^{N-k} > 2K$ for $k = N-1$, hence there exists a smallest integer $k_1 \geq 0$ such that $S_0 d + S_0 u^{k_1} d^{N-k_1} > 2K$. Since $S_0 u + S_0 u^{k+1} d^{N-k-1} > S_0 d + S_0 u^{N-1} d$ for $k = N-1$, there exists a smallest integer $k_2 \geq 0$ such that $S_0 u + S_0 u^{k_2+1} d^{N-k_2-1} > 2K$.

Therefore

$$\begin{aligned}
A_0 &= \sum_{k=k_1}^{N-1} \binom{N-1}{k} p^{*k} q^{*N-k} (S_0 d + S_0 u^k d^{N-k} - 2K) \\
&= (S_0 d - 2K) q^* \sum_{k=k_1}^{N-1} \binom{N-1}{k} p^{*k} q^{*N-1-k} \\
&\qquad + S_0 d q^* \sum_{k=k_1}^{N-1} \binom{N-1}{k} (up^*)^k (dq^*)^{N-1-k} \\
&= (S_0 d - 2K) q^* \Psi(k_1, N-1, p^*) + (1+i)^N S_0 d q^* \Psi(k_1, N-1, \hat{p}),
\end{aligned}$$

and

$$\begin{aligned}
A_1 &= p^* \sum_{k=k_2}^{N-1} \binom{N-1}{k} p^{*k} q^{*N-1-k} (S_0 u + S_0 u^{k+1} d^{N-1-k} - 2K) \\
&= (S_0 u - 2K) p^* \sum_{k=k_2}^{N-1} \binom{N-1}{k} p^{*k} q^{*N-1-k} \\
&\qquad + S_0 u p^* \sum_{k=k_2}^{N-1} \binom{N-1}{k} (up^*)^k (dq^*)^{N-1-k} \\
&= (S_0 u - 2K) p^* \Psi(k_2, N-1, p^*) + (1+i)^N S_0 u p^* \Psi(k_2, N-1, \hat{p}).
\end{aligned}$$

21. See Table D.2.

TABLE D.2: Variation of P with u, d, p, and K.

K	u	d	p	P	K	u	d	p	P
10	1.20	.80	.45	.04	30	1.20	.80	.45	.01
10	1.18	.80	.45	.02	30	1.18	.80	.45	.00
10	1.20	.82	.45	.13	30	1.20	.82	.45	.04
10	1.20	.80	.55	.62	30	1.20	.80	.55	.38
10	1.18	.80	.55	.46	30	1.18	.80	.55	.24
10	1.20	.82	.55	.82	30	1.20	.82	.55	.62

23. From $(s-K)^+ - (K-s)^+ = s - K$ we have

$$\begin{aligned}
(1+i)^N (C_0 - P_0) &= \sum_{j=0}^{N} \binom{N}{j} \left(S_0 u^j d^{N-j} - K\right) p^{*j} q^{*N-j} \\
&= \sum_{j=0}^{N} \binom{N}{j} S_0 (p^*/u)^j (q^*/d)^{N-j} - \sum_{j=0}^{N} \binom{N}{j} K p^{*j} q^{*N-j} \\
&= S_0 (1+i)^N - K.
\end{aligned}$$

25. Recalling that $S_n = S_0(u/d)^{U_n} d^n$, we have

$$\begin{aligned}\mathbb{E}^*(S_1S_2\cdots S_N)^{\frac{1}{N}} &= S_0 d^{\frac{N+1}{2}}\mathbb{E}^* w^{U_1+U_2+\cdots+U_N}\\ &= S_0 d^{\frac{N+1}{2}}\mathbb{E}^*\, w^{(NX_1+(N-1)X_2+\cdots+2X_{N-1}+X_N)}\\ &= S_0 d^{\frac{N+1}{2}}\left(\mathbb{E}^*\, w^{NX_1}\right)\left(\mathbb{E}^*\, w^{(N-1)X_2}\right)\cdots\left(\mathbb{E}^*\, w^{X_N}\right)\\ &= S_0 d^{\frac{N+1}{2}}\left(q^*+p^*w^N\right)\left(q^*+p^*w^{N-1}\right)\ldots\left(q^*+p^*w\right)\\ &= S_0 d^{\frac{N+1}{2}}(u-d)^{-N}\prod_{j=1}^{N}\left(u-a+(a-d)w^j\right).\end{aligned}$$

Since there is no cost to enter into a forward, $V_0 = (1+i)^{-N}\mathbb{E}^*H = 0$, hence

$$K = \frac{S_0 d^{\frac{N+1}{2}}}{(u-d)^N}\prod_{j=1}^{N}\left[u-a+(a-d)(u/d)^{j/N}\right], \quad a := 1+i.$$

Chapter 8

1. Conditioning on $\mathcal{G}$ we have

$$\mathbb{E}(XY) = \mathbb{E}\left[\mathbb{E}(XY \mid \mathcal{G})\right] = \mathbb{E}\left[X\mathbb{E}(Y \mid \mathcal{G})\right] = \mathbb{E}\left[X\mathbb{E}(Y)\right] = \mathbb{E}(X)\mathbb{E}(Y).$$

3. Set $Z = X+Y$ and $q = 1-p$. Then

$$p_{X|Z}(x \mid z) = \frac{p_X(x)p_Y(z-x)}{qp_Y(z)+pp_Y(z-1)} = \begin{cases}\dfrac{qp_Y(z)}{qp_Y(z)+pp_Y(z-1)} & \text{if } x=0,\\ \dfrac{pp_Y(z-1)}{qp_Y(z)+pp_Y(z-1)} & \text{if } x=1.\end{cases}$$

and

$$\sum_x xp_{X|Z}(x \mid z) = \frac{pp_Y(z-1)}{qp_Y(z)+pp_Y(z-1)} = \begin{cases}1 & z=2,\\ 1/2 & z=1,\\ 0 & z=0.\end{cases}$$

$$= \mathbf{1}_{\{2\}}(z) + \tfrac{1}{2}\mathbf{1}_{\{1\}}(z).$$

5. Set $Z = X+Y$. Then

$$p_{X|Z}(x \mid z) = \frac{p_X(x)p_Y(z-x)}{\sum_{x'} p_X(x')p_Y(z-x')}$$

and

$$\mathbb{E}(X \mid Z) = \sum_x x\,p_{X|Z}(x \mid Z).$$

7. Mimic the proof of (8.4).

Chapter 9

1. By the multistep property, $\mathbb{E}(M_n|\mathcal{F}_0) = M_0$, $n = 1, 2, \ldots$. Taking expectations of both sides produces the desired result.

3. Set $b := (pe^a + qe^{-a})$. Then $M_{n+1} = b^{-1}M_n e^{aX_{n+1}}$ and $\mathbb{E}\left(e^{aX_{n+1}}\right) = b$, hence

$$\mathbb{E}(M_{n+1}|\mathcal{F}_n^X) = b^{-1}M_n\mathbb{E}\left(e^{aX_{n+1}}|\mathcal{F}_n^X\right) = b^{-1}M_n\mathbb{E}\left(e^{aX_{n+1}}\right) = M_n.$$

5. Since $A_{n+1}B_{n+1} = (X_{n+1} + A_n)(Y_{n+1} + B_n)$,

$$\begin{aligned}
&\mathbb{E}(A_{n+1}B_{n+1} - A_nB_n|\mathcal{F}_n) \\
&\quad = \mathbb{E}(X_{n+1}Y_{n+1}|\mathcal{F}_n) + \mathbb{E}(B_nX_{n+1}|\mathcal{F}_n) + \mathbb{E}(A_nY_{n+1}|\mathcal{F}_n) \\
&\quad = \mathbb{E}(X_{n+1}Y_{n+1}) + B_n\mathbb{E}(X_{n+1}) + A_n\mathbb{E}(Y_{n+1}) \\
&\quad = \mathbb{E}(X_{n+1})\mathbb{E}(Y_{n+1}) = 0.
\end{aligned}$$

7. For $0 \le k \le m \le n$,

$$\begin{aligned}
\mathbb{E}[(M_n - M_m)M_k] &= \mathbb{E}\{\mathbb{E}[(M_n - M_m)M_k|\mathcal{F}_k]\} \\
&= \mathbb{E}\{M_k\mathbb{E}[(M_n - M_m)|\mathcal{F}_k]\} \\
&= \mathbb{E}(M_k(M_k - M_k)) = 0.
\end{aligned}$$

9. Because (δ_n) is predictable,

$$\mathbb{E}^*(S_{n+1}|\mathcal{F}_n) = (1 - \delta_{n+1})S_n\mathbb{E}^*(Z_{n+1}|\mathcal{F}_n^S) = (1 + i)(1 - \delta_{n+1})S_n.$$

Dividing both sides of this equation by $(1 + i)^{n+1}\xi_{n+1}$ establishes the assertion.

11. The expression for Z follows upon noting that

$$\mathbb{P}(\omega) = p^{Y_N(\omega)}q^{N-Y_N(\omega)} \quad \text{and} \quad \mathbb{P}^*(\omega) = p^{*Y_N(\omega)}q^{*N-Y_N(\omega)}.$$

To find $\mathbb{E}(Z|\mathcal{F}_n^S)$ set $a = \dfrac{p^*}{p}$, $b = \dfrac{q^*}{q}$, and $c = \dfrac{a}{b}$, so that

$$Z_n = a^{Y_n}b^{n-Y_n} = b^nc^{Y_n}.$$

By the factor and independence properties,

$$\begin{aligned}
\mathbb{E}(Z_m|\mathcal{F}_n^S) &= b^m\mathbb{E}(c^{Y_n}c^{Y_m-Y_n}|\mathcal{F}_n^S) = b^mc^{Y_n}\mathbb{E}(c^{Y_m-Y_n}) \\
&= b^mc^{Y_n}(pc + q)^{m-n} = b^nc^{Y_n}(bpc + bq)^{m-n}.
\end{aligned}$$

Since $bpc + bq = p^* + q^* = 1$, $\mathbb{E}(Z_m|\mathcal{F}_n^S) = Z_n$.

Chapter 10

1. For $n = 1, 2, \ldots, N$, let U_n denote any of the $\mathcal{F}^S_{n-1}$ random variables

$$\frac{1}{n}(S_0+S_1+\cdots+S_{n-1}),\ \max\{S_0, S_1, \cdots, S_{n-1}\},\ \min\{S_0, S_1, \cdots, S_{n-1}\},$$

and set $A_n = \{S_n > U_n\}$. Then $A_n \in \mathcal{F}^S_n$, and the functions in (a), (b) and (c) are of the form

$$\tau(\omega) = \begin{cases} \min\{n \mid \omega \in A_n\} & \text{if } \{n \mid \omega \in A_n\} \neq \emptyset, \\ N & \text{otherwise.} \end{cases}$$

Therefore,

$$\{\tau = n\} = A_1' A_2' \cdots A_{n-1}' A_n \in \mathcal{F}^S_n, \quad n < N,$$

and

$$\{\tau = N\} = A_1' A_2' \cdots A_{N-1}' \in \mathcal{F}^S_{N-1} \subseteq \mathcal{F}^S_N.$$

3. We show by induction on k that

$$v_k(S_k(\omega)) = f(S_k(\omega)) = 0 \tag{$\dagger$}$$

for all $k \geq n$ $(= \tau_0(\omega))$. By definition of τ_0, ($\dagger$) holds for $k = n$. Suppose ($\dagger$) holds for arbitrary $k \geq n$. Since

$$v_k(S_k(\omega)) = \max\big(f(S_k(\omega)), a v_{k+1}(S_k(\omega)u) + b v_{k+1}(S_k(\omega)d)\big)$$

and all terms comprising the expression on the right of this equation are nonnegative, ($\dagger$) implies that

$$v_{k+1}\left(S_k(\omega)u\right) = v_{k+1}\left(S_k(\omega)d\right) = 0,$$

that is, $v_{k+1}(S_{k+1}(\omega)) = 0$. But

$$v_{k+1}(S_{k+1}(\omega)) \\ = \max\big(f(S_{k+1}(\omega)), a v_{k+2}(S_{k+1}(\omega)u) + b v_{k+2}(S_{k+1}(\omega)d)\big),$$

which implies that $f(S_{k+1}(\omega)) = 0$. Therefore ($\dagger$) holds for $k + 1$.

Chapter 11

1. (a) $x^{-2}\,dx = \sin t\,dt \Rightarrow x^{-1} = \cos t + c \Rightarrow x = (\cos t + c)^{-1}$. Also, $x(0) = 1/3 \Rightarrow c = 2$. Therefore, $x(t) = (\cos t + 2)^{-1}$, $-\infty < t < \infty$.

(b) $x(0) = 2 \Rightarrow c = -1/2 \Rightarrow x(t) = (\cos t - 1/2)^{-1}$, $-\pi/3 < t < \pi/3$.

(c) $2x\,dx = (2t + \cos t)\,dt \Rightarrow x^2 = t^2 + \sin t + c$. $x(0) = 1 \Rightarrow c = 1 \Rightarrow x(t) = \sqrt{t^2 + \sin t + 1}$, valid for all t (positive root because $x(0) > 0$).

(d) $(x+1)^{-1}\,dx = \cot t\,dt \Rightarrow \ln|x+1| = \ln|\sin t| + c \Rightarrow x + 1 = \pm e^c \sin t$; $x(\pi/6) = 1/2 \Rightarrow x + 1 = \pm 3\sin t$. Positive sign is chosen because $x(\pi/6) + 1 > 0$. Therefore $x(t) = 3\sin t - 1$.

3. Use the partitions $\mathcal{P}_n$ described in the example to construct Riemann-Stieltjes sums that diverge.

5. $W(t) - W(s)$ and $2W(s)$ are independent, with $W(t) - W(s) \sim N(0, t-s)$ and $2W(s) \sim N(0, 4s)$. Since $W(s) + W(t) = 2W(s) + W(t) - W(s)$, the first assertion follows from Example 3.8.2. For the second assertion use the identity

$$W(s) + W(t) + W(r) = 3W(r) + 2[W(s) - W(r)] + [W(t) - W(s)]$$

and argue similarly.

7. By Theorem 11.3.3 $X_t = \int_0^t F(s)\,dW(s)$ has mean zero and variance

$$\mathbb{V}\,X_t = \int_0^t \mathbb{E}\left(F^2(s)\right)\,ds.$$

(a) Since $\mathbb{E}\left(sW^2(s)\right) = s^2$, $\mathbb{V}\,X_t = \int_0^t s^2\,ds = t^3/3$.

(b) Since $W(s) \sim N(0, s)$,

$$\mathbb{E}\,\exp(2W_s^2) = \frac{1}{\sqrt{2\pi s}} \int_{-\infty}^{\infty} e^{2x^2} e^{-x^2/2s}\,dx = \frac{1}{\sqrt{2\pi s}} \int_{-\infty}^{\infty} e^{-\alpha x^2/2}\,dx,$$

where $\alpha = s^{-1} - 4$. If $s \geq 1/4$ then $\alpha \leq 0$ and the integral diverges. Therefore, $\mathbb{V}\,Y_t = +\infty$ for $t \geq 1/4$. If $s \leq t < 1/4$ then, making the substitution $y = \sqrt{\alpha}x$, we have

$$\mathbb{E}\,\exp(2W_s^2) = \frac{1}{\sqrt{2\pi s\alpha}} \int_{-\infty}^{\infty} e^{-y^2/2}\,dy = \frac{1}{\sqrt{s\alpha}} = (1 - 4s)^{-1/2},$$

so that

$$\mathbb{V}\,X_t = \int_0^t (1 - 4s)^{-1/2}\,ds = \tfrac{1}{2}\left(1 - \sqrt{1 - 4t}\right).$$

(c) For $s > 0$

$$\begin{aligned}\mathbb{E}\,|W(s)| &= \frac{1}{\sqrt{2\pi s}} \int_{-\infty}^{\infty} |x| e^{-x^2/2s}\,dx = \frac{2}{\sqrt{2\pi s}} \int_0^{\infty} x e^{-x^2/2s}\,dx \\ &= \sqrt{\frac{2s}{\pi}} \int_0^{\infty} e^{-y}\,dy = \sqrt{\frac{2s}{\pi}}.\end{aligned}$$

Therefore,

$$\mathbb{V}\, X_t = \sqrt{\frac{2}{\pi}} \int_0^t \sqrt{s}\, ds = \frac{2}{3}\sqrt{\frac{2}{\pi}}\, t^{3/2}.$$

9. (a) Use Version 1 with $f(x) = e^x$.

   (b) From Version 2 with $f(t, x) = tx^2$,

$$d(tW^2) = 2tW\, dW + (W^2 + t)\, dt.$$

   (c) Use Version 4 with $f(t, x, y) = x/y$. Since $f_t = 0$, $f_x = 1/y$, $f_y = -x/y^2$, $f_{xx} = 0$, $f_{xy} = -1/y^2$, and $f_{yy} = 2x/y^3$, we have

$$d\left(\frac{X}{Y}\right) = \frac{dX}{Y} - \frac{X}{Y^2}\, dY + \frac{X}{Y^3}\, (dY)^2 - \frac{1}{Y^2} dX \cdot dY.$$

   Factoring out $\frac{X}{Y}$ gives the desired result.

11. Taking expectations in (11.20) gives

$$\mathbb{E}\, X_t = e^{-\beta t}\left(\mathbb{E}\, X_0 + \frac{\alpha}{\beta}(e^{\beta t} - 1)\right) = e^{-\beta t}\left(\mathbb{E}\, X_0 - \frac{\alpha}{\beta}\right) + \frac{\alpha}{\beta}.$$

   Since $\beta > 0$, $\lim_{t\to\infty} \mathbb{E}\, X_t) = \frac{\alpha}{\beta}$.

Chapter 12

1. From $(s - K)^+ - (K - s)^+ = s - K$ and (12.3.2) we have

$$e^{r(T-t)}(C_t - P_t) = e^{-\frac{1}{2}\sigma^2(T-t)} S_t \int_{-\infty}^{\infty} e^{\sigma\sqrt{T-t}\, y}\varphi(y)\, dy - K$$

   Now use $\int_{-\infty}^{\infty} e^{\alpha x}\varphi(x)\, dx = e^{\alpha^2/2}$ (Exercise 13).

3. By (12.14) and Theorems 12.4.1, 12.4.2,

$$\frac{\partial P}{\partial s} = \frac{\partial C}{\partial s} - 1 = \Phi(d_1) - 1 < 0,$$

$$\lim_{s\to\infty} P = 0, \quad \text{and} \quad \lim_{s\to 0+} P = Ke^{-r\tau}.$$

5. Taking $f(z) = \alpha \mathbf{1}_{(K,\infty)}(z)$ in Theorem 12.3.2 yields

$$V_t = e^{-r(T-t)} G(t, S_t),$$

where, as in the proof of Corollary 12.3.3

$$\begin{aligned}G(t,s) &= \int_{-\infty}^{\infty} A\mathbf{1}_{(K,\infty)}\left(s\exp\left[y\sigma\sqrt{T-t}+(r-\sigma^2/2)(T-t)\right]\right)\varphi(y)\,dy\\ &= A\Phi\big(d_2(T-t,s,K,\sigma,r)\big).\end{aligned}$$

7. Since

$$V_T = S_T\mathbf{1}_{(K_1,\infty)}(S_T) - S_T\mathbf{1}_{[K_2,\infty)}(S_T),$$

V_0 is seen to be the difference in the prices of two asset-or-nothing options.

9. Clearly,

$$V_t = C(T-t,S_t,F) + (F-K)e^{-r(T-t)}$$

and, in particular,

$$V_0 = C_0 + (F-K)e^{-rT},$$

where C_0 is the cost of a call option on the stock with strike price F. Setting $V_0 = 0$ and solving for K gives $K = F + e^{rT}C_0$.

11. Clearly, $S_T > K$ iff $\sigma W_T + (\mu - \sigma^2/2)T > \ln(K/S_0)$, hence the desired probability is

$$1-\Phi\left(\frac{\ln(K/S_0)-(\mu-\sigma^2/2)T}{\sigma\sqrt{T}}\right) = \Phi\left(\frac{\ln(S_0/K)+(\mu-\sigma^2/2)T}{\sigma\sqrt{T}}\right).$$

13. The expression for E_C follows from Theorem 12.4.1(i) and the Black-Scholes formula. To verify the limits write

$$E_C^{-1} = 1-\alpha\frac{\Phi(d_2)}{s\Phi(d_1)}, \qquad \alpha := Ke^{-rT}$$

and note that (a) follows from $\lim_{s\to\infty}\Phi(d_{1,2}) = 1$. For (b), apply L'Hôpital's Rule to obtain

$$\begin{aligned}\alpha(1-E_C^{-1})^{-1} &= \lim_{s\to 0+}\frac{s\Phi(d_1)}{\Phi(d_2)} = \lim_{s\to 0+}\frac{s\varphi(d_1)(\beta s)^{-1}+\Phi(d_1)}{\varphi(d_2)(\beta s)^{-1}}\\ &= \lim_{s\to 0+}\frac{s\varphi(d_1)}{\varphi(d_2)}\left[1+\beta\frac{\Phi(d_1)}{\varphi(d_1)}\right], \qquad \beta := \sigma\sqrt{T}.\end{aligned}$$

Since $d_2^2 - d_1^2 = (d_2-d_1)(d_2+d_1) = -\beta(d_2+d_1) = 2[\ln(K/s)-rT]$,

$$s\frac{\varphi(d_1)}{\varphi(d_2)} = s\exp\left[\tfrac{1}{2}(d_2^2-d_1^2)\right] = s\exp\left[\ln(K/s)-rT\right] = \alpha,$$

hence

$$\left(1 - E_C^{-1}\right)^{-1} = \alpha^{-1} \lim_{s\to 0+} \frac{s\Phi(d_1)}{\Phi(d_2)} = 1 + \beta \lim_{s\to 0+} \frac{\Phi(d_1)}{\varphi(d_1)}.$$

By L'Hôpital's Rule,

$$\lim_{s\to 0+} \frac{\Phi(d_1)}{\varphi(d_1)} = \lim_{s\to 0+} \frac{\varphi(d_1)(\beta s)^{-1}}{\varphi(d_1)(-d_1)(\beta s)^{-1}} = -\lim_{s\to 0+} \frac{1}{d_1} = 0.$$

Therefore $\lim_{s\to 0+} \left(1 - E_C^{-1}\right)^{-1} = 1$, which implies (b).

15. Make the substitution

$$z = s \exp\left\{y\sigma\sqrt{T-t} + (r - \tfrac{1}{2}\sigma^2)(T-t)\right\}$$

to obtain

$$G(t,s) = \frac{1}{\sigma\sqrt{2\pi\tau}} \int_0^\infty f(z) \exp\left(-\frac{1}{2}\left\{\frac{\ln(z/s) - (r - \sigma^2/2)\tau}{\sigma\sqrt{\tau}}\right\}^2\right) \frac{dz}{z}.$$

Chapter 13

1. Since W is a martingale,

$$\mathbb{E}(W_s W_t) = \mathbb{E}\left[\mathbb{E}(W_s W_t \mid \mathcal{F}_s^W)\right] = \mathbb{E}\left[W_s \mathbb{E}(W_t \mid \mathcal{F}_s^W)\right] = \mathbb{E}(W_s^2) = s.$$

3. Let

$$A = \{(u,v) \mid v \le y, u + v \le x\}$$

and

$$f(x,y) = \frac{1}{\sqrt{s(t-s)}} \varphi\left(\frac{x}{\sqrt{t-s}}\right) \varphi\left(\frac{y}{\sqrt{s}}\right).$$

By independent increments,

$$\begin{aligned}
\mathbb{P}(W_t \le x, W_s \le y) &= \mathbb{P}\big((W_t - W_s, W_s) \in A\big) \\
&= \iint_A f(u,v)\, du\, dv \\
&= \int_{-\infty}^{y} \int_{-\infty}^{x-v} f(u,v)\, du\, dv \\
&= \int_{-\infty}^{y} \int_{-\infty}^{x} f(u-v,v)\, du\, dv.
\end{aligned}$$

Differentiating with respect to x and y yields the desired result.

5. Since

$$M_s^{-1}\mathbb{E}\left(M_t - M_s \mid \mathcal{F}_s\right) = \mathbb{E}\left(e^{\alpha[W(t)-W(s)]+h(t)-h(s)} - 1 \mid \mathcal{F}_s\right)$$
$$= e^{h(t)-h(s)}\mathbb{E}\left(e^{\alpha[W(t)-W(s)]} \mid \mathcal{F}_s\right) - 1, \quad s \le t,$$

we see that M is a martingale iff

$$\mathbb{E}\left(e^{\alpha[W(t)-W(s)]} \mid \mathcal{F}_s\right) = e^{h(s)-h(t)}, \quad 0 \le s \le t.$$

By independence of increments and Exercise 6.14

$$\mathbb{E}\left(e^{\alpha[W(t)-W(s)]} \mid \mathcal{F}_s\right) = \mathbb{E}\left(e^{\alpha[W(t)-W(s)]}\right) = e^{\alpha^2(t-s)/2}.$$

Therefore, M is a martingale iff $h(t) - h(s) = \alpha^2(s-t)/2$. Setting $s = 0$ gives the desired result.

7. By Example 13.1.1(b) and iterated conditioning,

$$\mathbb{E}(W_t^2 - t \mid W_s) = \mathbb{E}[\mathbb{E}(W_t^2 - t \mid \mathcal{F}_s^W) \mid W_s] = \mathbb{E}(W_s^2 - s \mid W_s) = W_s^2 - s,$$

hence $\mathbb{E}(W_t^2 \mid W_s) = W_s^2 + t - s$. For $\mathbb{E}(W_t^3 \mid W_s)$ proceed similarly: By Exercise 6

$$\mathbb{E}(W_t^3 - 3tW_t \mid W_s) = \mathbb{E}[(\mathbb{E}(W_t^3 - 3tW_t \mid \mathcal{F}_s^W) \mid W_s)]$$
$$= \mathbb{E}(W_s^3 - 3sW_s \mid W_s)$$
$$= W_s^3 - 3sW_s,$$

hence, from Exercise 1

$$\mathbb{E}(W_t^3 \mid W_s) = 3t\mathbb{E}(W_t \mid W_s) + W_s^3 - 3sW_s = W_s^3 + 3(t-s)W_s.$$

9. For any x,

$$\mathbb{P}^*(X \le x) = \mathbb{E}^*\, \mathbf{1}_{(-\infty,x]}(X)$$
$$= e^{-\frac{1}{2}\alpha^2 T}\mathbb{E}\left(\mathbf{1}_{(-\infty,x]}(X)e^{-\alpha W_T}\right)$$
$$= e^{-\frac{1}{2}\alpha^2 T}\mathbb{E}(\mathbf{1}_{(-\infty,x]}(X))\mathbb{E}\left(e^{-\alpha W_T}\right)$$
$$= \mathbb{P}(X \le x),$$

the last equality from Exercise 6.14

Chapter 14

1. Let

$$f(t,x) = \exp\left[\sigma x + (r_d - r_e - \sigma^2)t\right] \quad \text{and}$$
$$g(t,x) = \exp\left[-\sigma x - (r_d - r_e - \sigma^2)t\right].$$

Then $f_t = (r_d - r_e - \sigma^2)f$, $f_x = \sigma f$, $f_{xx} = \sigma^2 f$, $g_t = -(r_d - r_e - \sigma^2)g$, $g_x = -\sigma g$, and $g_x = \sigma^2 g$, so, by the Ito-Doeblin formula,

$$dQ = df(t, W_t^*) = \sigma^2 Q\, dW^* + (r_d - r_e)Q\, dt$$

and

$$dQ^{-1} = dg(t, W_t^*) = \sigma^2 Q^{-1}\, dW^* + (r_e - r_d + \sigma^2)Q^{-1}\, dt.$$

3. $\left(e^{(\delta - r)t} S_t\right)$ is the discount of the process $(\hat{S}_t)$.

5. Since $V_T = (S_T - S_\tau)^+ + S_\tau - K$, the payoff of a shout option is that of a portfolio long in a call with strike price S_τ and a bond with face value $S_\tau - K$. At time τ, S_τ is known, hence the value at time $t \geq \tau$ is

$$V_t = C(T - t, S_t, S_\tau, \sigma, r) + e^{-r(T-t)}(S_\tau - K).$$

Chapter 15

1. The compound option has payoff $(P(S_{T_0}) - K_0)^+$, where $P(S_{T_0})$ is the value of the underlying put at time T_0. The expression for $P(s)$ comes from the put-call parity relation $P(s) = C(s) - s + Ke^{-r(T-T_0)}$ and the Black-Scholes formula. By Corollary 13.3.2 with $f(x) = \left(P(x) - K_0\right)^+$, the cost of the compound option is

$$V_0^{cp} = e^{-rT_0} \int_{-\infty}^{\infty} \left[P\left(g(y)\right) - K_0\right]^+ \varphi(y)\, dy.$$

Since $\left[P\left(g(y)\right) - K_0\right]^+$ is decreasing in y (Exercise 12.3),

$$\left[P\left(g(y)\right) - K_0\right]^+ = \left[P\left(g(y)\right) - K_0\right] \mathbf{1}_{(-\infty, y_1)}.$$

3. Let $A = \{m^S \geq c\}$. Then (almost surely)

$$C_T^{do} + C_T^{di} = (S_T - K)^+ \mathbf{1}_A + (S_T - K)^+ \mathbf{1}_{A'} = (S_T - K)^+.$$

The conclusion now follows from the law of one price.

5. Since $C_T^{do} - P_T^{do} = (S_T - K)^+ \mathbf{1}_A - (K - S_T)^+ \mathbf{1}_A = (S_T - K)\mathbf{1}_A$, the assertion follows from Exercise 4 and the law of one price.

7. Use

$$(K - m^S)^+ = K - \min\left(m^S, K\right)$$

together with

$$\min\left(m^S, K\right) = \min\left(\min\left(m_t^S, K\right), m_{t,T}^S\right)$$

and proceed as in the text, replacing m_t^S by $\min\left(m_t^S, K\right)$.

9. (a) The payoff for an up-and-in call option is $C_T^{ui} = (S_T - K)^+ \mathbf{1}_A$, where $A = \{M^S \geq c\}$. This may be written as

$$C_T^{ui} = (S_T - K)\mathbf{1}_B, \quad \text{where} \quad B = \{S_T \geq K,\ M^S \geq c\}.$$

Since $S_T \geq K$ iff $\hat{W}_T \geq a$, and $M^S \geq c$ iff $M^{\hat{W}} \geq b$, B has the required representation.

(b) By risk-neutral pricing the cost of the option is

$$C_0^{ui} = e^{-rT}\mathbb{E}^*\big[(S_T - K)\mathbf{1}_B\big],$$

which may be written exactly as in (15.6).

(c) By Exercise 8 and Lemma 15.1.2,

$$\begin{aligned}
\hat{\mathbb{E}}\left(e^{\lambda \hat{W}_T}\mathbf{1}_B\right) &= -\iint_D e^{\lambda x}\hat{g}_m(x,y)\,dA, \quad D := \{(x,y) \mid y \geq b, y \geq x \geq a\}, \\
&= -\int_a^b \int_b^\infty e^{\lambda x}\hat{g}_m(x,y)\,dy\,dx - \int_b^\infty \int_x^\infty e^{\lambda x}\hat{g}_m(x,y)\,dy\,dx \\
&= -I_\lambda(\infty; a, b) + I_\lambda(b; a, b) - I_\lambda(\infty; b, \infty) + I_\lambda(0; b, \infty) \\
&= e^{2b\lambda + \lambda^2 T/2}\left\{\Phi\left(\frac{2b - a + \lambda T}{\sqrt{T}}\right) - \Phi\left(\frac{b + \lambda T}{\sqrt{T}}\right)\right\} \\
&\qquad + e^{\lambda^2 T/2}\Phi\left(\frac{-b + \lambda T}{\sqrt{T}}\right).
\end{aligned}$$

(d) Setting λ to the appropriate values and using (b) and (c) verifies the assertion.

11. The steps leading up to (15.12) are valid for the new β and result in the expression given in part (c) of the problem.

13. Follows immediately from $\lim_{n\to\infty} a_n = 1/3$.

15. The desired probability is $1 - \mathbb{P}(C)$, where

$$C := \{m^S \geq c\} = \{m^W \geq b\}, \quad b := \sigma^{-1}\ln(c/S_0).$$

To find $\mathbb{P}(C)$, recall that the measures $\mathbb{P}^*$, $\hat{\mathbb{P}}$ and the processes W^*, $\hat{W}$ are defined by

$$\begin{aligned}
d\mathbb{P}^* &= e^{-\alpha W_T - \frac{1}{2}\alpha^2 T}\,d\mathbb{P}, \\
d\hat{\mathbb{P}} &= e^{-\beta W_T^* - \frac{1}{2}\beta^2 T}\,d\mathbb{P}^*, \\
W_T^* &= W_T + \alpha T, \quad \alpha := \frac{\mu - r}{\sigma}, \quad \text{and} \\
\hat{W}_T &= W_T^* + \beta T = W_T + (\alpha + \beta)T, \quad \beta := \frac{r}{\sigma} - \frac{\sigma}{2}.
\end{aligned}$$

It follows that $d\mathbb{P} = U d\hat{\mathbb{P}}$, where

$$U := e^{\lambda \hat{W}_T - \frac{1}{2}\lambda^2 T}, \quad \lambda := \alpha + \beta = \frac{\mu}{\sigma} - \frac{\sigma}{2}.$$

Therefore

$$\mathbb{P}(C) = \hat{\mathbb{E}}(\mathbf{1}_C U) = e^{-\lambda^2 T/2} \iint_D e^{\lambda x} \hat{g}_m(x,y)\, dA,$$

where

$$D = \{(x,y) \mid b \le y \le 0, x \ge y\}$$

(see (15.8)). This is the region of integration described in Figure 15.1 so by (15.11)

$$\mathbb{P}(C) = \Phi\left(\frac{-b + \lambda T}{\sqrt{T}}\right) - e^{2b\lambda}\Phi\left(\frac{b + \lambda T}{\sqrt{T}}\right).$$

Since

$$\frac{\pm b + \lambda T}{\sqrt{T}} = \frac{\pm \ln(c/S_0) + (\mu - \sigma^2/2)T}{\sigma\sqrt{T}}$$

and

$$2b\lambda = \left(\frac{2\mu}{\sigma^2} - 1\right)\ln(c/S_0),$$

the desired probability is

$$1 - \left\{\Phi(d_1) - \left(\frac{c}{S_0}\right)^{\frac{2\mu}{\sigma^2} - 1} \Phi(d_2)\right\},$$

where

$$d_{1,2} = \frac{\pm \ln(S_0/c) + (\mu - \sigma^2/2)T}{\sigma\sqrt{T}}.$$

Bibliography

[1] Bellalah, M., 2009, *Exotic Derivatives*, World Scientific, London.

[2] Bingham, N. H. and R. Kiesel, 2004, *Risk-Neutral Valuation*, Springer, New York.

[3] Etheridge, A., 2002, *A Course in Financial Calculus*, Cambridge University Press, Cambridge.

[4] Elliot, R. J. and P. E. Kopp, 2005, *Mathematics of Financial Markets*, Springer, New York.

[5] Grimmett, G. R. and D. R. Stirzaker, 1992, *Probability and Random Processes*, Oxford Science Publications, Oxford.

[6] Hida, T., 1980, *Brownian Motion*, Springer, New York.

[7] Hull, J. C., 2000, *Options, Futures, and Other Derivatives*, Prentice-Hall, Englewood Cliffs, N.J.

[8] Karatzas, I. and S. Shreve, 1998, *Methods of Mathematical Finance*, Springer, New York.

[9] Junghenn, H., 2015, *A Course in Real Analysis*, CRC Press, Boca Raton.

[10] Junghenn, H., 2018, *Principles of Analysis: Measure, Integration, Functional Analysis, and Applications*, CRC Press, Boca Raton.

[11] Kuo, H., 2006, *Introduction to Stochastic Integration*, Springer, New York.

[12] Kwok, Y., 2008, *Mathematical Models of Financial Derivatives*, Springer, New York.

[13] Lewis, M., 2010, *The Big Short*, W. W. Norton, New York.

[14] Musiela, M. and M. Rutowski, 1997, *Mathematical Models in Financial Modelling*, Springer, New York.

[15] Myneni, R., 1997, The pricing of the American option, *Ann. Appl. Prob.* 2, 1–23.

[16] Ross, S. M., 2011, *An Elementary Introduction to Mathematical Finance*, Cambridge University Press, Cambridge.

[17] Rudin, W., 1976, *Principles of Mathematical Analysis*, McGraw-Hill, New York.

[18] Shreve, S. E., 2004, *Stochastic Calculus for Finance I*, Springer, New York.

[19] Shreve, S. E., 2004, *Stochastic Calculus for Finance II*, Springer, New York.

[20] Steele, J. M., 2001, *Stochastic Calculus and Financial Applications*, Springer, New York.

[21] Yeh, J., 1973, *Stochastic Processes and the Wiener Integral*, Marcel Dekker, New York.

Index

For Product Safety Concerns and Information please contact our EU
representative GPSR@taylorandfrancis.com
Taylor & Francis Verlag GmbH, Kaufingerstraße 24, 80331 München, Germany

www.ingramcontent.com/pod-product-compliance
Ingram Content Group UK Ltd.
Pitfield, Milton Keynes, MK11 3LW, UK
UKHW021833240425
457818UK00006B/183

* 9 7 8 0 3 6 7 2 0 8 8 2 0 *